STATISTICS

··

For Engineering and the Sciences

FOURTH EDITION

William Mendenhall
University of Florida

Terry Sincich
University of South Florida

PRENTICE HALL
Englewood Cliffs, New Jersey 07632

© 1995 by PRENTICE-HALL, INC.
A Simon & Schuster Company
Englewood Cliffs, NJ 07632

10 9 8 7 6 5

ISBN 0-02-312718-X

Printed in the United States of America

Contents

· ·

Preface

· ·

This solutions manual is designed to accompany the text *Statistics for Engineering and the Sciences*, Fourth Edition, by William Mendenhall and Terry Sincich (1995). It provides answers to most odd-numbered exercises for each chapter in the text. Other methods of solution may also be appropriate; however, the author has presented one that she believes to be most instructive to the beginning statistics student. The student should first attempt to solve the assigned exercises without help from this manual. Then, if unsuccessful, the solution in the manual will clarify points necessary to the solution. The student who successfully solves an exercise should still refer to the manual's solution. Many points are clarified and expanded upon to provide maximum insight into and benefit from each exercise.

Instructors will also benefit from the use of this manual. It will save time in preparing presentations of the solutions and possibly provide another point of view regarding their meaning.

Some of the exercises are subjective in nature and thus omitted from the Answer Key at the end of *Statistics for Engineering and the Sciences,* Fourth Edition. The subjective decisions regarding these exercises have been made and are explained by the author. Solutions based on these decisions are presented; the solution to this type of exercise is often most instructive. When an alternative interpretation of an exercise may occur, the author has often addressed it and given justification for the approach taken.

I would like to thank Kelly Evans for creating the art work and Brenda Dobson for her assistance and for typing this work.

Nancy S. Boudreau
Bowling Green State University
Bowling Green, Ohio

CHAPTER ONE

· ·

Introduction

1.1   a.   The population of interest is the thion levels of all possible daily ambient air specimens that can be collected at the orchard.

         b.   The sample of interest is the thion levels for the 13 ambient air specimens actually measured at the orchard.

1.3   a.   The population of interest is the powerload status (high neutral current or not) at all U.S. sites with computer powers systems.

         b.   The sample of interest is the powerload status at the 146 U.S. sites selected.

         c.   The sample indicated that less than 10% of the selected sites had high neutral status. We would infer that less than 10% of all sites in the U.S. have high neutral status.

1.5   The population of interest would be the times required for a CT scanner to project images. To collect the necessary sample data, the times necessary for the CT scanner to project an image would be recorded for 'n' images.

1.7   "Ratio of neutral current to full load current" would be quantitative. A ratio is a numerical value.

       "Type of load" would be qualitative. There are two categories for type: line-to-line and line-to-neutral. These categories are not numeric.

       "Computer system vendor" would be qualitative. A vendor is a company that sells computer systems. A vendor is not numeric.

1.9   a.   Chip discharge rate (number of chips discarded per minute) is quantitative. The number of chips is a numerical value.

         b.   Drilling depth (millimeters) is quantitative. The depth is a numerical value.

         c.   Oil velocity (millimeters per second) is quantitative. The velocity is a numerical value.

         d.   Type of drilling (single-edge, BTA, or ejector) is qualitative. The type of drilling is not a numerical value.

         e.   Quality of hole surface is qualitative. The quality can be judged as poor, good, excellent, etc., which are categories and are not numerical values.

1.11   a.   The population of interest is the lifelengths of all hardware components.

         b.   The sample is the lifelengths of the 100 computer components selected.

         c.   Lifelengths are measured on a numerical scale. Thus, they are quantitative.

         d.   We could calculate the average lifelength of the 100 sampled components. We could use this average to estimate the average lifelength of all computer components.

1.13   a.   The population of interest is the status of "fluid statics coverage in their undergraduate engineering program" (yes or no) at all colleges and universities. The sample of interest is the status of "fluid statics coverage in their undergraduate engineering program" at the 100 selected colleges.

         b.   "Fluid statics coverage" is qualitative. There are two possible responses: yes or no. No numbers are associated with these responses.

         c.   From the sample, 66 out of 100 (or 66%) of the colleges covered fluid statics in their undergraduate engineering program. We would infer that 66% of all colleges covered fluid statics in their undergraduate engineering program.

1.15   The variables and their type are listed below:

Model name–qualitative. The model names are NSX, Colt, etc., which are not numbers.

Manufacturer–qualitative. The manufacturers are Acura, Dodge, etc., which are not numbers.

Transmission–qualitative. The types of transmissions are Automatic and Manual. These are not numbers.

Engine size–quantitative. The values for engine size are 3.0, 1.5, etc., which are numbers.

Number of cylinders–quantitative. The values for number of cylinders are 6, and 4, which are numbers.

Estimated city miles/gallon–quantitative. The values for estimated city miles/gallon are 18, 32, etc., which are numbers.

Estimated highway miles/gallon–quantitative. The values for estimated highway miles/gallon are 23, 40, etc., which are numbers.

CHAPTER TWO

Descriptive Statistics

2.1 a. From the pie chart, more than three-fourths (77.6%) of all the automobile tires that are scrapped in the U.S. are dumped. Only 10.7% of the tires are burned for fuel, while 6.7% are recycled into new products. The remaining 5% are exported.

b. To convert the pie chart into a relative frequency bar chart, we first need to convert the percents to relative frequencies by dividing the percents by 100%. The relative frequency table is:

Tire Fate	Percent	Relative Frequency
Dumped	77.6	.776
Burned for fuel	10.7	.107
Recycled	6.7	.067
Exported	5.0	.050

The relative frequency bar chart is:

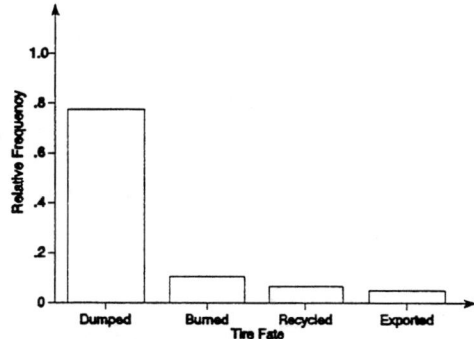

c. To find the frequencies for each of the categories, we multiply the relative frequencies by the total sample size, 242 million. The frequency table is:

Tire Fate	Relative Frequency	Frequency (millions)
Dumped	.776	187.792
Burned for fuel	.107	25.894
Recycled	.067	16.214
Exported	.050	12.100
		242.000

The frequency bar chart is:

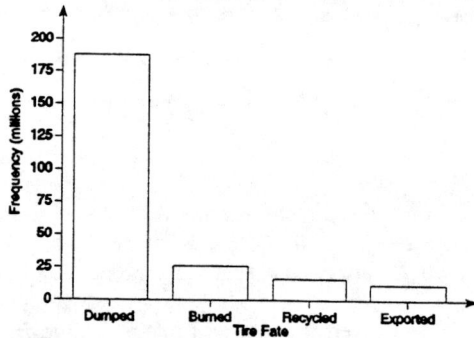

Frequency (millions)

200
175
150
125
100
75
50
25
0

Dumped   Burned   Recycled   Exported

Tire Fate

2.3   The Pareto diagram is:

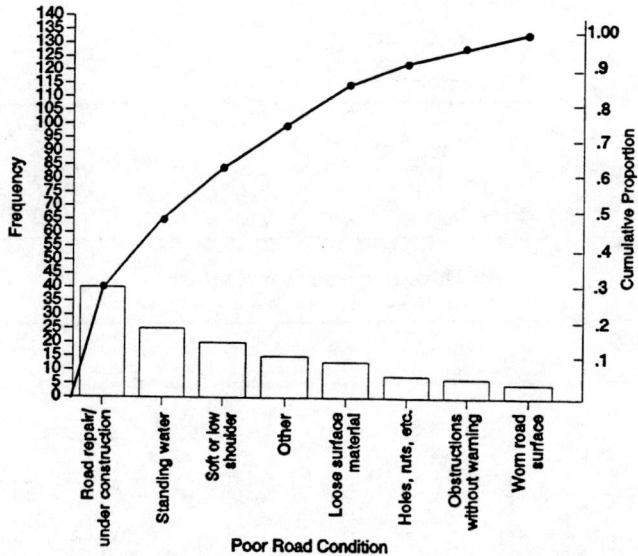

Frequency

140
135
130
125
120
115
110
105
100
95
90
85
80
75
70
65
60
55
50
45
40
35
30
25
20
15
10
5
0

Cumulative Proportion

1.00
.9
.8
.7
.6
.5
.4
.3
.2
.1

Road repair/ under construction   Standing water   Soft or low shoulder   Other   Loose surface material   Holes, nuts, etc.   Obstructions without warning   Worn road surface

Poor Road Condition

The poor road condition that caused the most accidents was "road repairs/under construction" with 39 accidents. The poor road condition that caused the next most accidents was "standing water" with 25 accidents. The poor road condition that caused the least number of accidents was "worn road surface" with 6 accidents.

2.5

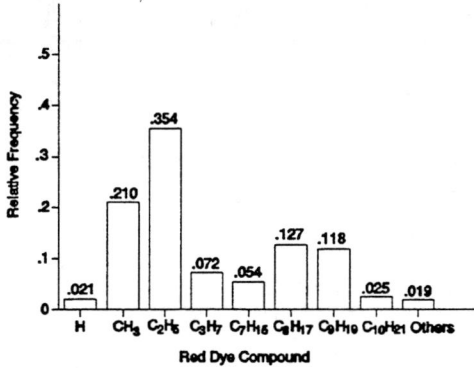

Compound C_2H_5 has the highest relative abundance (.354), with compound CH_3 the next highest with .210. Compounds H, C_3H_7, C_7H_{15}, $C_{10}H_{21}$, and others all have relative abundance less than .1.

2.7   a.   To get the relative frequencies, divide the percentages by 100%.

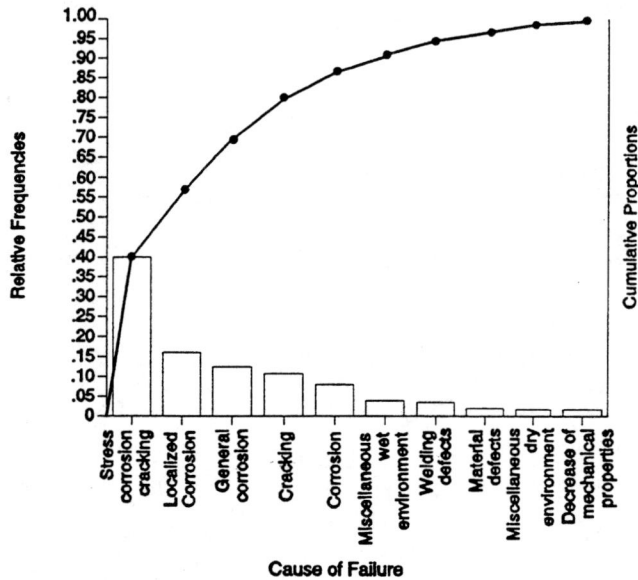

   b.   Yes. The researchers claimed that stress corrosion cracking is the greatest single cause of steel alloy failure. From the chart, it is quite evident that stress corrosion cracking is the greatest single cause of failure because stress corrosion cracking has the highest percentage (39.9%).

2.9   The most common undergraduate major of those in the graduate program in hazardous waste management is Chemical Engineering with 23.3%. The next most common undergraduate major is Biology with 14.8%. The next three most common undergraduate majors are Civil

Engineering with 14.3%, Chemistry with 13.0% and Geology with 7.2%. All the rest of the undergraduate majors are lumped together into the "Other" category and make up 27.4%.

2.11   a.   To convert a frequency histogram to a relative frequency histogram, we must first divide each of the frequencies by the sum of all the frequencies, which is $1 + 1 + 3 + 4 + 10 + 7 + 5 + 4 + 3 + 2 + 3 + 2 + 3 + 2 = 50$. The relative frequency table is:

Class Interval	Frequency	Relative Frequency
5.5–15.5	1	1/50 = .02
15.5–25.5	1	1/50 = .02
25.5–35.5	3	3/50 = .06
35.5–45.5	4	4/50 = .08
45.5–55.5	10	10/50 = .20
55.5–65.5	7	7/50 = .14
65.5–75.5	5	5/50 = .10
75.5–85.5	4	4/50 = .08
85.5–95.5	3	3/50 = .06
95.5–105.5	2	2/50 = .04
145.5–155.5	3	3/50 = .06
165.5–175.5	2	2/50 = .04
175.5–185.5	3	3/50 = .06
195.5–205.5	2	2/50 = .04

The relative frequency histogram is:

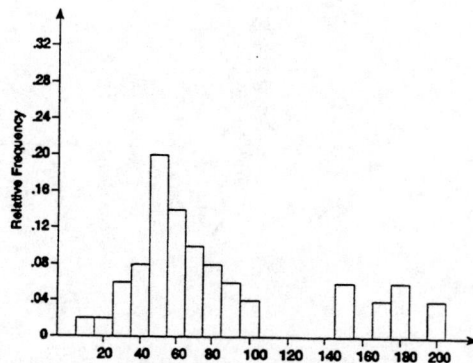

   b.   It would be very unusual to observe a drill chip with a length of at least 190 mm. There are only 2 out of 50 drill chips that are 190 mm or longer. The proportion of drill chips with lengths of at least 190 mm is .04.

2.13   a.   The stems are the left most digits of the numbers in the table and have values 6 through 9. The leaves are the right most digits of the numbers in the table and have values 0 through 9.

b. From the graph, 7 of the 91 observations have values less than 86. Thus, $91 - 7 = 84$ of the 91 observations have values of 86 or higher. Thus, the proportion is 84/91 = .923.

c. The inspection score of 70 is the second number from the top of the stem-and-leaf display.

2.15 a. From the graph, 16% of the sites have a load capacity between 20% and 30%. Dividing 16% by 100% gives a proportion of .16.

b. From the graph, $12\% + 5\% + 6\% + 2\% + 1\% = 26\%$ of the sites have a load capacity of 50% or higher. Dividing 26% by 100% gives a proportion of .26.

2.17 a. From the frequency histogram, the number of copper particles with diameters ranging from 5 to 7 nanometers is 130.

b. To convert a frequency histogram to a relative frequency histogram, we must first divide each of the frequencies by the sum of all the frequencies, which is $35 + 130 + 70 + 15 + 5 = 255$. The relative frequency table is:

Class Interval	Frequency	Relative Frequency
3–5	35	35/255 = .137
5–7	130	130/255 = .510
7–9	70	70/255 = .275
9–11	15	15/255 = .059
11–13	5	5/255 = .020

The relative frequency histogram is:

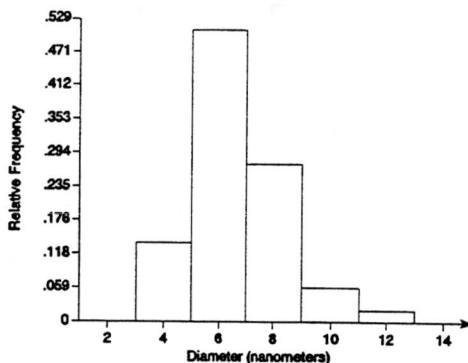

c. The proportion of copper particles exceeding 9 nanometers in diameter is approximately $20/255 = .078$.

2.19 a. First, rank the data in order: 3, 4, 5, 8, 10

mean: $\bar{y} = \dfrac{\sum y}{n} = \dfrac{30}{5} = 6$

median: m = middle (3rd) measurement = 5

mode: Most frequently occurring value

Since all values occur only once, all are modes.

b. First, rank the data in order: 2, 4, 4, 5, 6, 6, 9, 12

mean: $\bar{y} = \dfrac{\sum y}{n} = \dfrac{48}{8} = 6$

median: m = average of the middle (4th and 5th) measurements
$$= \frac{5 + 6}{2} = \frac{11}{2} = 5.5$$

mode: Most frequently occurring value.
     Mode = 4 and 6 (both occur twice)

2.21   The mean or average score is 91.044. The median or middle score, the score with half of the observations above it and half below it, is 92.000.

2.23   First arrange the data in order: .112, .205, .225, .239, .241, .270, .270, .330, .375, .523, .591, .618

mean: $\bar{y} = \dfrac{\sum y}{n} = \dfrac{3.999}{12} = .333$

median: m = average of the middle two (6th and 7th) measurements
$$= \frac{.270 + .270}{2} = .270$$

mode: Most frequently occurring value.
     Mode = .270

Although the data set is slightly skewed to the right, the mean would probably be preferred to the median and mode because it has nicer properties.

2.25   a. The population is the percentage of iron in all possible specimens of Chilean iron ore.

b. One possible objective of the sampling procedure would be to determine the average percentage of iron in all possible specimens.

c. First, calculate the range by subtracting the smallest observation from the largest

Range = 64.34 − 61.68 = 2.66

Next, decide on the number of classes. Suppose we pick 8 classes. To find the class width, we divide the range by the number of classes:

Class width = 2.66/8 = .3325 ≈ .34

The first class will begin at 61.675, just below the smallest percentage. The resulting eight class intervals are shown as follows.

Class	Class Interval	Tally	Class Frequency	Class Relative Frequency
1	61.675–62.015	IIII	4	.061
2	62.015–62.355	HH III	8	.121
3	62.355–62.695	HH I	6	.091
4	62.695–63.035	HH HH HH III	18	.273
5	63.035–63.375	HH HH HH	15	.227
6	63.375–63.715	HH IIII	9	.136
7	63.715–64.055	III	3	.045
8	64.055–64.395	III	3	.045
		Totals $n = 66$		.999

For each class, count the number of observations. This gives the class frequency. The class relative frequency is then calculated by dividing the class frequency by the number of observations, $n = 66$. For class 1, the class relative frequency is $4/66 = .061$. The other class relative frequencies are computed in a similar fashion and are listed in the table. The data above are then used to construct the relative frequency histogram.

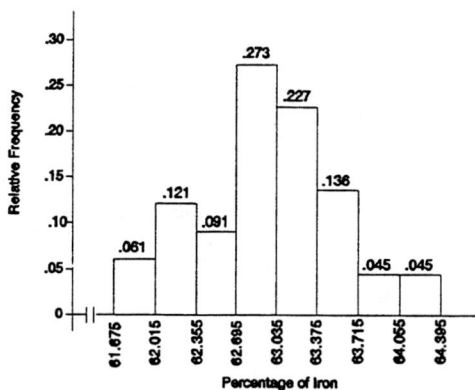

d. mean: $\bar{y} = \dfrac{\sum y}{n} = \dfrac{4155.59}{66} = 62.963$

variance: $s^2 = \dfrac{\sum y^2 - \dfrac{(\sum y)^2}{n-1}}{n-1} = \dfrac{261674.4983 - \dfrac{(4155.59)^2}{66}}{65} = \dfrac{24.0703}{65} = .3703$

standard deviation: $s = \sqrt{.3703} = .6085$

e. $\bar{y} \pm 2s \Rightarrow 62.963 \pm 2(.6085) \Rightarrow 62.963 \pm 1.217 \Rightarrow (61.746, 64.180)$

Yes. The percentage of observations that appear in this interval is $64/66 \times 100\% = 96.97\%$. This is close to the 95% given by the Empirical Rule.

2.27   a.   From the printout, $\bar{y} = 2.425$ and $s = 1.259$

      b. $\bar{y} \pm s \Rightarrow 2.425 \pm 1.259 \Rightarrow (1.166, 3.684)$

            $\bar{y} \pm 2s \Rightarrow 2.425 \pm 2(1.259) \Rightarrow 2.425 \pm 2.518 \Rightarrow (-0.093, 4.943)$

            $\bar{y} \pm 3s \Rightarrow 2.425 \pm 3(1.259) \Rightarrow 2.425 \pm 3.777 \Rightarrow (-1.352, 6.202)$

      c.   24 observations fall in the interval $\bar{y} \pm s$ or $24/40 = .60$. The Empirical Rule says there should be approximately .68 of the measurements within 1 standard deviation of the mean. This is fairly close.

          38 observations fall in the interval $\bar{y} \pm 2s$ or $38/40 = .95$. The Empirical Rule says there should be approximately .95 of the measurements within 2 standard deviations of the mean. This agrees with the Empirical Rule.

          40 observations fall in the interval $\bar{y} \pm 3s$ or $40/40 = 1.00$. The Empirical Rule says that approximately all of the measurements should fall within 3 standard deviations of the mean. Again, this agrees with the Empirical Rule.

2.29   New Location: $\bar{y} = 9.422$ and $s = .479$

        $\bar{y} \pm s \Rightarrow 9.422 \pm 0.479 \Rightarrow (8.943, 9.901)$. The Empirical Rule says that approximately 68% of the observations should fall in this interval.

        $\bar{y} \pm 2s \Rightarrow 9.422 \pm 2(0.479) \Rightarrow 9.422 \pm 0.958 \Rightarrow (8.464, 10.380)$. The Empirical Rule says that approximately 95% of the observations should fall in this interval.

        $\bar{y} \pm 3s \Rightarrow 9.422 \pm 3(0.479) \Rightarrow 9.422 \pm 1.437 \Rightarrow (7.985, 10.859)$. The Empirical Rule says that approximately all of the observations should fall in this interval.

    Old Location: $\bar{y} = 9.804$ and $s = .541$

        $\bar{y} \pm s \Rightarrow 9.804 \pm 0.541 \Rightarrow (9.263, 10.345)$. The Empirical Rule says that approximately 68% of the observations should fall in this interval.

        $\bar{y} \pm 2s \Rightarrow 9.804 \pm 2(0.541) \Rightarrow 9.804 \pm 1.082 \Rightarrow (8.722, 10.886)$. The Empirical Rule says that approximately 95% of the observations should fall in this interval.

        $\bar{y} \pm 3s \Rightarrow 9.804 \pm 3(0.541) \Rightarrow 9.804 \pm 1.623 \Rightarrow (8.181, 11.427)$. The Empirical Rule says that approximately all of the observations should fall in this interval.

2.31 From Exercise 2.23, $\bar{y} = .333$

$$\text{variance: } s^2 = \frac{\sum y^2 - \frac{(\sum y)^2}{n-1}}{n-1} = \frac{1.620455 - \frac{3.999^2}{12}}{12-1} = \frac{.28778825}{11} = .026162568$$

standard deviation: $s = \sqrt{.026162568} = .162$

$\bar{y} \pm s \Rightarrow 0.333 \pm 0.162 \Rightarrow (0.171, 0.495)$. The Empirical Rule says that approximately 68% of the observations should fall in this interval.

$\bar{y} \pm 2s \Rightarrow 0.333 \pm 2(0.162) \Rightarrow 0.333 \pm 0.324 \Rightarrow (0.009, 0.657)$. The Empirical Rule says that approximately 95% of the observations should fall in this interval.

$\bar{y} \pm 3s \Rightarrow 0.333 \pm 3(0.162) \Rightarrow 0.333 \pm 0.486 \Rightarrow (-0.153, 0.819)$. The Empirical Rule says that approximately all of the observations should fall in this interval.

2.33 The mean cyanide concentration is 84.0 and the median is 28.8. Since the mean is much greater than the median, the data are skewed to the right. Since the data are not mound-shaped, the Empirical Rule does not apply. The variance is 6,400, so the standard deviation is 80.

The upper quartile is 88.5. Thus, 75% of all the measurements are less than 88.5.

2.35 a. The 50th percentile is the same as the median. Thus, the 50th percentile is 7.15.

b. Tchebysheff's theorem says that at least 3/4 of the observations will fall in the interval $\bar{y} \pm 2s$. Thus, an upper bound on the 75th percentile is $\bar{y} + 2s$ or $24.36 + 2(98.38) = 24.36 + 196.76 = 221.12$.

2.37 a. For the old location, $\bar{y} = 9.804$ and $s = .541$. The z-score for 10.50 is

$$z = \frac{10.50 - 9.804}{.541} = 1.29$$

b. For the new location, $\bar{y} = 9.422$ and $s = .479$. The z-score for 10.50 is

$$z = \frac{10.50 - 9.422}{.479} = 2.25$$

c. The closer the z-score is to 0, the more likely it is to occur. Thus, a voltage reading of 10.50 is more likely to occur at the old location because the z-score is closer to 0.

2.39 a. From Exercise 2.27, median = 2.000, $Q_L = 1.250$, and $Q_U = 3.000$. The interquartile range is IQR = $Q_U - Q_L = 3.000 - 1.250 = 1.750$. The inner and outer fences are located a distance of 1.5 (IQR) = 1.5(1.75) = 2.625 and 3(IQR) = 3(1.75) = 5.25 below Q_L and above Q_U, respectively. Values between the inner and outer fences are suspect outliers and are designated by *. Highly suspect outliers lie outside the outer fences and are designated by o. The closest points to the inner fences which are still inside the inner fences are marked by x and whiskers are drawn between these points and the box. The box plot is shown below:

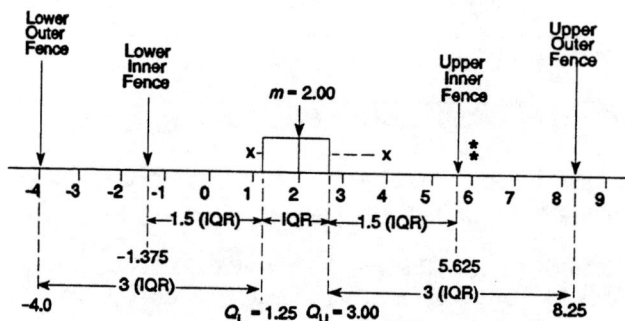

b.   From Exercise 2.27, $\bar{y} = 2.425$ and $s = 1.259$. The z-score corresponding to the suspect outliers is

$$z = \frac{6 - 2.425}{1.259} = 2.84$$

Since this value is fairly close to 3, the two observations with the value 6 are suspect outliers.

2.41   Two methods for detecting outliers are discussed in this chapter. The first uses z-scores and the second uses box plots.

In order to use the z-score method, we must first calculate the mean and standard deviation of the data.

mean: $\bar{y} = \dfrac{\sum y}{n} = \dfrac{120.6}{20} = 6.03$

variance: $s^2 = \dfrac{\sum y^2 - \dfrac{(\sum y)^2}{n}}{n - 1} = \dfrac{1983.06 - \dfrac{120.6^2}{20}}{20 - 1} = \dfrac{1255.842}{19} = 66.09694737$

standard deviation: $s = \sqrt{66.09694737} = 8.13$

The largest value in the data set is 36. The z-score corresponding to 36 is:

$$z = \frac{36.0 - 6.03}{8.13} = 3.69$$

Since the z-score is greater than 3, 36 is an outlier.

The next largest value in the data set is 20. The z-score corresponding to 20 is:

$$z = \frac{20.0 - 6.03}{8.13} = 1.72$$

Since the z-score is not greater than 3, 20 is not an outlier.

The second method for finding outliers is using a box plot. To construct a box plot, we first arrange the data in order from smallest to largest:

1.6, 1.8, 1.8, 2.0, 2.1, 2.5, 2.5, 3.0, 3.1, 3.1, 3.5, 4.1, 4.6, 4.7, 6.8, 6.9, 7.2, 20.0, 36.0

The median is the average of the middle two numbers and is $m = \dfrac{3.1 + 3.3}{2} = 3.2$

Next, we need to compute Q_L and Q_U. L = $1/4(n + 1) = 1/4(20 + 1) = 5.25 \approx 5$. Thus, Q_L is the 5th observation and is 2.1. U = $3/4(n + 1) = 3/4(20 + 1) = 15.75 \approx 16$. Thus, Q_U is the 16th observation and is 6.8. The interquartile range is IQR = $Q_U - Q_L = 6.8 - 2.1 = 4.7$. The inner and outer fences are located a distance of 1.5 (IQR) = 1.5(4.7) = 7.05 and 3(IQR) = 3(4.7) = 14.1 below Q_L and above Q_U, respectively. Values between the inner and outer fences are suspect outliers and are designated by *. Highly suspect outliers lie outside the outer fences and are designated by o. The closest points to the inner fences which are still inside the inner fences are marked by x and whiskers are drawn between these points and the box. The box plot is shown below:

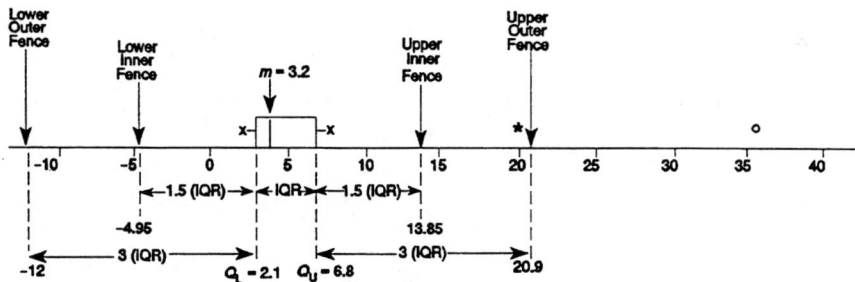

From the box plot, the observation 36 is an outlier because it lies outside the outer fence and the observation 20 is a suspect outlier because it lies between the inner and outer fences.

2.43   a.   In order to construct a box plot, we first need to calculate the median, Q_L, and Q_U. Using SAS, the summary statistics are:

Univariate Procedure

Variable=WEIGHT

Moments

N	144	Sum Wgts	144
Mean	1049.715	Sum	151159
Std Dev	376.5461	Variance	141787
Skewness	0.500552	Kurtosis	0.368447
USS	1.7895E8	CSS	20275537
CV	35.87126	Std Mean	31.37884
T:Mean=0	33.45296	Pr>¦T¦	0.0001
Num ^=0	144	Num > 0	144
M(Sign)	72	Pr>=¦M¦	0.0001
Sgn Rank	5220	Pr>=¦S¦	0.0001

Quantiles(Def=5)

100% Max	2302	99%	2061	Range	2129
75% Q3	1260.5	95%	1717	Q3-Q1	455.5
50% Med	1000	90%	1524	Mode	886
25% Q1	805	10%	549		
0% Min	173	5%	514		
		1%	353		

Extremes

Lowest	Obs	Highest	Obs
173(	143)	1763(	31)
353(	80)	1770(	24)
358(	140)	2006(	36)
393(	47)	2061(	93)
441(	45)	2302(	33)

Thus, median = 1000, Q_L = 805, and Q_U = 1260.5. The interquartile range is IQR = $Q_U - Q_L$ = 1260.5 − 805 = 455.5. The inner and outer fences are located a distance of 1.5(IQR) = 1.5(455.5) = 683.25 and 3(IQR) = 3(455.5) = 1366.5 below Q_L and above Q_U, respectively. Values between the inner and outer fences are suspect outliers and are designated by *. Highly suspect outliers lie outside the outer fences and are designated by o. The closest points to the inner fences which are still inside the inner fences are marked by x and whiskers are drawn between these points and the box. The box plot is shown below:

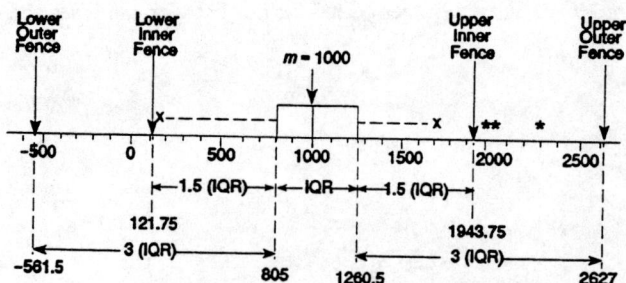

There are 3 suspect outliers, data points lying between the inner and outer fences. They have values 2006, 2061, and 2302. There are no points outside the outer fences, so there are no highly suspect outliers.

b. From the printout, $\bar{y} = 1049.715$ and $s = 376.5461$

For data point 2302, $z = \dfrac{2302 - 1049.715}{376.5461} = 3.33$

For data point 2061, $z = \dfrac{2061 - 1049.715}{376.5461} = 2.69$

Using the method of z-scores, only one data point has a z-score of more than 3. Thus, there is only one outlier.

2.45 a. To construct a stem and leaf display for the PCB levels of rural soil samples, we will use the digits to the left of the decimal point as the stems and the digits to the right of the decimal point as the leaves.

Stems	Leaves	Frequency	Relative Frequency
1	0 5 6 8	4	.286
2		0	.000
3	5	1	:071
4		0	.000
5	3	1	.071
6			
7			
8	1 2	2	.143
9	0 7 8	3	.214
10			
11			
12	0	1	.071
13			
14			
15	0	1	.071
⋮			
23	0	1	.071
	Totals $n = 14$		.998

b. For the urban samples, the stem-and-leaf display is:

Stems	Leaves	Frequency	Relative Frequency
11	0 0	2	.133
12	0	1	.067
13	0	1	.067
14			
15			
16	0	1	.067
17		0	.000
18	0 0	2	.133
19			
20			
21	0	1	.067
22	0	1	.067
23		0	.000
24	0	1	.067
⋮			
29	0	1	.067
⋮			
49	0	1	.067
⋮			
94	0	1	.067
⋮			
107	0	1	.067
⋮			
141	0	1	.067
	Totals: $n = 15$		1.003

c.

Stems	Leaves	Frequency	Relative Frequency
1	0 5 6 8	4	.138
2		0	.000
3	5	1	.034
4		0	.000
5	3	1	.034
6			
7			
8	1 2	2	.069
9	0 7 8	3	.103
10			
11	⓪⓪	2	.069
12	⓪ 0	1	.069
13	⓪	1	.034
14			.000
15	0		.034
16	⓪	1	.034
17		0	.000
18	⓪⓪	2	.069
19			
20			
21	⓪	1	.034
22	⓪	1	.034
23	0	0	.034
24	⓪	1	.034
⋮			
29	⓪	1	
⋮			
49	⓪	1	.034
⋮			
94	⓪	1	.034
⋮			
107	⓪	1	.034
⋮			
141	⓪	1	.034

Totals: $n = 29$ .993

The graph supports the researchers' claim that a significant difference exists between the PCB levels for urban and rural areas. The rural PCB levels are almost all less than the lowest reading for the urban areas.

2.47 a.   From the stem-and-leaf display, most of the reactor units had between 0 and 7 scrams in a recent year. The box plot indicates that there are two suspect outliers or toe reactor units which had unusually large number of scrams. The two units had 12 and 13 scrams. The mean number of scrams per year per reactor unit is 4.036 and the median is 3.00. This indicates that the data are skewed to the right. The standard deviation is 3.027. Since the data are skewed to the right, the Empirical Rule is not appropriate here.

b.   A score of 11 has a z-score of $z = \dfrac{11 - 4.036}{3.027} = 2.30$. Most of the observations will fall within 2 standard deviations of the mean. Since 11 has a z-score of 2.30, it is not a very likely value to observe.

2.49 a.   Large Firms

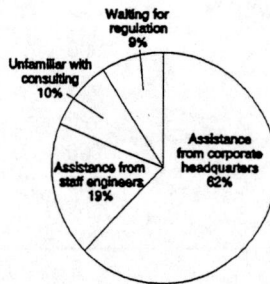

b.   Small Firms

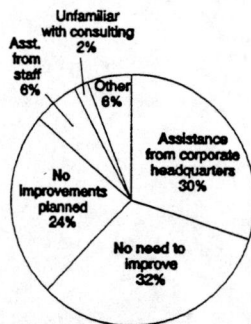

c.   Yes. The pie chart for large firms shows that over half (62%) say they do not need consulting engineering services because they receive assistance from corporate headquarters. However, only 30% of the small firms gave the same answer. Of the small firms, 56% mentioned they did not need consulting assistance because they either had no waste or were not planning improvements. None of the large firms gave these reasons.

2.51   Using the leftmost digit as the stem and the first digit to the right of the decimal point as the leaf, the stem-and-leaf display is:

Stem	Leaf
2	7899
3	158
4	688
5	113
6	188
7	46
8	7

From the stem-and-leaf display, the data are skewed to the right. There are more smaller values of injury rates than larger values. However, the data appear to be fairly uniform from 3 to 6.

2.53   Because the data are qualitative, we can use a bar chart or a pie chart to describe the data. A bar chart will be constructed with the relative frequency on the vertical axis and the databases on the horizontal axis.

Of the 10 most accessed databases, the relative frequency of access of Medline was a little more than one-third (.343). The relative frequency of access for Agricola, Biosis, and CAB were .149, .123, and .106, respectively. Of the remaining top 10 accessed databases, none had relative frequencies of access greater than .10.

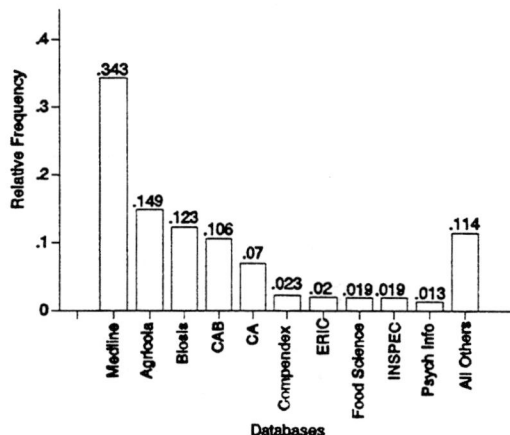

2.55   a.   The source contribution the most to the total IPM is Secondary sulfate, which contributes 58%.

         b.   The percentage of the total IPM due to industry, oil burning, and motor vehicles is 6% + 6% + 10% = 22%.

2.57   Using MINITAB, the numerical description of the data is:

N	MEAN	MEDIAN	TRMEAN	STDEV	SEMEAN
52	0.812	0.275	0.547	1.505	0.209

MIN	MAX	Q1	Q3
0.036	8.788	0.136	0.595

The mean of the data is .812 while the median is .275. This indicates that the data are skewed to the right. The lower quartile is .136 while the upper quartile is .595. The standard deviation is 1.505. Since the data are not mound shaped, it is not appropriate to use the Empirical Rule.

Using MINITAB, the stem-and-leaf display is:

```
Stem-and-leaf of TIME          N = 52
Leaf unit = 0.10

   (42)   0  000000000111111111122222222233333344555669
    10    1  00268
     5    2
     5    3  089
     2    4  1
     1    5
     1    6
     1    7
     1    8  7
```

Again, from the stem-and-leaf display, it is very apparent that the data are skewed to the right. Forty-seven of the 52 observations are between 0 and 2, while there are only five observations greater than 2.

CHAPTER THREE

$\cdots\cdots\cdots\cdots\cdots\cdots\cdots\cdots\cdots\cdots\cdots\cdots\cdots\cdots\cdots\cdots\cdots\cdots$

Probability

3.1   a.   There are five simple events, each corresponding to a classification:

BB, TG, GG, S, and *G*

b.   Reasonable probabilities would be probabilities that correspond to the proportions. Thus, the probabilities are:

Simple Event	Probability
BB	.28
TG	.11
GG	.11
S	.26
G	.24

c. $P(BB \text{ or } G) = .28 + .24 = .52$

d.   Those that support environmentalism in some fashion are the True-blue greens, the Greenback greens, and the Sprouts. The probability is:

$P(TG, GG, \text{ or } S) = .11 + .11 + .26 = .48$

3.3   a.   Define event *A* as {Extremity getting caught in spoke}.

$$P(A) = \frac{21}{100} = .21$$

b.   Define event *B* as {Child falling out of seat or seat falling off bike}.

$$P(B) = \frac{39}{100} + \frac{6}{100} = \frac{45}{100} = .45$$

c.   Define event *C* as {Injury did not occur as a result of an accident with a car}. Event *C* contains all simple events except "accident with car."

$$P(C) = \frac{39}{100} + \frac{24}{100} + \frac{6}{100} + \frac{21}{100} = \frac{90}{100} = .90$$

3.5   a.   The experiment consists of selecting a rule and recording its type. The simple events are the list of all possible types:

Batch scheduling, JES queue space, C-to-C links,
Hardware errors, SMF management, Quiesce and IPL,
Performance, and Background monitor.

b.   Since there are 548 different rules and differing number of rules per simple event, we will assign probabilities to each simple event corresponding to the number of rules per simple event divided by the total number of rules, 548. Thus,

$$P(\text{Batch schedule}) = 139/548,$$
$$P(\text{JES}) = 104/548,$$
$$P(\text{C-to-C links}) = 68/548,$$
$$P(\text{Hardware errors}) = 87/548,$$
$$P(\text{SMF management}) = 25/548,$$
$$P(\text{Quiesce and IPL}) = 52/548,$$
$$P(\text{Performance}) = 41/548, \text{ and}$$
$$P(\text{Background monitor}) = 32/548.$$

c.   Define event A as {C-to-C link or Hardware error}

$$P(A) = 68/548 + 87/548 = 155/548$$

d.   Define event B as {Not performance}. Event B contains all the simple events except performance. Thus,

$$P(B) = 139/548 + 104/548 + 68/548 + 87/548 + 25/548 + 52/548 + 32/548$$
$$= 507/548$$

3.7   a.   Let A be the event that a dragonfly species inhabits a dragonfly hotspot. Then $P(A) = .92$.

b.   Let B be the event that a dragonfly species inhabits a dragonfly hotspot. Then $P(B) = .92$.

c.   Since 1.00 or 100% of all butterfly species inhabit bird hotspots, then the butterfly hotspots have to be part of the bird hotspots. Thus, all butterfly hotspots are contained in bird hotspots.

3.9   A simple event for this experiment consists of the outcomes (dry well or oil gusher) for the 6 oil wells. Using the multiplicative rule, there are a total of $2 \times 2 \times 2 \times 2 \times 2 \times 2 = 64$ simple events. If we let D represent a dry well and G represent an oil gusher, a simple event will be of the form $DDGDGG$, where each position corresponds to one of the six wells. If each simple event is equally likely, then each simple event has a probability of 1/64. Define event A as the event that at least 1 oil gusher will be discovered. Of the 64 simple events, only 1 ($DDDDDD$) is not in A. Thus,

$$P(A) = 1 - P(A^c) = 1 - 1/64 = 63/64 = .984$$

3.11   a.   Using the multiplicative rule, the total number of ways to throw two dice is $6 \times 6 = 36$. Of the 36 outcomes, there are 6 {(1, 6), (2, 5), (3, 4), (4, 3), (5, 2), (6, 1)} ways to get a 7 and 2 {(5, 6), (6, 5)} ways to get an 11. The total number of ways to get a natural is $6 + 2 = 8$. The probability of getting a natural is $8/36 = 2/9$.

       b.   There is 1 way to get a 2 {(1, 1)}, 2 ways to get a 3 {(1, 2), (2, 1)}, and 1 way to get a 12 {(6, 6)}. Thus, the total number of ways to get craps is $1 + 2 + 1 = 4$. The probability of getting craps is 4/36. If A is the event {Throw craps}, then A^c is the event {Does not throw craps}.

$$P(A^c) = 1 - P(A) = 1 - 4/36 = 32/36 = 8/9$$

3.13   a.   Define the following events:

           A: {Cloudy conditions}
           B: {Unmanned system detects an intruder}

From the table,

$$P(A) = \frac{234}{692}, \; P(B) = \frac{21 + 228 + 226 + 7 + 185}{692} = \frac{667}{692}, \text{ and}$$

$$P(A \cap B) = \frac{228}{692}$$

$$P(B \mid A) = \frac{P(A \cap B)}{P(A)} = \frac{228/692}{234/692} = .974$$

       b.   Define the following events:

           C: {Snowy conditions}

From the table,

$$P(C) = \frac{10}{692}$$

$$P(B^c) = 1 - P(B) = 1 - \frac{667}{692} = \frac{25}{692}, \text{ and } P(B^c \cap C) = \frac{3}{692}$$

$$P(C \mid B^c) = \frac{P(B^c \cap C)}{P(B^c)} = \frac{3/692}{25/692} = .12$$

3.15   a.   Refer to the solution given in Exercise 3.11. Of the 36 total outcomes, half or 18 will have a sum that is odd and 18 will have a sum that is even. Define event A as {Throwing craps} and event B as {Sum is odd}. The probability of B is $P(B) = 18/36 = 1/2$. The event $A \cap B$ is the event the sum is odd and craps is thrown. There are 2 simple events in $A \cap B$, {(2, 1) and (1, 2)}. $P(A \cap B) = 2/36$. The probability of throwing craps, given the sum is odd is

$$P(A \mid B) = \frac{P(A \cap B)}{P(B)} = \frac{2/36}{18/36} = 1/9 = .111$$

. . . .

b. Let event C be {Player throws a double}. Event C consists of 6 simple events {(1, 1), (2, 2), (3, 3), (4, 4), (5, 5), (6, 6)}. $P(C) = 6/36$. The event $A^c \cap C$ is the event a player does not throw craps and he throws a double. There are 4 simple events in $A^c \cap C$, {(2, 2), (3, 3), (4, 4), (5, 5)}. $P(A^c \cap C) = 4/36$. The probability a player throws a double given he does not throw craps is

$$P(C \mid A^c) = \frac{P(A^c \cap C)}{P(A^c)} = \frac{4/36}{32/36} = \frac{1}{8} = .125$$

3.17 Define the following events:

  A: {System has high selectivity}
  B: {System has high fidelity}

From the problem, $P(A) = .72$, $P(B) = .59$, and $P(A \cap B) = .33$.

Thus, $P(A \mid B) = \dfrac{P(A \cap B)}{P(B)} = \dfrac{.33}{.59} = .559$

3.19 Define the following events:

  A: {Less than 1 latent cancer fatality/year}
  B: {Core melt occurs during a year}

From the problem, $P(A \mid B) = .00005$ and $P(B) = 1/100,000 = .00001$. The probability that at least 1 latent cancer fatality will occur as a result of a core melt is $P(A^c \cap B) = P(A^c \mid B)P(B)$. Since $P(A \mid B) = .00005$,

$$P(A^c \mid B) = 1 - P(A \mid B) = 1 - .00005 = .99995$$

Thus, $P(A^c \cap B) = .99995(.00001) = .0000099995 \approx .00001$

3.21 a. Define the following events:

  A_i: {Shuttle mission i does not result in a critical-item failure}

$P(A_i) = 1 - 1/63 = 62/63$

Then the event that none of the 8 shuttle flights results in a critical-item failure can be written as $A_1 \cap A_2 \cap A_3 \cap A_4 \cap A_5 \cap A_6 \cap A_7 \cap A_8$. The event that at least one of the 8 shuttle flights results in a critical-item failure is $(A_1 \cap A_2 \cap A_3 \cap A_4 \cap A_5 \cap A_6 \cap A_7 \cap A_8)^c$. Thus,

$$\begin{aligned}
&P(A_1 \cap A_2 \cap A_3 \cap A_4 \cap A_5 \cap A_6 \cap A_7 \cap A_8)^c \\
&= 1 - P(A_1 \cap A_2 \cap A_3 \cap A_4 \cap A_5 \cap A_6 \cap A_7 \cap A_8) \\
&= 1 - P(A_1)P(A_2)P(A_3)P(A_4)P(A_5)P(A_6)P(A_7)P(A_8) \\
&= 1 - (62/63)(62/63)(62/63)(62/63)(62/63)(62/63)(62/63)(62/63) \\
&= 1 - .880 = .120 \text{ (assuming the missions are independent of each other)}
\end{aligned}$$

b. Using the same logic as above, the probability that at least one of the 40 shuttle missions will result in a critical-item failure is:

$$P(A_1 \cap A_2 \cap A_3 \cap \cdots \cap A_{40})^c$$
$$= 1 - P(A_1 \cap A_2 \cap A_3 \cap \cdots \cap A_{40})$$
$$= 1 - P(A_1)P(A_2)P(A_3) \cdots P(A_{40})$$
$$= 1 - (62/63)^{40}$$
$$= 1 - .527 = .473$$

3.23 a. The probability that a single oil well prospect will result in no more than 100,000 barrels of oil is

$$P(0) + P(50,000) + P(100,000) = .60 + .10 + .15 = .85$$

b. The probability that a single oil well prospect will strike oil is the complement of the event that a single oil well prospect will not strike oil and is

$$1 - P(0) = 1 - .60 = .40$$

c. If two wells are drilled, the total number of outcomes is $5 \times 5 = 25$ using the multiplicative rule. The simple events are (in thousands):

(0, 0)   (0, 50)   (0, 100)   (0, 500)   (0, 1,000)
(50, 0)   (50, 50)   (50, 100)   (50, 500)   (50, 1,000)
(100, 0)   (100, 50)   (100, 100)   (100, 500)   (100, 1,000)
(500, 0)   (500, 50)   (500, 100)   (500, 500)   (500, 1,000)
(1,000, 0) (1,000, 50) (1,000, 100) (1,000, 500) (1,000, 1,000)

d. We are given in the problem that $P(0) = .6$, $P(50) = .1$, $P(100) = .15$, $P(500) = .1$, and $P(1,000) = .05$. The event (0, 0) is really the event $0 \cap 0$.

Since the outcomes of the two wells are independent,

$$P(0, 0) = P(0 \cap 0) = P(0)P(0) = .6(.6) = .36$$

Similarly, $P(0, 50) = P(0)P(50) = .6(.1) = .06$

The probabilities to all the simple events are listed below:

$$P(0, 0) = .6(.6) = .36 \qquad P(0, 50) = .6(.1) = .06$$
$$P(0, 100) = .6(.15) = .09 \qquad P(0, 500) = .6(.1) = .06$$
$$P(0, 1{,}000) = .6(.05) = .03 \qquad P(50, 0) = .1(.6) = .06$$
$$P(50, 50) = .1(.1) = .01 \qquad P(50, 100) = .1(.15) = .015$$
$$P(50, 500) = .1(.1) = .01 \qquad P(50, 1{,}000) = .1(.05) = .005$$
$$P(100, 0) = .15(.6) = .09 \qquad P(1{,}000, 50) = .15(.1) = .015$$
$$P(100, 100) = .15(.15) = .0225 \qquad P(1{,}000, 500) = .15(.1) = .015$$
$$P(100, 1{,}000) = .15(.05) = .0075 \qquad P(500, 0) = .1(.6) = .06$$
$$P(500, 50) = .1(.1) = .01 \qquad P(500, 100) = .1(.15) = .015$$
$$P(500, 500) = .1(.1) = .01 \qquad P(500, 1{,}000) = .1(.05) = .005$$
$$P(1{,}000, 0) = .05(.6) = .03 \qquad P(1{,}000, 50) = .05(.1) = .005$$
$$P(1{,}000, 100) = .05(.15) = .0075 \qquad P(1{,}000, 500) = .05(.1) = .005$$
$$P(1{,}000, 1{,}000) = .05(.05) = .0025$$

e. Define the event:

   A: {At least 1 of the 2 oil prospects strikes oil}

   Then A^c is the event that neither of the two oil prospects strikes oil.

   $$P(A^c) = P(0, 0) = .36$$

   Thus, $P(A) = 1 - P(A^c) = 1 - .36 = .64$

3.25 a. Define the following events:

   A_1: {Particle 1 is reflected}
   A_2: {Particle 2 is reflected}

   From the problem, $P(A_1) = .16$ and $P(A_2) = .16$. The event both particles are reflected is $A_1 \cap A_2$. If we assume the events A_1 and A_2 are independent,

   $$P(A_1 \cap A_2) = P(A_1)P(A_2) = .16(.16) = .0256$$

b. Let A_3, A_4, and A_5 be defined similarly to A_1 and A_2 above for particles 3, 4, and 5. The event all 5 particles will be absorbed is $A_1^c \cap A_2^c \cap A_3^c \cap A_4^c \cap A_5^c$. Again, assuming the events are independent,

   $$P(A_1^c \cap A_2^c \cap A_3^c \cap A_4^c \cap A_5^c)$$
   $$= P(A_1^c)P(A_2^c)P(A_3^c)P(A_4^c)P(A_5^c)$$
   $$= [1 - P(A_1)][1 - P(A_2)][1 - P(A_3)][1 - P(A_4)][1 - P(A_5)]$$
   $$= (1 - .16)(1 - .16)(1 - .16)(1 - .16)(1 - .16) = .84^5 = .418$$

c. We must assume the simple events are independent.

3.27 a.   Define the following events:

> A: {Failure in earthworks}
> B: {Failure in earth-retaining structures and excavations}.

From the problem, $P(A) = .01$ and $P(B) = .001$. The event failure in either the earthworks or earth-retaining structures and excavations is $A \cup B$.

$$P(A \cup B) = P(A) + P(B) - P(A \cap B)$$

If A and B are independent events,

$$P(A \cup B) = P(A)P(B) = .01(.001) = .00001$$

Then, $P(A \cup B) = .01 + .001 - .00001 = .01099$

b.   Define event C as {Failure in foundations}. The event failure in all 3 design areas is $A \cap B \cap C$.

If the events are independent,

$$P(A \cap B \cap C) = P(A)P(B)P(C) = .01(.001)(.0001) = .000000001$$

3.29 a.   Define the following event:

> A: {Particular 1-mile stretch of interstate highway eligible to be posted at 65 mph is actually posted at 65 mph}

$P(A) = .97$

b.   Define the following event:

> B: {Particular 1-mile stretch of interstate highway eligible to be posted at 65 mph is rural}

$P(B^c) = 1 - P(B) = 1 - .96 = .04$

c.   No. It is possible that a particular 1-mile stretch of highway eligible to be posted at 65 mph is actually posted at 65 mph and is not rural. We know from part b that .04 or 4% of the miles are not rural. From part a, we know that .97 or 97% of the miles are posted at 65 mph and 3% of the miles are not posted at 65 mph. Thus, at least 1% and not more than 4% of the miles are both non-rural and posted at 65 mph, or $.01 \le P(A \cap B^c) \le .04$. Events A and B^c are not mutually exclusive.

3.31 a.   From the problem, $P_0 = P(D \cap A^c)$ and $P_1 = P(D) = P(D \cap A) + P(D \cap A^c)$.

$$\text{Thus, } P(A \mid D) = \frac{P(D \cap A)}{P(D)} = \frac{P(D) - P(D \cap A^c)}{P_1} = \frac{P_1 - P_0}{P_1}$$

b. Since A and B cannot occur at the same time,

$$P(D \cap A) = P(D \cap A \cap B^c) = P_1,$$
$$P(D \cap B) = P(D \cap A^c \cap B) = P_2, \text{ and}$$
$$P(D) = P(D \cap A \cap B^c) + P(D \cap A^c \cap B) = P_1 + P_2$$

Thus, $P(A \mid D) = \dfrac{P(D \cap A)}{P(D)} = \dfrac{P_1}{P_1 + P_2}$ and $P(B \mid D) = \dfrac{P(D \cap B)}{P(D)} = \dfrac{P_2}{P_1 + P_2}$

c. From the problem, $P_1 = P(D \cap A) = P(D \cap A \cap B^c) + P(D \cap A \cap B)$ and
$$P_2 = P(D \cap B) = P(D \cap A^c \cap B) + P(D \cap A \cap B)$$

Since A and B are independent, $P_1 P_2 = P(D \cap A)P(D \cap B) = P(D \cap A \cap B)$

$$
\begin{aligned}
P(D) &= P(D \cap A \cap B^c) + P(D \cap A^c \cap B) + P(D \cap A \cap B) \\
&= P(D \cap A \cap B^c) + P(D \cap A \cap B) + P(D \cap A^c \cap B) + P(D \cap A \cap B) \\
&\quad - P(D \cap A \cap B) \\
&= P_1 + P_2 - P_1 P_2
\end{aligned}
$$

Thus, $P(A \mid D) = \dfrac{P(D \cap A)}{P(D)} = \dfrac{P_1}{P_1 + P_2 - P_1 P_2}$

and $P(B \mid D) = \dfrac{P(D \cap B)}{P(D)} = \dfrac{P_2}{P_1 + P_2 - P_1 P_2}$

3.33 From the problem, $P(CC \mid 275\text{–}300) = .775$, $P(CC \mid 305\text{–}325) = .77$, and $P(CC \mid 330\text{–}350) = .86$. Also $P(275\text{–}300) = .52$, $P(305\text{–}325) = .39$, and $P(330\text{–}350) = .09$

We want to find $P(275\text{–}300 \mid CC)$. First, we must find $P(CC \cap 275\text{–}300)$

$$P(CC \cap 275\text{–}300) = P(CC \mid 275\text{–}300)P(275\text{–}300) = .775(.52) = .403$$

Next, we must find $P(CC) = P(CC \cap 275\text{–}300) + P(CC \cap 305\text{–}325) + P(CC \cap 330\text{–}350)$

$$P(CC \cap 305\text{–}325) = P(CC \mid 305\text{–}325)P(305\text{–}325) = .77(.39) = .3003$$
$$P(CC \cap 330\text{–}350) = P(CC \mid 330\text{–}350)P(330\text{–}350) = .86(.09) = .0774$$

Thus, $P(CC) = P(CC \cap 275\text{–}300) + P(CC \cap 305\text{–}325) + P(CC \cap 330\text{–}350)$
$$= .403 + .3003 + .0774 = .7807$$

Finally, $P(275\text{–}300 \mid CC) = \dfrac{P(CC \cap 275\text{–}300)}{P(CC)} = \dfrac{.403}{.7807} = .516$

3.35 Define the following events:

A: {Computer system shuts down}   B: {Hardware failure}
C: {Software failure}   D: {Power failure}

From the problem, we know $P(A \mid B) = .73$, $P(A \mid C) = .12$, $P(A \mid D) = .88$, $P(B) = .01$, $P(C) = .05$, and $P(D) = .02$. The probability the current shutdown is due to hardware failure is $P(B \mid A) = P(B \cap A)/P(A)$.

Using Bayes' Rule,
$$\begin{aligned}
P(A) &= P(A \cap B) + P(A \cap C) + P(A \cap D) \\
&= P(A \mid B)P(B) + P(A \mid C)P(C) + P(A \mid D)P(D) \\
&= .73(.01) + .12(.05) + .88(.02) \\
&= .0309
\end{aligned}$$

and $P(B \cap A) = P(A \mid B)P(B) = .73(.01) = .0073$

Then, $P(B \mid A) = P(B \cap A)/P(A) = .0073/.0309 = .236$

The probability the current shutdown is due to software failure is

$$P(C \mid A) = P(C \cap A)/P(A)$$

Using Bayes' Rule, $P(C \cap A) = P(A \mid C)P(C) = .12(.05) = .0060$

Then, $P(C \mid A) = P(C \cap A)/P(A) = .0060/.0309 = .194$

The probability the current shutdown is due to power failure is $P(D \mid A) = P(D \cap A)/P(A)$

Using Bayes' Rule, $P(D \cap A) = P(D \mid C)P(C) = .88(.02) = .0176$

Then, $P(D \mid A) = P(D \cap A)/P(A) = .0176/.0309 = .570$

3.37   a.   There are a total of $3 \times 3 = 9$ different scenarios possible.

       b.   Let O correspond to "optimist," M correspond to "moderate," and P correspond to "pessimist." The scenarios are represented by pairs of values, where the first value corresponds to the perspective on abatement and the second value corresponds to the perspective on climate change damage. The scenarios are:

           OO, OM, OP, MO, MM, MP, PO, PM, PP

       c.   We will be picking 5 scenarios from 9 possible scenarios. The number of combinations of 5 scenarios from 9 scenarios is

$$\binom{9}{5} = \frac{9!}{5!(9 - 5)!} = \frac{9 \cdot 8 \cdot 7 \cdot 6}{4 \cdot 3 \cdot 2 \cdot 1} = 126$$

3.39   a.   There are a total of $4 \times 4 = 16$ metal-support combinations possible.

       b.   For each of the metals, there are $P_4^4 = \dfrac{4!}{(4 - 4)!} = 4 \cdot 3 \cdot 2 \cdot 1 = 24$ different orderings of the four metals.

3.41   a.   To find the number of different responses to the questionnaires, the engineers must evaluate 2 building parameters for each of 3 parts. Using the multiplicative rule, there are $2 \times 3 = 6$ parameter-part combinations to evaluate. Each of the 6 parameter-part combinations is to be rated with one of 3 responses. Again, using the multiplicative rule, there are a total of

$$3 \times 3 \times 3 \times 3 \times 3 \times 3 = 3^6 = 729$$

different possible responses to the questionnaire.

      b.   The engineers want to select the 3 parameter-part combinations with the highest ratings from the 6 possible combinations and rank them in order. Using the permutation rule, there are a total of

$$P_3^6 = \frac{6!}{(6-3)!} = \frac{6 \times 5 \times 4 \times 3 \times 2 \times 1}{3 \times 2 \times 1} = 120$$

3.43   a.   The total number of machining conditions possible is $2 \times 2 \times 6 \times 7 = 168$.

      b.   If the 8 conditions actually employed were randomly selected, the total number of ways of selecting 8 conditions from the 168 is a combination of 168 things taken 8 at a time or

$$\binom{168}{8} = \frac{168!}{8!(168-8)!} = \frac{168!}{8!160!}$$

If there is one and only one machining condition that will detect a flaw in the system, that one condition must be one of the eight chosen. The other seven conditions can be randomly chosen from the remaining 167 conditions. The total number of ways of selecting 7 conditions from the 167 is a combination of 167 things taken 7 at a time or

$$\binom{167}{7} = \frac{167!}{7!(167-7)!} = \frac{167!}{7!160!}$$

Thus, the probability that the experiment conducted will detect the system flaw is

$$\frac{\frac{167!}{7!160}}{\frac{168!}{8!168!}} = \frac{167!8!}{168!7!} = \frac{8}{168}$$

      c.   The system flaw occurs when drilling steel material with a .25 inch drill size at a speed of 2,500 rpm. There are actually 7 conditions out of the 168 that meet the above criteria (the above criteria with each of the 7 feed rates). Of these 7 conditions, 2 of them are in the experiment. Thus, the probability that the experiment will detect the system flaw is 2/7.

3.45   a.   Since order does not matter, the number of ways to choose 6 numbers out of 49 is a combination of 49 numbers taken 6 at a time. The number of ways is:

$$\binom{49}{6} = \frac{49!}{6!(49-6)!} = \frac{49!}{6!43!} = 13,983,816$$

The probability of matching all 6 numbers is $\dfrac{1}{13,983,816} = .0000000715$

   b.   The number of tickets you would need to buy would be a combination of 7 numbers taken 6 at a time or

$$\binom{7}{6} = \frac{7!}{6!(7-6)!} = \frac{7!}{6!1!} = 7$$

The 7 sets of numbers would be:

2, 7, 18, 23, 30, 32   2, 7, 18, 23, 30, 39
2, 7, 18, 23, 32, 39   2, 7, 18, 30, 32, 39
2, 7, 23, 30, 32, 39   2, 18, 23, 30, 32, 39
7, 18, 23, 30, 32, 39

   c.   The probability of winning is $\dfrac{7}{13,983,816} = .0000005$

This increases the odds of winning. However, one has to buy 7 tickets to get the higher odds.

   d.   Even though this set of 6-numbers contains a neighboring pair, the probability of winning is only $\dfrac{1}{13,983,816} = .0000000715$, the same as that in part a.

3.47   a.   To find the probability of the dealer drawing blackjack, we must find the total number of ways to get blackjack and the total number of ways to draw two cards. The total number of ways to draw 2 cards from a deck of 52 cards is found using the combination rule and is

$$\binom{52}{2} = \frac{52!}{2!(52-2)!} = \frac{52!}{2!50!} = \frac{52 \times 51}{2 \times 1 \times 50 \times 49} = 1326$$

To find the total number of ways to get blackjack, we must multiply the number of ways to get 1 ace and the number of ways to get 1 card with a value of 10. The number of ways to get 1 ace is found using the combination rule and is

$$\binom{4}{1} = \frac{4!}{1!(4-1)!} = \frac{4!}{1!3!} = \frac{4 \times 3 \times 2 \times 1}{1 \times 3 \times 2 \times 1} = 4$$

The number of ways to get 1 card with a value of 10 (there are 16 cards with a value of 10) is found using the combination rule and is

$$\begin{pmatrix} 16 \\ 1 \end{pmatrix} = \frac{16!}{1!(16-1)!} = \frac{16!}{1!15!} = \frac{16 \times 15 \times \cdots \times 1}{1 \times 15 \times 14 \times \cdots \times 1} = 16$$

The total number of ways to draw blackjack is $4 \times 16 = 64$

The probability of drawing blackjack is $64/1326$

b. Define the following events:

   B: {Player draws Blackjack}
   D: {Dealer draws Blackjack}

The event we want to find is $P(B \cap D^c)$.

We know that $P(B \cap D^c) = P(B)P(D^c \mid B)$.

From part a, $P(B) = \dfrac{64}{1326}$

Now we will find $P(D \mid B)$. If the player has already been dealt blackjack, then there are only 50 cards left from which the dealer draws 2. Thus, the total number of ways of selecting 2 cards from 50 is

$$\begin{pmatrix} 50 \\ 2 \end{pmatrix} = \frac{50!}{2!(50-2)!} = \frac{50!}{2!48!} = \frac{50 \cdot 49}{2} = 1225$$

To find the total number of ways the dealer can get blackjack given the player has already been dealt blackjack, we must multiply the number of ways to get 1 ace given the player has 1 ace and the number of ways to get 1 card with a value of 10 given the player has 1 card with a value of 10. The number of ways to get 1 ace given the player has 1 ace is found using the combination rule and is

$$\begin{pmatrix} 3 \\ 1 \end{pmatrix} = \frac{3!}{1!(3-1)!} = \frac{3!}{1!2!} = 3$$

The number of ways to get 1 card with a value of 10 given the player has 1 card with a value of 10 is found using the combination rule and is

$$\begin{pmatrix} 15 \\ 1 \end{pmatrix} = \frac{15!}{1!(15-1)!} = \frac{15!}{1!14!} = 15$$

The total number of ways to draw blackjack given the player has drawn blackjack is $3 \times 15 = 45$.

$$P(D \mid B) = \frac{45}{1225} \text{ and } P(D^c \mid B) = 1 - P(D \mid B) = 1 - \frac{45}{1225} = \frac{1180}{1225}$$

Finally, $P(B \cap D^c) = P(B)P(D^c \mid B) = \dfrac{64}{1326} \cdot \dfrac{1180}{1225} = \dfrac{75,520}{1,624,350} = .0465$

3.49 Define the following events:

A_i: {Diskette i is not defective} $P(A_i) = .99$

If we assume that the claim is true and all the diskettes are independent, then the probability that none of the diskettes are defective is

$$P(A_1 \cap A_2 \cap A_3 \cap A_4) = P(A_1)P(A_2)P(A_3)P(A_4) = .99^4 = .961$$

Thus, the probability that at least 1 diskette is defective is the probability of the complement of the above event or $1 - .961 = .039$. If the claim is true, it would be very unlikely that at least one of the next 4 diskettes manufactured will be defective (probability .039). We would infer that the probability that a diskette is defective is greater than 1 in 100.

3.51 a. Define the following events:

A_1: {1st antiaircraft shell strikes within 30 feet of target}
A_2: {2nd antiaircraft shell strikes within 30 feet of target}
A_3: {3rd antiaircraft shell strikes within 30 feet of target}

From the text, $P(A_i) = p = .45$ for $i = 1, 2, 3$

$$\begin{aligned}
P(\text{All 3 shells miss their targets}) &= P(A_1^c \cap A_2^c \cap A_3^c) = P(A_1^c)P(A_2^c)P(A_3^c) \\
&= [1 - P(A_1)][1 - P(A_2)][1 - P(A_3)] \\
&= (1 - .45)(1 - .45)(1 - .45) = .166
\end{aligned}$$

Assume the simple events are independent. Since this is not a very small probability, there is no evidence to conclude that in battle conditions, p differs from .45.

b. Define events $A_4, A_5, \ldots, A_{10}$ as in part a for shells 4 through 10.

$$\begin{aligned}
P(\text{all 10 miss their targets}) &= P(A_1^c \cap A_2^c \cdots \cap A_{10}^c) \\
&= P(A_1^c)P(A_2^c) \cdots P(A_{10}^c) \\
&= (1 - .45)^{10} = .0025
\end{aligned}$$

Since this is a very small probability, there is evidence to conclude that in battle conditions, p differs from .45.

3.53 a. Let A = {No mandatory seat belt law}

$$P(A) = \frac{63 + 42 + 38 + 38}{387} = \frac{181}{387} = .468$$

b. Let B = {Uses seat belts infrequently}

$$P(B) = \frac{18 + 23 + 38}{387} = \frac{79}{387} = .204$$

c. Let C = {Pending mandatory seat belt law}
   D = {Never uses seat belts}

$$P(C \cap D) = \frac{8}{387} = .021$$

d. Let E = {Mandatory seat belt law}
   F = {Always uses seat belts}

$$P(E \cup F) = \frac{67 + 24 + 18 + 19 + 27 + 63}{387} = \frac{218}{387} = .563$$

e. $P(D) = \frac{19 + 8 + 38}{387} = \frac{65}{387}$

$$P(A \cap D) = \frac{38}{387}$$

$$P(A \mid D) = \frac{P(A \cap D)}{P(D)} = \frac{38/387}{65/387} = \frac{38}{65} = .585$$

f. Let G = {Frequently uses seat belts}

$$P(C) = \frac{27 + 20 + 23 + 8}{387} = \frac{78}{387}$$

$$P(G \cap C) = \frac{20}{387}$$

$$P(G \mid C) = \frac{P(G \cap C)}{P(C)} = \frac{20/387}{78/387} = \frac{20}{78} = .257$$

g. If events E and B are independent, then $P(E \cap B) = P(E)P(B)$.

$$P(E \cap B) = \frac{18}{387} = .0465, \quad P(E) = \frac{128}{387}, \text{ and } P(B) = \frac{79}{387}$$

$$P(E)P(B) = \frac{128}{387} \cdot \frac{79}{387} = .0675$$

Since $.0465 \neq .0675$, events B and E are not independent.

3.55 a. The total number of tablets selected was 6 per lot × 30 lots = 180. Eight measurements were taken on each tablet, so the total number of measurements is

$$180 \times 8 = 1440$$

b. An average measurement was computed for each lot (30) at each time period (8). The total number of averages computed is

$$30 \times 8 = 240$$

3.57 a. From the problem, $P(A) = .15$

b. $P(B \mid A)$ = probability the person will gladly switch to mass transit given he/she regularly drives his/her own car = .80. This is given in the problem.

c. Event C^c is the event the person lives more than 3 miles from the center of the city. It is given that $P(C^c) = .4$.

$$P(C) = 1 - P(C^c) = 1 - .4 = .6$$

d. Two events are mutually exclusive if they have no simple events in common or if the probability of their intersection is 0. Events A and B are not mutually exclusive because $P(A \cap B) = P(B \mid A)P(A) = .80(.15) = .12 \neq 0$. Events A and C are not mutually exclusive since there are workers who live within 3 miles of the center of the city and who regularly drive their own cars to work. Events B and C are not mutually exclusive since there are workers who live within 3 miles of the center of the city and who would gladly switch to mass transit.

3.59 a. Define the following events:

A: {1st system breaks down}
B: {2nd system breaks down}

From the problem, $P(A) = .2$ and $P(B \mid A) = .3$. The event the system is not working is $A \cap B$.

$$P(A \cap B) = P(B \mid A)P(A) = .2(.3) = .06$$

b. The event the system is working is the complement of event $A \cap B$.

$$P(A \cap B)^c = 1 - P(A \cap B) = 1 - .06 = .94$$

3.61 Define the following events:

A: {Component fails when first used}
B: {Component lasts for 1 year}

From the problem, $P(A) = .1$ and $P(B \mid A^c) = .99$. The event the component will last 1 year is the event $B \cap A^c$.

$$P(A^c) = 1 - P(A) = 1 - .1 = .9$$
$$P(B \cap A^c) = P(B \mid A^c)P(A^c) = .99(.9) = .891$$

3.63 There are five suppliers and a company will choose at least 2 suppliers. Therefore, we must find the number of ways to select 2, 3, 4, and 5 suppliers from the 5 suppliers. The total number of options is

$$\binom{5}{2} + \binom{5}{3} + \binom{5}{4} + \binom{5}{5} = \frac{5!}{2!(5-2)!} + \frac{5!}{3!(5-3)!} + \frac{5!}{4!(5-4)!} + \frac{5!}{5!(5-5)!}$$

$$= \frac{5 \times 4 \times 3 \times 2 \times 1}{2 \times 1 \times 3 \times 2 \times 1} + \frac{5 \times 4 \times 3 \times 2 \times 1}{3 \times 2 \times 1 \times 2 \times 1}$$

$$+ \frac{5 \times 4 \times 3 \times 2 \times 1}{4 \times 3 \times 2 \times 1 \times 1} + \frac{5 \times 4 \times 3 \times 2 \times 1}{5 \times 4 \times 3 \times 2 \times 1 \times 1}$$

$$= 10 + 10 + 5 + 1 = 26$$

3.65 Define the following events:

   A: {Part is supplied by company A}
   B: {Part is supplied by company B}
   C: {Part is defective}

From the problem, $P(A) = .8$, $P(B) = .2$, $P(C \mid A) = .05$, and $P(C \mid B) = .03$. We know the given part is defective. We want to find the probability it came from company A and the probability it came from company B or $P(A \mid C)$ and $P(B \mid C)$. First, we find

$P(A \cap C) = P(C \mid A)P(A) = .05(.8) = .04,$

$P(B \cap C) = P(C \mid B)P(B) = .03(.2) = .006,$ and

$P(C) = P(A \cap C) + P(B \cap C) = .04 + .006 = .046$

$P(A \mid C) = P(A \cap C)/P(C) = .04/.046 = .87,$ and

$P(B \mid C) = P(B \cap C)/P(C) = .006/.046 = .13$

Thus, it is more likely the defective part came from company A (probability is .87) than from company B (probability is .13).

3.67 a. There are 3 contracts to be awarded. Since no company can be awarded more than one contract, there are five possible companies to which to award the first contract, only four companies to which to award the second contract, and three companies to which to award the third contract. The total number of ways to award the contracts is found using the permutation of five companies taken 3 at a time or

$$P_3^5 = \frac{5!}{(5-3)!} = \frac{5 \times 4 \times 3 \times 2 \times 1}{2 \times 1} = 60$$

b. If company 2 is awarded the first contract, then there are only 4 companies to which to award the second contract and only 3 companies to which to award the third contract. Thus, the total number of ways to award contracts so company 2 receives the first contract is the permutation of 4 companies taken 2 at a time or

$$P_4^2 = \frac{4!}{(4-2)!} = \frac{4 \times 3 \times 2 \times 1}{2 \times 1} = 12$$

Company 2 could also be awarded the second or third contract. There are also 12 ways company 2 could receive the second contract and 12 ways company 2 could receive the third contract. Thus, the total number of ways company 2 could be awarded a contract is $12 + 12 + 12 = 36$.

$$P(\text{Company 2 awarded a contract}) = 36/60 = .6$$

c. There are 6 ways companies 4 and 5 can be awarded 2 of the contracts. One way is for company 4 to receive the 1st and company 5 the 2nd. The remaining ways are: company 4 the 1st and company 5 the 3rd, company 4 the 2nd and company 5 the 3rd, company 5 the 1st and company 4 the 2nd, company 5 the 1st and company 4 the 3rd, and company 5 the 2nd and company 4 the 3rd. For any of the above mentioned ways that companies 4 and 5 can receive contracts, there are 3 companies left from which to select 1 to fill in the remaining contract. Thus, the total number of ways companies 4 and 5 can receive contracts is $6 \times 3 = 18$.

$$P(\text{Company 4 and company 5 awarded contracts}) = 18/60 = .3$$

3.69 a. A flush is 5 cards of the same suit. The total number of ways to get a flush in one suit is a combination of 13 cards taken 5 at a time or

$$\binom{13}{5} = \frac{13!}{5!(13-5)!} = 1287$$

Since there are four suits, the total number of ways to draw a flush is $4(1287) = 5148$. The total number of ways to draw 5 cards from a deck of 52 is

$$\binom{52}{5} = \frac{52!}{5!(52-5)!} = 2,598,960$$

Thus, $P(A) = 5148/2,598,960 = .0019808$

b. A straight is 5 cards in sequence, regardless of suit. The total number of ways to get a straight starting with an ace is

$$\binom{4}{1}\binom{4}{1}\binom{4}{1}\binom{4}{1}\binom{4}{1}$$

(combination of 4 aces taken 1 at a time times a combination of 4 twos taken one at a time, etc.)

$$= \frac{4!}{1!(4-1)!} \times \frac{4!}{1!(4-1)!} \times \frac{4!}{1!(4-1)!} \times \frac{4!}{1!(4-1)!} \times \frac{4!}{1!(4-1)!}$$
$$= 4^5 = 1024$$

However, a straight can start with an ace, 2, 3, ... , 10. The total number of ways to draw a straight is $1024(10) = 10,240$.

$$P(B) = 10,240/2,598,960 = .00394$$

c.   There are 10 ways to get a straight flush in one suit (straight starting with an ace, 2, 3, ... , 10). Since there are four suits, the total number of ways to draw a straight flush is 10(4) = 40.

$$P(A \cap B) = 40/2,598,960 = .0000154$$

3.71   Define the following events:

A: {No Soviet attack is made}
B: {Warning system send signal}
C: {Launch is made}

From the problem, $P(A^c) = .01$, thus, $P(A) = 1 - P(A^c) = 1 - .01 = .99$, $(B \mid A) = .02$, and $P(C \mid A \cap B) = .9$

We want to find $P(A \cap B \cap C) = (C \mid A \cap B)P(B \mid A)(A) = .9(.02)(.99) = .01782$

CHAPTER FOUR

....

Discrete Random Variables

4.1   a.   The two requirements for a valid discrete probability distribution are:

1. $0 \le p(y) \le 1$

2. $\displaystyle\sum_{\text{all } y} p(y) = 1$

For this distribution, none of the rectangles are higher than 1 and none are below 0. Thus, $0 \le p(y) \le 1$ holds.

From the graph, $p(7) = .10$, $p(8) = .10$, $p(9) = .15$, $p(10) = .20$, $p(11) = .20$, $p(12) = .15$, and $p(13) = .10$

$$\sum_{\text{all } y} p(y) = .10 + .10 + .15 + .20 + .20 + .15 + .10 = 1.00$$

Therefore, this is a valid probability distribution.

b.

y	$p(y)$
7	.10
8	.10
9	.15
10	.20
11	.20
12	.15
13	.10

$\displaystyle\sum_{\text{all } y} p(y) = 1.00$

c. $P(y = 9) = p(9) = .15$

d. $P(y < 12) = p(7) + p(8) + p(9) + p(10) + p(11)$
$= .10 + .10 + .15 + .20 + .20 = .75$

4.3   The total number of ways to select 2 cornrakes from 12 cornrakes is

$$\binom{12}{2} = \frac{12!}{2!(12 - 2)!} = \frac{12!}{2!10!} = \frac{12 \cdot 11}{2 \cdot 1} = 66$$

If y = the number of the captured cornrakes that are capable of mating, then y can take on the values 0, 1, and 2. Thus, we must find $p(0)$, $p(1)$, and $p(2)$.

If $y = 0$, then 0 of the captured cornrakes come from the 8 fertile cornrakes and 2 of the captured cornrakes come from the infertile cornrakes. The number of ways to get 2 infertile cornrakes is

$$\binom{4}{2}\binom{8}{0} = \frac{4!}{2!(4-2)!}\frac{8!}{0!(8-0)!} = \frac{4!}{2!2!}\frac{8!}{0!8!} = \frac{4\cdot 3}{2\cdot 1} = 6$$

Thus, $P(y = 0) = p(0) = \dfrac{6}{66}$

If $y = 1$, then 1 of the captured cornrakes come from the 8 fertile cornrakes and 1 of the captured cornrakes come from the infertile cornrakes. The number of ways to get 1 fertile cornrake is

$$\binom{4}{1}\binom{8}{1} = \frac{4!}{1!(4-1)!}\frac{8!}{1!(8-1)!} = \frac{4!}{1!3!}\frac{8!}{1!7!} = 4\cdot 8 = 32$$

Thus, $P(y = 1) = p(1) = \dfrac{32}{66}$

If $y = 2$, then 2 of the captured cornrakes come from the 8 fertile cornrakes and 0 of the captured cornrakes come from the infertile cornrakes. The number of ways to get 2 fertile cornrakes is

$$\binom{4}{0}\binom{8}{2} = \frac{4!}{0!(4-0)!}\frac{8!}{2!(8-2)!} = \frac{4!}{0!4!}\frac{8!}{2!6!} = \frac{8\cdot 7}{2\cdot 1} = 28$$

Thus, $P(y = 2) = p(2) = \dfrac{28}{66}$

The probability distribution of y is

y	0	1	2
$p(y)$	6/66	32/66	28/66

4.5   There are 5 firing pins, of which 3 are defective. Let y = number of firing pins tested until the first defective is found.

Let D_i represent a defective firing pin on the ith trial and G_i represent a good firing pin on the ith trial. The sample space for this experiment is

$D_1, \ G_1D_2, \ G_1G_2D_3$

$P(D_1) = 3/5 = P(y = 1)$
$P(G_1D_2) = P(D_2 \mid G_1)P(G_1) = 3/4(2/5) = 3/10 = P(y = 2)$
$P(G_1G_2D_3) = P(D_3 \mid G_1G_2)P(G_2 \mid G_1)P(G_1) = 3/3(1/4)(2/5) = 1/10 = P(y = 3)$

The probability distribution of y is

y	1	2	3
$p(y)$	3/5	3/10	1/10

4.7   a.   Define the following event:

A: {Consumer is environmentalist}

$P(A) = P(TG, GG, \text{ or } S) = .11 + .11 + .26 = .48$

$P(A^c) = 1 - P(A) = 1 - .48 = .52$

Let $y =$ number of consumers sampled until the first environmentalist is found. Then y can take on values, 1, 2, 3, 4, ...

$P(y = 1) = P(A) = .48$

$P(y = 2) = P(A^c \cap A) = P(A^c)P(A) = .52(.48) = .2496$

$P(y = 3) = P(A^c \cap A^c \cap A) = P(A^c)P(A^c)P(A) = .52(.52)(.48) = .1298$

$P(y = 4) = P(A^c \cap A^c \cap A^c \cap A) = P(A^c)P(A^c)P(A^c)P(A)$
$\qquad\qquad = .52(.52)(.52)(.48)$
$\qquad\qquad = .0675$

$P(y = 5) = P(A^c \cap A^c \cap A^c \cap A^c \cap A) = P(A^c)P(A^c)P(A^c)P(A^c)P(A)$
$\qquad\qquad = .52(.52)(.52)(.52)(.48)$
$\qquad\qquad = .0351$

$P(y = 6) = P(A^c \cap A^c \cap A^c \cap A^c \cap A^c \cap A)$
$\qquad\qquad = P(A^c)P(A^c)P(A^c)P(A^c)P(A^c)P(A)$
$\qquad\qquad = .52(.52)(.52)(.52)(.52)(.48)$
$\qquad\qquad = .0182$

etc.

y	1	2	3	4	5	6	...
$p(y)$	.4800	.2496	.1298	.0675	.0351	.0182	...

b. $p(y) = .48(.52)^{y-1}$

4.9   From Exercise 4.2, the probability distribution of y is

y	0	1	2	3
$p(y)$	1/8	3/8	3/8	1/8

$$E(y) = \sum_{\text{all } y} yp(y) = 0(1/8) + 1(3/8) + 2(3/8) + 3(1/8) = 12/8 = 1.5$$

$$\sigma^2 = E(y - \mu^2) = E(y^2) - \mu^2 = \sum_{\text{all } y} y^2 p(y) - \mu^2$$

$$= 0^2(1/8) + 1^2(3/8) + 2^2(3/8) + 3^2(1/8) - 1.5^2$$

$$= 24/8 - 2.25 = 3 - 2.25 = .75$$

4.11
$$E(y) = \sum_{\text{all } y} yp(y) = -50,000(.60) - 20,000(.10) + 30,000(.15) + 430,000(.10)$$

$$+ 950,000(.05)$$

$$= -30,000 - 2,000 + 4,500 + 43,000 + 47,500$$

$$= 63,000$$

$$\sigma^2 = V(y) = E[(y - \mu)^2] = \sum_{\text{all } y} (y - \mu)^2 p(y)$$

$$= (-50,000 - 63,000)^2(.60) + (-20,000 - 63,000)^2(.10)$$

$$+ (30,000 - 63,000)^2(.15) + (430,000 - 63,000)^2(.10)$$

$$+ (950,000 - 63,000)^2(.05)$$

$$= 7,661,400,000 + 688,900,000 + 163,350,000 + 13,468,900,000$$

$$+ 39,338,450,000$$

$$= 61,321,000,000$$

4.13   Let y = time to evacuate. From the probability table,

$$P(y \le 14) = .04 + .25 = .29$$

4.15   From Exercise 4.8, $\mu = 10.15$ and $\sigma^2 = 3.1275$.

Let x = amount company earns = $15,000y$

Thus, $\mu_x = E(x) = E(15,000y) = 15,000\mu = 15,000(10.15) = \$152,250$

$$\sigma_x^2 = E(x - \mu_x)^2 = E(15,000y - 15,000\mu)^2 = E[(15,000)^2(y - \mu)^2]$$
$$= 15,000^2\sigma^2 = 15,000^2(3.1275) = 703,687,500$$

4.17   From Exercise 4.3, the probability distribution of y is

y	0	1	2
$p(y)$	6/66	32/66	28/66

Using Theorem 4.4, $\sigma^2 = E(y^2) - \mu^2$

$$\mu = E(y) = \sum_{\text{all } y} yp(y) = 0(6/66) + 1(32/66) + 2(28/66)$$

$$= 0 + 32/66 + 56/66 = 88/66 = 4/3$$

$$E(y^2) = \sum_{\text{all } y} y^2 p(y) = 0^2(6/66) + 1^2(32/66) + 2^2(28/66)$$

$$= 0 + 32/66 + 112/66 = 144/66 = 24/11$$

$$\sigma^2 = E(y^2) - \mu^2 = \frac{24}{11} - \left[\frac{4}{3}\right]^2 = \frac{24}{11} - \frac{16}{9} = \frac{216 - 176}{99} = \frac{40}{99} = .404$$

4.19   Theorem 4.1 is: $E(c) = c$

$$E(c) = \sum_{\text{all } y} cp(y) = c \sum_{\text{all } y} p(y) = c(1) = c$$

4.21   Theorem 4.3 is: $E[g_1(y) + g_2(y) + \cdots + g_k(y)] = E[(g_1(y)] + E[(g_2(y)] + \cdots + E[(g_k(y)]$

From Definition 4.5,

$$E[g_1(y) + g_2(y) + \cdots + g_k(y)]$$

$$= \sum_{\text{all } y} [g_1(y) + g_2(y) + \cdots + g_k(y)]p(y)$$

$$= \sum_{\text{all } y} [g_1(y)p(y) + g_2(y)p(y) + \cdots + g_k(y)p(y)]$$

$$= \sum_{\text{all } y} g_1(y)p(y) + \sum_{\text{all } y} g_2(y)p(y) + \cdots + \sum_{\text{all } y} g_k(y)p(y)$$

$$= E[g_1(y)] + E[g_2(y)] + \cdots + E[g_k(y)]$$

4.23   a.   Using Table 1 with $n = 10$ and $p = .1$:

$$p(0) = .3487$$
$$p(1) = .7361 - .3487 = .3874$$
$$p(2) = .9298 - .7361 = .1937$$
$$p(3) = .9872 - .9298 = .0574$$
$$p(4) = .9984 - .9872 = .0112$$
$$p(5) = .9999 - .9984 = .0015$$
$$p(6) = 1 - .9999 = .0001$$
$$p(7) = p(8) = p(9) = p(10) = 1 - 1 = 0$$

    b.   Using Table 1 with $n = 10$ and $p = .5$:

$$p(0) = .0010$$
$$p(1) = .0107 - .0010 = .0097$$
$$p(2) = .0547 - .0107 = .0440$$
$$p(3) = .1719 - .0547 = .1172$$
$$p(4) = .3770 - .1719 = .2051$$
$$p(5) = .6230 - .3770 = .2460$$

$p(6) = .8281 - .6230 = .2051$
$p(7) = .9453 - .8281 = .1172$
$p(8) = .9893 - .9453 = .0440$
$p(9) = .9990 - .9893 = .0097$
$p(10) = 1 - .9990 = .0010$

c.   Using Table 1 with $n = 10$ and $p = .9$:

$p(0) = 0$
$p(1) = p(2) = p(3) = 0 - 0 = 0$
$p(4) = .0001 - 0 = .0001$
$p(5) = .0016 - .0001 = .0015$
$p(6) = .0128 - .0016 = .0112$
$p(7) = .0702 - .0128 - .0574$
$p(8) = .2639 - .0702 = .1937$
$p(9) = .6513 - .2639 = .3874$
$p(10) = 1 - .6513 = .3487$

d.   The graph for part a is:

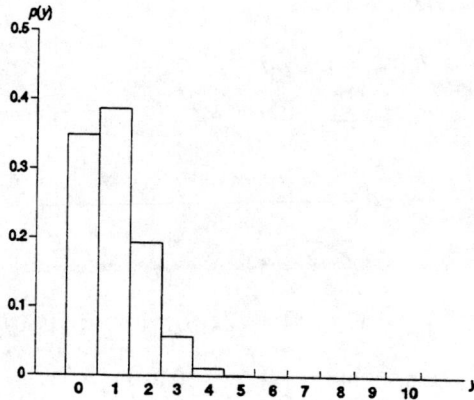

The graph for part b is:

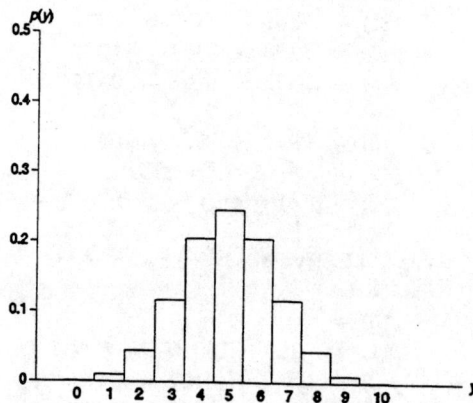

The graph for part **c** is:

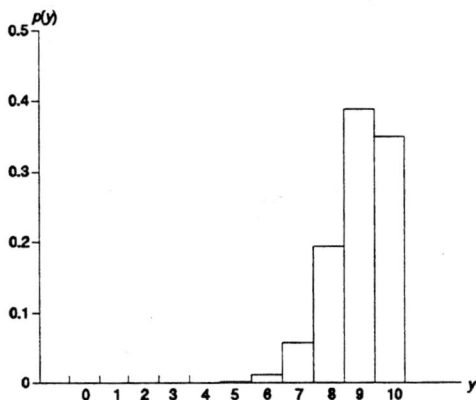

4.25 The experiment consists of $n = 10$ trials. Each trial results in an S (drywall installer is a woman) or an F (drywall installer is not a woman). The probability of success, p, is .01 and $q = 1 - p = 1 - .01 = .99$. We assume the trials are independent. Let $y =$ number of drywall installers who are women in 10 trials. Thus, y has a binomial distribution with $n = 10$ and $p = .01$.

Using Table 1, $P(y \leq 1) = .9957$

4.27 The experiment consists of $n = 3$ trials. Each trial results in an S(swordfish piece has level of mercury above FDA minimum amount) or an F (swordfish piece does not have level of mercury above FDA minimum amount). The probability of success, p, is .40 and $q = 1 - .40 = .60$. We assume the trials are independent. Let $y =$ number of swordfish pieces out of 3 that have level of mercury above FDA minimum amount. Thus, y has a binomial distribution with $n = 3$ and $p = .40$.

a. $P(y = 3) = \begin{pmatrix} 3 \\ 3 \end{pmatrix} .4^3 (.6)^{3-3} = \dfrac{3!}{3!(3-3)!} .4^3 = .064$

b. $P(y = 1) = \begin{pmatrix} 3 \\ 1 \end{pmatrix} .4^1 (.6)^{3-1} = \dfrac{3!}{1!(3-1)!} .4(.6)^2 = 3(.4)(.36) = .432$

c. $P(y \leq 1) = P(y = 0) + P(y = 1) = \begin{pmatrix} 3 \\ 0 \end{pmatrix} .4^0 (.6)^{3-0} + .432$

$= \dfrac{3!}{0!(3-0)!} .6^3 + .432 = .216 + .432 = .648$

4.29 The experiment consists of $n = 10$ trials. Each trial results in an S (intruder detected) or an F (intruder not detected). The probability of success, p, is .5 and $q = 1 - .5 = .5$. We assume the trials are independent. Therefore, y has a binomial distribution with $n = 10$ and $p = .5$.

a. Using Table 1 with $n = 10$ and $p = .5$:

$P(y \geq 7) = 1 - P(y \leq 6) = 1 - .8281 = .1719$

b. If the probability of intruder detection in snowy conditions is only .5, it would not be unusual for the system to detect at least 7 intruders ($p = .1719$). The sample size of 10 is rather small to use to estimate probabilities.

4.31 This experiment consists of $n = 1000$ trials. Each trial results in an S (person develops cancer) or an F (person does not develop cancer). The probability of success, p, is .0000389. We will assume the trials are independent. Let y = number of people who develop cancer in 1000 trials. Then y has a binomial distribution with $n = 1000$ and $p = .0000389$.

a. $\mu = np = 1000(.0000389) = .0389$
$\sigma^2 = npq = 1000(.0000389)(.9999611) = .0388985$

b. Using Tchebysheff's Theorem, at least $1 - \dfrac{1}{2^2} = \dfrac{3}{4}$ of the observations are within 2 standard deviations of the mean.

$\sigma = \sqrt{npq} = \sqrt{.0388985} = .1972$

$\mu \pm 2\sigma \Rightarrow .0389 \pm 2(.1972) \Rightarrow .0389 \pm .3944 \Rightarrow (-.3555, .4333)$

Since 1 is not contained in the interval, we would not expect to see 1 male with cancer among the 1000 males.

4.33 $(q + p)^n = \begin{bmatrix} n \\ 0 \end{bmatrix} q^n + \begin{bmatrix} n \\ 1 \end{bmatrix} q^{n-1} p + \begin{bmatrix} n \\ 2 \end{bmatrix} q^{n-2} p^2 + \dots + \begin{bmatrix} n \\ n \end{bmatrix} p^n$

$= p(0) + p(1) + p(2) + \dots + p(n)$

$= \sum_{y=0}^{n} p(y) = 1$

4.35 $E[y(y - 1)] = E(y^2 - y) = E(y^2) - E(y) = E(y^2) - \mu$

From Exercise 4.34, $E[y(y - 1)] = npq + \mu^2 - \mu$

$\Rightarrow npq + \mu^2 - \mu = E(y)^2 - \mu$

$\Rightarrow E(y^2) = npq + \mu^2$

4.37 $P(y_1, y_2, y_3) = \dfrac{n!}{y_1! y_2! y_3!} (p_1)^{y_1}(p_2)^{y_2}(p_3)^{y_3}$

a. $P(3, 1, 1) = \dfrac{5!}{3!1!1!} (.2)^3(.5)^1(.3)^1 = .024$

b. $P(0, 5, 0) = \dfrac{5!}{0!5!0!} (.2)^0(.5)^5(.3)^0 = .03125$

c. $P(1, 3, 1) = \dfrac{5!}{1!3!1!} (.2)^1(.5)^3(.3)^1 = .15$

4.39   This experiment consists of 8 identical trials. There are 3 possible outcomes on each trial: brighter side up with probability .65, darker side up with probability .15, and brighter and darker sides aligned with probability .20. Assuming the trials are independent, this is a multinomial experiment with $n = 8$, $k = 3$, $p_1 = .65$, $p_2 = .15$, and $p_3 = .20$.

a. $P(8, 0, 0) = \dfrac{8!}{8!0!0!}(.65)^8(.15)^0(.20)^0 = .0319$

b. $P(4, 3, 1) = \dfrac{8!}{4!3!1!}(.65)^4(.15)^3(.20)^1 = .0337$

c. $\mu_1 = np_1 = 8(.65) = 5.2$

4.41   This experiment consists of 10 identical trials. There are 3 paths available on each trial with path probabilities of .25, .30, and .45. Assuming the trials are independent, this is a multinomial experiment with $n = 10$, $k = 3$, $p_1 = .25$, $p_2 = .30$, $p_3 = .45$.

a. $P(2, 4, 4) = \dfrac{10!}{2!4!4!}(.25)^2(.30)^4(.45)^4 = .06539$

b. $E(y_2) = n \cdot p_2 = 10(.30) = 3$

$$\sigma_2^2 = V(y_2) = n \cdot p_i(1 - p_i) = n \cdot p_2(1 - p_2) = 10(.30)(.70) = 2.1$$

$$\sigma_2 = \sqrt{\sigma_2^2} = \sqrt{2.1} = 1.45$$

We expect y_2, the number of times path two is used, to fall within two standard deviations of its mean.

$$\mu \pm 2\sigma \Rightarrow 3 \pm 2(1.45) \Rightarrow 3 \pm 2.90 \Rightarrow .10 \text{ to } 5.90$$

4.43 $E(c) = E(4y_1 + y_2) = E(4y_1) + E(y_2)$

$y_1 = $ the number of bits with one defect
$y_2 = $ the number of bits with two or more defects

$E(y_1) = n \cdot p_1$ where p_1 is the probability of observing a lot with exactly one defect.

$E(y_2) = n \cdot p_2$ where p_2 is the probability of observing a lot with more than one defect.

$E(c) = 4n \cdot p_1 + n \cdot p_2 = n(4p_1 + p_2)$

4.45 $[a + (b + c)]^2 = \begin{bmatrix} 2 \\ 0 \end{bmatrix} a^2 + \begin{bmatrix} 2 \\ 1 \end{bmatrix} a^1(b + c) + \begin{bmatrix} 2 \\ 2 \end{bmatrix} (b + c)^2$

$$= \begin{bmatrix} 2 \\ 0 \end{bmatrix} a^2 + \begin{bmatrix} 2 \\ 1 \end{bmatrix} ab + \begin{bmatrix} 2 \\ 1 \end{bmatrix} ac + \begin{bmatrix} 2 \\ 2 \end{bmatrix} \left[\begin{bmatrix} 2 \\ 0 \end{bmatrix} b^2 + \begin{bmatrix} 2 \\ 1 \end{bmatrix} bc + \begin{bmatrix} 2 \\ 2 \end{bmatrix} c^2 \right]$$

$$= \frac{2!}{2!0!} a^2 + \frac{2!}{1!1!} a^1 b^1 + \frac{2!}{1!1!} a^1 c^1 + \frac{2!}{0!2!} b^2 + \frac{2!}{1!1!} bc + \frac{2!}{2!0!} c^2$$

$$= \frac{2!}{2!0!0!} a^2 b^0 c^0 + \frac{2!}{1!1!0!} a^1 b^1 c^0 + \frac{2!}{1!0!1!} a^1 b^0 c^1 + \frac{2!}{0!2!0!} a^0 b^2 c^0$$

$$+ \frac{2!}{0!1!1!} a^0 b^1 c^1 + \frac{2!}{0!0!2!} a^0 b^0 c^0$$

Substituting $a = p_1$, $b = p_2$, $c = p_3$ yields:

$$= P(2, 0, 0) + P(1, 1, 0) + P(1, 0, 1) + P(0, 2, 0) + P(0, 1, 1) + P(0, 0, 2) = 1$$

4.47 a. For $p = .6$ and $r = 3$

$$P(y = 6) = \begin{bmatrix} 6 - 1 \\ 3 - 1 \end{bmatrix} .6^3 .4^{6-3} = \frac{5!}{2!(5 - 2)!} .6^3 .4^3 = .13824$$

$$P(y = 7) = \begin{bmatrix} 7 - 1 \\ 3 - 1 \end{bmatrix} .6^3 .4^{7-3} = \frac{6!}{2!(6 - 2)!} .6^3 .4^4 = .08294$$

$$P(y = 8) = \begin{bmatrix} 8 - 1 \\ 3 - 1 \end{bmatrix} .6^3 .4^{8-3} = \frac{7!}{2!(7 - 2)!} .6^3 .4^5 = .04645$$

$$P(y = 9) = \begin{bmatrix} 9 - 1 \\ 3 - 1 \end{bmatrix} .6^3 .4^{9-3} = \frac{8!}{2!(8 - 2)!} .6^3 .4^6 = .02477$$

b. To construct a probability histogram, we must find probabilities for $y = 3, 4, 5$.

$$P(y = 3) = \begin{bmatrix} 3 - 1 \\ 3 - 1 \end{bmatrix} .6^3 .4^{3-3} = \frac{2!}{2!(2 - 2)!} .6^3 .4^0 = .21600$$

$$P(y = 4) = \begin{bmatrix} 4 - 1 \\ 3 - 1 \end{bmatrix} .6^3 .4^{4-3} = \frac{3!}{2!(3 - 2)!} .6^3 .4^1 = .25920$$

$$P(y = 5) = \begin{bmatrix} 5 - 1 \\ 3 - 1 \end{bmatrix} .6^3 .4^{5-3} = \frac{4!}{2!(4 - 2)!} .6^3 .4^2 = .20736$$

The histogram is:

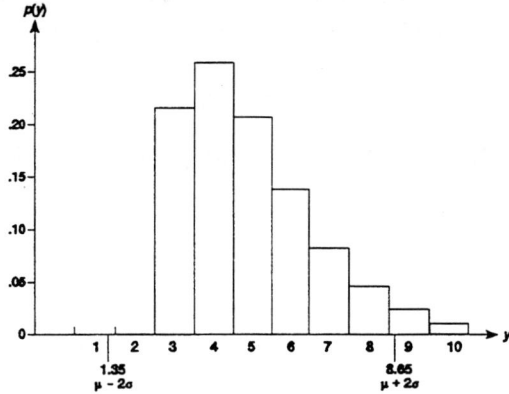

c. $\mu = \dfrac{r}{p} = \dfrac{3}{.6} = 5,$ $\sigma^2 = \dfrac{rq}{p^2} = \dfrac{3(.4)}{.6^2} = 3.33$

$\sigma = \sqrt{\sigma^2} = \sqrt{3.33} = 1.825$

d. $\mu \pm 2\sigma \Rightarrow 5 \pm 2(1.825) \Rightarrow 5 \pm 3.65 \Rightarrow 1.35$ to 8.65

These points are marked above.

$P(1.35 \leq y \leq 8.65) = P(2 \leq y \leq 8) = .95019$

4.49   a.   For $p = .544$ and $r = 1$

$P(y = 3) = \begin{bmatrix} 3 - 1 \\ 1 - 1 \end{bmatrix} (.544)^1(.456)^{3-1} = \dfrac{2!}{0!2!}(.544)(.456)^2 = .1131$

b. $P(y \leq 2) = p(1) + p(2)$

$= \begin{bmatrix} 1 - 1 \\ 1 - 1 \end{bmatrix} (.544)^1(.456)^0 + \begin{bmatrix} 2 - 1 \\ 1 - 1 \end{bmatrix} (.544)^1(.456)^1 = .544 + .248 = .792$

c. $P(y > 2) = 1 - P(y \leq 2) = 1 - .792 = .208$

4.51   The probability of getting an environmentalist is $.11 + .11 + .26 = .48$. Since we are looking for the first success, we can use the geometric distribution.

a. $\mu = \dfrac{1}{p} = \dfrac{1}{.48} = 2.083$

$\sigma^2 = \dfrac{q}{p^2} = \dfrac{.52}{.48^2} = 2.2569$

$\sigma = \sqrt{2.2569} = 1.502$

b. We know from Tchebysheff's Theorem that at least 3/4 of the observations are within 2 standard deviations of the mean.

$$\mu \pm 2\sigma \Rightarrow 2.083 \pm 2(1.502) \Rightarrow 2.083 \pm 3.004 \Rightarrow (-.921, 5.087)$$

Since we know y cannot be negative, the interval should be (0, 5.087)

4.53 Using the geometric distribution with $p = .3$:

a. $P(y \le 3) = p(1) + p(2) + p(3)$
$$= (.3)(.7)^0 + (.3)(.7)^1 + (.3)(.7)^2$$
$$= .3 + .21 + .147 = .657$$

b. $\mu = \dfrac{1}{p} = \dfrac{1}{.3} = 3.33$

$\sigma^2 = \dfrac{q}{p^2} = \dfrac{.7}{(.3)^2} = 7.78$

$\sigma = \sqrt{\sigma^2} = \sqrt{7.78} = 2.79$

c. $\mu \pm 2\sigma \Rightarrow 3.33 \pm 2(2.79) \Rightarrow 3.33 \pm 5.58 \Rightarrow -2.25$ to 8.91

Since this interval doesn't contain 10, it is unlikely y will exceed 10.

d. $P(y \le 7) = p(2) + p(3) + \cdots + p(7)$
$$= \begin{bmatrix} 2 - 1 \\ 2 - 1 \end{bmatrix}(.3)^2(.7)^0 + \begin{bmatrix} 3 - 1 \\ 2 - 1 \end{bmatrix}(.3)^2(.7) + \cdots + \begin{bmatrix} 7 - 1 \\ 2 - 1 \end{bmatrix}(.3)^2(.7)^5$$
$$= .6706$$

4.55 $E(w) = E(y - r) = E(y) - r = \dfrac{r}{p} - r = \dfrac{r - rp}{p} = \dfrac{r(1 - p)}{p} = \dfrac{rq}{p}$

$\sigma_w^2 = E\left[\left(w - \dfrac{rq}{p}\right)^2\right] = E\left[\left(y - r - \dfrac{rq}{p}\right)^2\right]$

$= E\left[\left(y - r\left(1 + \dfrac{q}{p}\right)\right)^2\right] = E\left[\left(y - r\left(\dfrac{p + q}{p}\right)\right)^2\right]$

$= E\left[\left(y - r\left(\dfrac{1}{p}\right)\right)^2\right] = E\left[\left(y - \dfrac{r}{p}\right)^2\right] = \sigma_y^2 = \dfrac{rq}{p^2}$

4.57 a. $p(1) = \dfrac{\begin{bmatrix} 4 \\ 1 \end{bmatrix}\begin{bmatrix} 5 - 4 \\ 3 - 1 \end{bmatrix}}{\begin{bmatrix} 5 \\ 3 \end{bmatrix}} = 0$

b. $p(3) = \dfrac{\begin{bmatrix}3\\3\end{bmatrix}\begin{bmatrix}10-3\\5-3\end{bmatrix}}{\begin{bmatrix}10\\5\end{bmatrix}} = \dfrac{\dfrac{3!}{3!0!} \cdot \dfrac{7!}{2!5!}}{\dfrac{10!}{5!5!}} = \dfrac{21}{252} = \dfrac{1}{12}$

c. $p(2) = \dfrac{\begin{bmatrix}2\\2\end{bmatrix}\begin{bmatrix}3-2\\2-2\end{bmatrix}}{\begin{bmatrix}3\\2\end{bmatrix}} = \dfrac{\dfrac{2!}{2!0!} \cdot \dfrac{1!}{0!1!}}{\dfrac{3!}{1!2!}} = \dfrac{1}{3}$

d. $p(0) = \dfrac{\begin{bmatrix}2\\0\end{bmatrix}\begin{bmatrix}4-2\\2-0\end{bmatrix}}{\begin{bmatrix}4\\2\end{bmatrix}} = \dfrac{\dfrac{2!}{0!2!} \cdot \dfrac{2!}{0!2!}}{\dfrac{4!}{2!2!}} = \dfrac{1}{6}$

4.59 a. $P(y = 3) = \dfrac{\begin{bmatrix}3\\3\end{bmatrix}\begin{bmatrix}15-3\\3-3\end{bmatrix}}{\begin{bmatrix}15\\3\end{bmatrix}} = \dfrac{\dfrac{3!}{3!0!} + \dfrac{12!}{0!12!}}{\dfrac{15!}{3!12!}} = \dfrac{1}{455} = .0022$

b. $P(y \geq 1) = 1 - P(y = 0) = 1 - \dfrac{\begin{bmatrix}3\\0\end{bmatrix}\begin{bmatrix}15-3\\3-0\end{bmatrix}}{\begin{bmatrix}15\\3\end{bmatrix}} = 1 - \dfrac{\dfrac{3!}{0!3!} + \dfrac{12!}{3!9!}}{\dfrac{15!}{3!12!}} = 1 - \dfrac{220}{455}$

$$= 1 - .4835 = .5165$$

4.61 For this exercise, the number of elements in the population is $N = 7$, and the sample size is $n = 3$.

a. For $r = 1$, $P(y = 0) = \dfrac{\begin{bmatrix}1\\0\end{bmatrix}\begin{bmatrix}7-1\\3-0\end{bmatrix}}{\begin{bmatrix}7\\3\end{bmatrix}} = \dfrac{\dfrac{1!}{0!1!} + \dfrac{6!}{3!3!}}{\dfrac{7!}{3!4!}} = \dfrac{20}{35} = \dfrac{4}{7}$

b. For $r = 3$, $P(y = 0) = \dfrac{\begin{bmatrix}3\\0\end{bmatrix}\begin{bmatrix}7-3\\3-0\end{bmatrix}}{\begin{bmatrix}7\\3\end{bmatrix}} = \dfrac{\dfrac{3!}{0!3!} + \dfrac{4!}{3!1!}}{\dfrac{7!}{3!4!}} = \dfrac{4}{35}$

4.63 For this exercise, the number of elements in the population is $N = 492$, and the sample size is $n = 4$.

a. For $r = 331$ and $N - r = 165$, $P(y = 4) = \dfrac{\dbinom{331}{4}\dbinom{165}{0}}{\dbinom{496}{4}} = \dfrac{\dfrac{331!}{4!327!} + \dfrac{165!}{0!165!}}{\dfrac{496!}{4!492!}}$

$= .197$

b. For $N = 492$, $r = 327$ and $N - r = 165$,

$$P(y = 0) = \dfrac{\dbinom{327}{0}\dbinom{165}{2}}{\dbinom{492}{2}} = \dfrac{\dfrac{327!}{0!327!} + \dfrac{165!}{2!163!}}{\dfrac{492!}{2!490!}} = .112$$

c. For $N = 492$, $r = 396$ and $N - r = 96$,

$$P(y = 0) = \dfrac{\dbinom{396}{0}\dbinom{96}{2}}{\dbinom{492}{2}} = \dfrac{\dfrac{396!}{0!396!} + \dfrac{96!}{2!96!}}{\dfrac{492!}{2!490!}} = .038$$

4.65 For a Poisson random variable, $p(y) = \dfrac{\lambda^y e^{-\lambda}}{y!}$

a. Letting $\lambda = 5.5$,

$$p(0) = \dfrac{5.5^0 e^{-5.5}}{0!} = .0041$$

$$p(1) = \dfrac{5.5 e^{-5.5}}{1!} = .0225$$

$$p(2) = \dfrac{5.5^2 e^{-5.5}}{2!} = .0618$$

$$p(3) = \dfrac{5.5^3 e^{-5.5}}{3!} = .1133$$

$$p(4) = \dfrac{5.5^4 e^{-5.5}}{4!} = .1558$$

$$p(5) = \dfrac{5.5^5 e^{-5.5}}{5!} = .1714$$

$$p(6) = \dfrac{5.5^6 e^{-5.5}}{6!} = .1571$$

$$p(7) = \dfrac{5.5^7 e^{-5.5}}{7!} = .1234$$

$$p(8) = \frac{5.5^8 e^{-5.5}}{8!} = .0849$$

$$p(9) = \frac{5.5^9 e^{-5.5}}{9!} = .0519$$

$$p(10) = \frac{5.5^{10} e^{-5.5}}{10!} = .0285$$

The graph of $p(y)$ is:

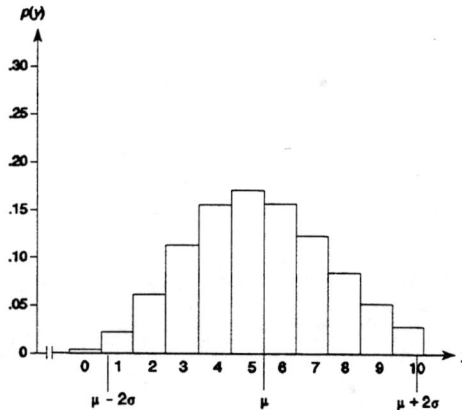

b. $\mu = \lambda = 5.5$

$\sigma^2 = \lambda = 5.5 \Rightarrow \sigma = \sqrt{5.5} = 2.35$

$\mu \pm 2\sigma \Rightarrow 5.5 \pm 2(2.35) \Rightarrow 5.5 \pm 4.70 \Rightarrow (.80, 10.20)$

c. $P(.80 \leq y \leq 10.20) = p(1) + p(2) + \cdots + p(10)$

$= .0225 + .0618 + .1133 + .1558 + .1714 + .1571 + .1234$

$+ .0849 + .0519 + .0285$

$= .9706$

4.67 a. Let y = number of independent spectral image exposures per asteroid. Then y has a Poisson distribution with $\lambda = \mu = 2.5$.

$$P(y = 1) = \frac{2.5^1 e^{-2.5}}{1!} = .2052$$

b. $P(y \leq 2) = p(0) + p(1) + p(2) = \dfrac{2.5^0 e^{-2.5}}{0!} + \dfrac{2.5^1 e^{-2.5}}{1!} + \dfrac{2.5^2 e^{-2.5}}{2!}$

$= .0821 + .2052 + .2565 = .5438$

c. $\mu = \lambda = 2.5$ and $\sigma^2 = \lambda = 2.5$ and $\sigma = \sqrt{2.5} = 1.581$

We know from Tchebysheff's Theorem that at least 3/4 of the observations are within 2 standard deviations of the mean.

$$\mu \pm 2\sigma \Rightarrow 2.5 \pm 2(1.581) \Rightarrow 2.5 \pm 3.162 \Rightarrow (-.662, 5.662)$$

Since 7 is not in the above interval, it would be unusual to observe 7 or more independent spectral image exposures during a main-belt asteroid sighting.

4.69   Let y = number of arrivals in a 1 minute interval. Then y has a Poisson distribution with $\lambda = 1$.

a. $P(y \geq 3) = 1 - P(y = 0) - P(y = 1) - P(y = 2)$

$$= 1 - \frac{1^0 e^{-1}}{0!} - \frac{1^1 e^{-1}}{1!} - \frac{1^2 e^{-1}}{2!} = 1 - .3679 - .3679 - .1839 = .0803$$

b. $P(y > 3) = 1 - P(y = 0) - P(y = 1) - P(y = 2) - P(y = 3)$

$$= .0803 - \frac{1^3 e^{-1}}{3!} = .0803 - .0613 = .019$$

Yes.   Since the probability of observing more than 3 arrivals is so small (.019), one can assure the engineer that the number of arrivals will rarely exceed 3 per minute.

4.71   a. Let y = number of unplanned scrams in a year. Then y has a Poisson distribution with $\lambda = \mu = 4$.

$$P(y \geq 10) = 1 - P(y \leq 9) = 1 - .9919 = .0081 \text{ (from Table 3, Appendix II)}$$

b. Since the probability of observing 10 or more unplanned scrams in a year is so small using $\mu = 4$, we would infer that μ is probably larger than 4.

4.73   a. $\mu = \lambda = 15$, $\sigma^2 = \lambda = 15$, $\sigma = \sqrt{15} = 3.873$

b. Since it would be very difficult to compute $P(y > 27)$, we will answer the question based on $\mu \pm 3\sigma$. (By Tchebysheff's Theorem, at least 8/9 of observations fall within 3 standard deviations of the mean.)

$$\mu \pm 3\sigma \Rightarrow 15 \pm 3(3.873) \Rightarrow 15 \pm 11.619 \Rightarrow (3.381, 26.619)$$

Since 27 falls outside this interval, the approximate probability that y exceeds 27 is 0.

c. If the variance is substantially less than 15, it is unreasonable to believe y follows a Poisson process. The probability in part b will decrease. If the variance is smaller, the width of the interval will decrease, thus making the probability smaller.

4.75 **a.** Show for Poisson random variable y that $0 \le p(y) \le 1$. The probability function for y is

$$p(y) = \frac{\lambda^y e^{-\lambda}}{y!}$$

For a Poisson random variable, $\lambda > 0$ and $y \ge 0$. Thus, $p(y) \ge 0$.

The Taylor expansion for e^λ is $1 + \dfrac{\lambda^1}{1!} + \dfrac{\lambda^2}{2!} + \dfrac{\lambda^3}{3!} + \dfrac{\lambda^4}{4!} + \cdots$

Thus, any 1 term of the Taylor expansion is less than e^λ, so a term of the expansion times $e^{-\lambda} < 1$. Thus, $0 \le p(y) \le 1$.

b. Show for Poisson random variable, y, $\sum\limits_{y=0}^{\infty} p(y) = 1$.

$$\sum_{y=0}^{\infty} p(y) = \frac{\lambda^0 e^{-\lambda}}{0!} + \frac{\lambda^1 e^{-\lambda}}{1!} + \frac{\lambda^2 e^{-\lambda}}{2!} + \frac{\lambda^3 e^{-\lambda}}{3!} + \cdots$$

$$= e^{-\lambda} \left[\frac{\lambda^0}{0!} + \frac{\lambda^1}{1!} + \frac{\lambda^2}{2!} + \frac{\lambda^3}{3!} + \cdots \right] = e^{-\lambda}(e^\lambda) = 1$$

As shown above, the terms inside the parentheses are the Taylor expansion for e^λ.

c. $E[y(y-1)] = \sum\limits_{y=0}^{\infty} y(y-1)p(y) = \sum\limits_{y=0}^{\infty} y(y-1)\dfrac{\lambda^y e^{-\lambda}}{y!}$

$$= \sum_{y=2}^{\infty} \frac{\lambda^y e^{-\lambda}}{(y-2)!} = \lambda^2 \sum_{y=2}^{\infty} \frac{\lambda^{y-2} e^{-\lambda}}{(y-2)!}$$

Let $z = y - 2$, $\Rightarrow \lambda^2 \sum\limits_{z=0}^{\infty} \dfrac{\lambda^z e^{-\lambda}}{z!} = \lambda^2$,

since $\sum\limits_{z=0}^{\infty} \dfrac{\lambda^z e^{-\lambda}}{z!} = 1$

$E[y(y-1)] = \lambda^2 = E(y^2) - E(y)$

$\Rightarrow E(y^2) = \lambda^2 + E(y) = \lambda^2 + \lambda$

4.77 $m(t) = E(e^{ty}) = \sum\limits_{y=0}^{\infty} e^{ty}\dfrac{\lambda^y e^{-\lambda}}{y!} = e^{-\lambda}\sum\limits_{y=0}^{\infty}\dfrac{(\lambda e^t)^y}{y!} = e^{-\lambda}e^{\lambda e^t}\sum\limits_{y=0}^{\infty}\dfrac{(\lambda e^t)^y e^{-\lambda e^t}}{y!}$

Let $\beta = \lambda e^t = e^{-\lambda}e^{\lambda e^t}\sum\limits_{y=0}^{\infty}\dfrac{\beta^y e^{-\beta}}{y!} = e^{-\lambda}e^{\lambda e^t} = e^{\lambda(e^t-1)} \Rightarrow m(t) = e^{\lambda(e^t-1)}$

4.79 $m(t) = \dfrac{pe^t}{1 - (1 - p)e^t} = \dfrac{pe^t}{1 - e^t + pe^t}$

$$\mu_1' = \dfrac{dm(t)}{dt}\Bigg]_{t=0} = \dfrac{pe^t(1 - e^t + pe^t) - pe^t(-e^t + pe^t)}{[1 - e^t + pe^t]^2}\Bigg]_{t=0}$$

$$= \dfrac{pe^t - p(e^t)^2 + p^2(e^t)^2 + p(e^t)^2 - p^2(e^t)^2}{[1 - e^t + pe^t]^2}\Bigg]_{t=0}$$

$$= \dfrac{pe^t}{[1 - e^t + pe^t]^2}\Bigg]_{t=0} = \dfrac{p}{p^2} = \dfrac{1}{p}$$

$$\mu_2' = \dfrac{d^2m(t)}{dt}\Bigg]_{t=0} = \left[\dfrac{pe^t[1 - e^t + pe^t]^2 - pe^t(2)[1 - e^t + pe^t](-e^t + pe^t)}{[1 - e^t + pe^t]^4}\right]_{t=0}$$

$$= \dfrac{p[1 - 1 + p]^2 - p(2)[1 - 1 + p)(-1 + p)]}{[1 - 1 + p]^4}$$

$$= \dfrac{p^3 + 2p^2(1 - p)}{[p]^4} = \dfrac{p^2[p + 2(1 - p)]}{p^4}$$

$$= \dfrac{p + 2 - 2p}{p^2} = \dfrac{2 - p}{p^2}$$

$$\sigma^2 = \mu_2' - (\mu_1')^2 = \dfrac{2 - p}{p^2} - \dfrac{1}{p^2} = \dfrac{1 - p}{p^2}$$

4.81   a.   Let y = loss next year.

For firm A,

$$E(y) = \sum_{\text{all } y} yp(y) = 0(.01) + 500(.01) + 1000(.01) + 1500(.02) + 2000(.35)$$
$$+ 2500(.30) + 3000(.25) + 3500(.02) + 4000(.01)$$
$$+ 4500(.01) + 5000(.01) = 2450$$

For firm B,

$$E(y) = \sum_{\text{all } y} yp(y) = 0(0) + 200(.01) + 700(.02) + 1200(.02) + 1700(.15)$$
$$+ 2200(.30) + 2700(.30) + 3200(.15) + 3700(.02)$$
$$+ 4200(.02) + 4700(.01) = 2450$$

b. For firm A,

$$\sigma^2 = E(y^2) - \mu^2 = \sum_{\text{all } y} y^2 p(y) - \mu^2$$

$$= 0^2(.01) + 500^2(.01) + 1000^2(.01) + 1500^2(.02) + 2000^2(.35)$$
$$+ 2500^2(.30) + 3000^2(.25) + 3500^2(.02) + 4000^2(.01) + 4500^2(.01)$$
$$+ 5000^2(.01) - 2450^2$$
$$= 6,440,000 - 6,002,500 = 437,500$$

$$\sigma = \sqrt{437,500} = 661.4378$$

For firm B,

$$\sigma^2 = E(y^2) - \mu^2 = \sum_{\text{all } y} y^2 p(y) - \mu^2$$

$$= 0^2(0) + 200^2(.01) + 700^2(.02) + 1200^2(.02) + 1700^2(.15) + 2200^2(.30)$$
$$+ 2700^2(.25) + 3200^2(.15) + 3700^2(.02) + 4200^2(.02) + 4700^2(.01)$$
$$- 2450^2$$
$$= 6,495,000 - 6,002,500 = 492,500$$

$$\sigma = \sqrt{492,500} = 701.7834$$

Since the standard deviation for firm B is larger, firm B faces the greater risk.

c. Since there is a chance of incurring an economic loss but no chance of gain for both data sets, part b was concerned with pure risk.

4.83 For the $n = 20$ trials, let S = plant discharges more than the EPA's suggested maximum and F = plant does not discharge more than the EPA's suggested maximum. Then $p = .05$ and $q = .95$. Let y = number of plants out of 20 that exceed the maximum. Then y has a binomial distribution with $n = 20$ and $p = .05$.

a. Using Table 1, with $n = 20$ and $p = .05$:

$$P(y < 1) = p(0) = .3585$$

b. $P(y \le 1) = .7358$

c. $P(y < 2) = P(y \le 1) = .7358$

d. $P(y > 1) = 1 - P(y \le 1) = 1 - .7358 = .2642$

e. $P(y \ge 3) = 1 - P(y \le 2) = 1 - .9245 = .0755$

Since the probability of observing a value of y greater than or equal to 3 is rather small (.0755), we have either seen a rare event or the executives' claim is incorrect.

4.85   The experiment consists of $n = 10$ trials. Each trial results in an S (person prefers factory shutdowns and lost jobs to waiving BAT standards) and F (person does not prefer factory shutdowns and lost jobs to waiving BAT standards). The probability of success, p, is .50 and $q = 1 - p = 1 - .50 = .50$. We assume the trials are independent. Let y = number of people who prefer factory shutdowns and lost jobs to waiving BAT standards in 10 trials. Thus y has a binomial distribution with $n = 10$ and $p = .50$.

a.   Using Table 1, $P(y = 0) = .0010$

b.   Using Table 1, $P(y \geq 5) = 1 - P(y \leq 4) = 1 - .3770 = .6230$

c.   Using Table 1, $P(y \geq 1) = 1 - P(y = 0) = 1 - .0010 = .9990$

4.87   Let y = number of shutdowns per month. Then y has a Poisson distribution with $\lambda = 6.5$.

a. $P(y \geq 5) = 1 - P(y = 0) - P(y = 1) - P(y = 2) - P(y = 3) - P(y = 4)$

$$= 1 - \frac{6.5^0 e^{-6.5}}{0!} - \frac{6.5^1 e^{-6.5}}{1!} - \frac{6.5^2 e^{-6.5}}{2!} - \frac{6.5^3 e^{-6.5}}{3!} - \frac{6.5^4 e^{-6.5}}{4!}$$

$$= 1 - .0015 - .0098 - .0318 - .0688 - .1118$$

$$= .7763$$

b. $P(y = 4) = \dfrac{6.5^4 e^{-6.5}}{4!} = .1118$

4.89 a. The sample space, values of x (in thousands) and associated probabilities are:

Sample Space	x	$p(x)$
−50, −50	−100	.6(.6) = .36
−50, −20	−70	.6(.1) = .06
−50, 30	−20	.6(.15) = .09
−50, 430	380	.6(.1) = .06
−50, 950	900	.6(.05) = .03
−20, −50	−70	.1(.6) = .06
−20, −20	−40	.1(.1) = .01
−20, 30	10	.1(.15) = .015
−20, 430	410	.1(.1) = .01
−20, 950	930	.1(.05) = .005
30, −50	−20	.15(.6) = .09
30, −20	10	.15(.1) = .015
30, 30	60	.15(.15) = .0025
30, 430	460	.15(.1) = .015
30, 950	980	.15(.05) = .0075
430, −50	380	.1(.6) = .06
430, −20	410	.1(.1) = .01
430, 30	460	.1(.15) = .015
430, 430	860	.1(.1) = .01
430, 950	1380	.1(.05) = .005
950, −50	900	.05(.6) = .03
950, −20	930	.05(.1) = .005
950, 30	980	.05(.15) = .0075
950, 430	1380	.05(.1) = .005
950, 950	1900	.05(.05) = .0025

The probability distribution of x is:

x	$p(x)$
-100,000	.36
-70,000	.12
-40,000	.01
-20,000	.18
10,000	.03
60,000	.0225
380,000	.12
410,000	.02
460,000	.03
860,000	.01
900,000	.06
930,000	.01
980,000	.015
1,380,000	.01
1,900,000	.0025

b. $E(x) = \sum_{\text{all } x} xp(x) = -100,000(.36) - 70,000(.12) - 40,000(.01) + \cdots$

$$+ 1,900,000(.0025) = 126,000$$

$V(x) = E(x^2) - \mu^2 = \sum_{\text{all } x} x^2 p(x) - \mu^2$

$$= (-100,000)^2(.36) + (-70,000)^2(.12) + (-40,000)^2(.01) + \cdots$$
$$+ 1,900,000^2(.0025) - 126,000^2 = 122,642,000,000$$

c. The probability of doubling the $100,000 investment is

$$P(x \geq 200,000) = P(x = 380,000) + P(x = 410,000) + \cdots + P(x = 1,900,000)$$
$$= .12 + .02 + .03 + .01 + .06 + .01 + .015 + .01 + .0025$$
$$= .2775$$

d. The probability of 2 dry holes is $P(x = -100,000) = .36$

From Exercise 4.11, $P(y = -50,000) = .6$

Thus, the probability of one dry hole is much greater than the probability of two dry holes.

4.91 a. Let y = number of trials until the first misread price is observed. The probability of a misread price is .001. Then y has a geometric distribution with p = .001. The probability distribution for y is

$$p(y) = .001(.999)^{y-1} \qquad y = 1, 2, \ldots$$

b. $P(y > 5) = 1 - P(y \le 5) = 1 - p(1) - p(2) - p(3) - p(4) - p(5)$

$$= 1 - .001(.999)^0 - .001(.999)^1 - .001(.999)^2$$
$$- .001(.999)^3 - .001(.999)^4 = 1 - .005 = .995$$

c. $P(y \le 3) = p(1) + p(2) + p(3) = .001(.999)^0 + .001(.999)^1 + .001(.999)^2 = .003$

Since the probability of having the first misread price within the first 3 trials is very small, there is evidence the claim is too small.

4.93 Let y = number of vehicles who use less than one-third of the acceleration lane in 5 trials. Then y has a binomial distribution with n = 5 and p = 1/6.

a. $P(y = 0) = \binom{5}{0}\left(\frac{1}{6}\right)^0\left(\frac{5}{6}\right)^{5-0} = \frac{5!}{0!(5-0)!}\left(\frac{5}{6}\right)^5 = .4019$

b. $P(y = 2) = \binom{5}{2}\left(\frac{1}{6}\right)^2\left(\frac{5}{6}\right)^{5-2} = \frac{5!}{2!(5-2)!}\left(\frac{1}{6}\right)^2\left(\frac{5}{6}\right)^3 = .1608$

4.95 Let y = number of shifts before the robot breaks down two times. Then y has a negative binomial distribution with r = 2 and p = .2.

$P(y \le 5) = p(2) + p(3) + p(4) + p(5)$

$$= \binom{2-1}{2-1}.2^2.8^{2-2} + \binom{3-1}{2-1}.2^2.8^{3-2} + \binom{4-1}{2-1}.2^2.8^{4-2}$$

$$+ \binom{5-1}{2-1}.2^2.8^{5-2}$$

$$= \frac{1!}{1!(1-1)!}.2^2 + \frac{2!}{1!(2-1)!}.2^2.8 + \frac{3!}{1!(3-1)!}.2^2.8^2 + \frac{4!}{1!(4-1)!}.2^2.8^3$$

$$= .04 + .064 + .0768 + .08192 = .26272$$

4.97 a. Let y = number of breakdowns per 8-hour shift. Then y has a Poisson distribution with $\lambda = \mu = 1.5$.

$$P(y = 2) = \frac{1.5^2 e^{-1.5}}{2!} = .2510$$

b. $P(y < 2) = p(0) + p(1) = \dfrac{1.5^0 e^{-1.5}}{0!} + \dfrac{1.5^1 e^{-1.5}}{1!} = .2231 + .3347 = .5578$

c. The probability that there are no breakdowns in one 8-hour shift is

$$P(y = 0) = p(0) = \frac{1.5^0 e^{-1.5}}{0!} = .2231$$

Let x = number of shifts with no breakdowns in 3 trials. Then x is a binomial random variable with $n = 3$ and $p = .2231$.

$$P(x = 3) = \binom{3}{3} .2231^3 \, .7769^0 = .0111$$

4.99 $E(y) = p \sum_{y=1}^{\infty} y q^{y-1}$

Using $\dfrac{dq^y}{dq} = y q^{y-1}$, we get

$$E(y) = p \frac{d}{dq} \sum_{y=1}^{\infty} q^y$$

Using $\displaystyle\sum_{y=1}^{\infty} q^y = \frac{q}{1-q}$, we get

$$E(y) = p \frac{d}{dq} \left[\frac{q}{1-q} \right] = p \left[\frac{(1-q)+q}{(1-q)^2} \right] = p \left[\frac{1}{(1-q)^2} \right] = p \cdot \frac{1}{p^2} = \frac{1}{p}$$

CHAPTER FIVE

. . . .

Continuous Random Variables

5.1 **a.** We know $\int_0^2 cy^2 dy = 1$

$$\Rightarrow \left. \frac{cy^3}{3} \right]_0^2 = \frac{c(2^3)}{3} - \frac{c(0^3)}{3} = \frac{8c}{3} - 0 = 1$$

$$\Rightarrow 8c = 3 \Rightarrow c = \frac{3}{8}$$

b. $F(y) = \int_{-\infty}^y f(t)dt = \int_0^y \frac{3}{8}t^2 dt = \frac{3}{8} \cdot \left. \frac{t^3}{3} \right]_0^y = \frac{3y^3}{24} - \frac{3(0^3)}{24} = \frac{y^3}{8}$

c. $F(1) = \frac{1^3}{8} = \frac{1}{8}$

d. $F(.5) = \frac{.5^3}{8} = .015625$

e. $P(1 \le y \le 1.5) = F(1.5) - F(1) = \frac{1.5^3}{8} - \frac{1^3}{8} = \frac{3.375 - 1}{8} = \frac{2.375}{8} = .2969$

5.3 **a.** We know $\int_0^\infty ce^{-y} dy = 1$

$$\Rightarrow \left. -ce^{-y} \right]_0^\infty = -c(0) - \left(-ce^{-0}\right) = 1 = 1$$

$$\Rightarrow c = 1$$

b. $F(y) = \int_{-\infty}^y f(t)dt = \int_0^y = e^{-t} dt = \left. -e^{-t} \right]_0^y = -e^{-y} - \left(-e^{-0}\right) = -e^{-y} + 1 = 1 - e^{-y}$

c. $F(2.6) = 1 - e^{-2.6} = 1 - .07427 = .92573$

d. $F(0) = 1 - e^{-0} = 1 - 1 = 0$
$F(\infty) = 1 - e^{-\infty} = 1 - 0 = 1$

e. $P(1 \le y \le 5) = F(5) - F(1) = (1 - e^{-5}) - (1 - e^{-1})$
$\qquad\qquad\qquad\qquad = (1 - .0067) - (1 - .3679) = .3612$

. . . .

5.5   **a.**   We know $\int_{-1}^{0}(c + y)dy + \int_{0}^{1}(c - y)dy = 1$

$$\Rightarrow cy + \frac{y^2}{2}\Bigg]_{-1}^{0} + cy - \frac{y^2}{2}\Bigg]_{0}^{1}$$

$$= c(0) + \frac{0^2}{2} - \left[c(-1) + \frac{(-1)^2}{2}\right] + c(1) - \frac{1^2}{2} - \left[c(0) - \frac{(0)^2}{2}\right]$$

$$= c - \frac{1}{2} + c - \frac{1}{2} = 2c - 1 = 1 \Rightarrow c = 1$$

   b.   For $y < 0$,

$$F(y) = \int_{\infty}^{y}(c + t)dt = \int_{-1}^{y}(1 + t)dt = t + \frac{t^2}{2}\Bigg]_{-1}^{y} = y + \frac{y^2}{2} - \left[-1 + \frac{(-1)^2}{2}\right]$$

$$= y + \frac{y^2}{2} + 1 - \frac{1}{2} = \frac{1}{2}(y^2 + 2y + 1) = \frac{1}{2}(y + 1)^2 \text{ for } y < 0$$

     For $y \geq 0$,

$$F(y) = \int_{-1}^{0}(1 + t)dt + \int_{0}^{y}(1 - t)dt = t + \frac{t^2}{2}\Bigg]_{-1}^{0} + t - \frac{t^2}{2}\Bigg]_{0}^{y}$$

$$= 0 + \frac{0^2}{2} - \left[-1 + \frac{(-1)^2}{2}\right] + y - \frac{y^2}{2} - \left[0 - \frac{(0)^2}{2}\right]$$

$$= 1 - \frac{1}{2} + y - \frac{y^2}{2} = \frac{1}{2}(1 + 2y - y^2) = \frac{1}{2}[y(2 - y) + 1] \text{ for } y \geq 0$$

   c. $F(-.5) = \frac{1}{2}(-.5 + 1)^2 = \frac{1}{2}\left[\frac{1}{2}\right]^2 = .125$

   d. $P(0 \leq y \leq .5) = F(.5) - F(0)$

$$= \frac{1}{2}[.5(2 - .5) + 1] - \frac{1}{2}[0(2 - 0) + 1] = \frac{1}{2}(1.75) - \frac{1}{2} = .375$$

5.7 $\mu = E(y) = \int_{0}^{2}\frac{3}{8}y^3 dy = \frac{3}{8} \cdot \frac{y^4}{4}\Bigg]_{0}^{2} = \frac{3(2^4)}{32} - \frac{3(0^4)}{32} = \frac{3}{2} - 0 = 1.5$

$$E(y^2) = \int_{0}^{2}\frac{3}{8}y^4 dy = \frac{3}{8} \cdot \frac{y^5}{5}\Bigg]_{0}^{2} = \frac{3(2^5)}{40} - \frac{3(0^5)}{40} = 2.4 - 0 = 2.4$$

$$\sigma^2 = E(y^2) - \mu^2 = 2.4 - 1.5^2 = .15$$

$\sigma = \sqrt{.15} = .387$

$\mu \pm 2\sigma \Rightarrow 1.5 \pm 2(.387) \Rightarrow 1.5 \pm .774 \Rightarrow (.726, 2.274)$

$$P(.726 < y < 2.274) = \int_{.726}^{2} \frac{3}{8}y^2 dy = \frac{3}{8}\frac{y^3}{3}\Big]_{.726}^{2} = \frac{3(2^3)}{24} - \frac{3(.726^3)}{24} = 1 - .048$$

$$= .952$$

This is very close to the results of the Empirical Rule.

5.9 $\mu = E(y) = \int_0^\infty ye^{-y}dy = \frac{e^{-y}}{(-1)^2}(-y-1)\Big]_0^\infty$

$$= e^{-\infty}(-\infty - 1) - e^{-0}(0-1) = 0 - 1(-1) = 1$$

$E(y^2) = \int_0^\infty y^2 e^{-y}dy = \frac{y^2 e^{-y}}{-1}\Big]_0^\infty - \frac{2}{-1}\int_0^\infty ye^{-y}dy$

$$= \frac{(\infty)^2 e^{-\infty}}{(-1)} - \frac{0^2 e^{-0}}{-1} + 2\left[\frac{e^{-y}}{(-1)^2}(-y-1)\right]_0^\infty$$

$$= 0 - 0 + 2\left[\frac{e^{-\infty}}{(-1)^2}(-\infty - 1) - \frac{-e^{-0}}{(-1)^2}(0-1)\right] = 2\left[0 - \frac{1}{1}(-1)\right] = 2(1) = 2$$

$\sigma^2 = E(y^2) - \mu^2 = 2 - 1^2 = 1$ $\sigma = \sqrt{1} = 1$

$\mu \pm 2\sigma \Rightarrow 1 \pm 2(1) \Rightarrow 1 \pm 2 \Rightarrow (-1, 3)$

$P(-1 < y < 3) = \int_0^3 e^{-y}dy = -e^{-y}\Big]_0^3 = -e^{-3} - (-e^{-0}) = -.050 + 1 = .950$

This is very close to the results of the Empirical Rule.

5.11 $\mu = E(y) = \int_{-1}^0 (y + y^2)dy + \int_0^1 (y - y^2)dy = \frac{y^2}{2} + \frac{y^3}{3}\Big]_{-1}^0 + \frac{y^2}{2} - \frac{y^3}{3}\Big]_0^1$

$$= \frac{0^2}{2} + \frac{0^3}{3} - \left[\frac{(-1)^2}{2} + \frac{(-1)^3}{3}\right] + \frac{1^2}{2} - \frac{1^3}{3} - \left[\frac{0^2}{2} - \frac{0^3}{3}\right]$$

$$= \frac{-1}{2} + \frac{1}{3} + \frac{1}{2} - \frac{1}{3} = 0$$

$$E(y^2) = \int_{-1}^{0}(y^2 + y^3)dy + \int_{0}^{1}(y^2 - y^3)dy = \left.\frac{y^3}{3} + \frac{y^4}{4}\right]_{-1}^{0} + \left.\frac{y^3}{3} - \frac{y^4}{4}\right]_{0}^{1}$$

$$= \frac{0^3}{3} + \frac{0^4}{4} - \left[\frac{(-1)^3}{3} + \frac{(-1)^4}{4}\right] + \frac{1^3}{3} - \frac{1^4}{4} - \left[\frac{0^3}{3} - \frac{0^4}{4}\right]$$

$$= \frac{1}{3} - \frac{1}{4} + \frac{1}{3} - \frac{1}{4} = \frac{4 - 3 + 4 - 3}{12} = \frac{2}{12} = \frac{1}{6}$$

$$\sigma^2 = E(y^2) - \mu^2 = \frac{1}{6} - 0^2 = \frac{1}{6} \qquad \sigma = \sqrt{1/6} = .408$$

$$P(\mu - 2\sigma < y < \mu + 2\sigma) = P[0 - 2(.408) < y < 0 + 2(.408)] = P(-.816 < y < .816)$$

$$= \int_{-.816}^{0}(1 + y)dy + \int_{0}^{.816}(1 - y)dy = \left.y + \frac{y^2}{2}\right]_{-.816}^{0} + \left.y - \frac{y^2}{2}\right]_{0}^{.816}$$

$$= 0 + \frac{0^2}{2} - \left[-.816 + \frac{(-.816)^2}{2}\right] + .816 - \frac{.816^2}{2} - \left[0 - \frac{0^2}{2}\right]$$

$$= .816 - .333 + .816 - .333 = .966$$

5.13   a.   Show $\int_{-\infty}^{\infty} f(y)dy = 1$

$$\int_{\mu-c}^{\mu}\frac{(c - \mu) + y}{c^2}dy + \int_{\mu}^{\mu+c}\frac{(c + \mu) - y}{c^2}dy$$

$$= \frac{1}{c^2}\left[(c - \mu)y + \frac{y^2}{2}\right]\Bigg]_{\mu-c}^{\mu} + \frac{1}{c^2}\left[(c + \mu)y - \frac{y^2}{2}\right]\Bigg]_{\mu}^{\mu+c}$$

$$= \frac{1}{c^2}\left\{(c - \mu)\mu + \frac{\mu^2}{2} - \left[(c - \mu)(\mu - c) + \frac{(\mu - c)^2}{2}\right]\right\}$$

$$+ \frac{1}{c^2}\left\{(c + \mu)(\mu + c) - \frac{(\mu + c)^2}{2} - \left[(c + \mu)\mu - \frac{\mu^2}{2}\right]\right\}$$

$$= \frac{1}{c^2}\left\{c\mu - \mu^2 + \frac{\mu^2}{2} + c^2 - 2c\mu + \mu^2 - \frac{\mu^2}{2} + c\mu - \frac{c^2}{2}\right\}$$

$$+ \frac{1}{c^2}\left\{c^2 + 2c\mu + \mu^2 - \frac{\mu^2}{2} - \mu c - \frac{c^2}{2} - c\mu - \mu^2 + \frac{\mu^2}{2}\right\}$$

$$= \frac{1}{c^2}\left[\frac{c^2}{2}\right] + \frac{1}{c^2}\left[\frac{c^2}{2}\right] = \frac{1}{2} + \frac{1}{2} = 1$$

b. $\quad E(y) = \int_{\mu-c}^{\mu} \dfrac{(c-\mu)y + y^2}{c^2}dy + \int_{\mu}^{\mu+c} \dfrac{(c+\mu)y - y^2}{c^2}dy$

$$= \frac{1}{c^2}\left[\frac{(c-\mu)y^2}{2} + \frac{y^3}{3}\right]\Bigg]_{\mu-c}^{\mu} + \frac{1}{c^2}\left[\frac{(c+\mu)y^2}{2} - \frac{y^3}{3}\right]\Bigg]_{\mu}^{\mu+c}$$

$$= \frac{1}{c^2}\left\{\frac{(c-\mu)\mu^2}{2} + \frac{\mu^3}{3} - \left[\frac{(c-\mu)(\mu-c)^2}{2} + \frac{(\mu-c)^3}{3}\right]\right\}$$

$$+ \frac{1}{c^2}\left\{\frac{(c+\mu)(\mu+c)^2}{2} - \frac{(\mu+c)^3}{3} - \left[\frac{(c+\mu)\mu^2}{2} - \frac{\mu^3}{3}\right]\right\}$$

$$= \frac{1}{c^2}\left\{\frac{c\mu^2}{2} - \frac{\mu^3}{2} + \frac{\mu^3}{3} - \left[\frac{c^3 - 3c^2\mu + 3c\mu^2 - \mu^3}{2} + \frac{\mu^3 - 3\mu^2 c + 3\mu c^2 - c^3}{3}\right]\right\}$$

$$+ \frac{1}{c^2}\left\{\frac{c^3 + 3c^2\mu + 3c\mu^2 + \mu^3}{2} - \frac{c^3 + 3c^2\mu + 3c\mu^2 + \mu^3}{3} - \left[\frac{c\mu^2}{2} + \frac{\mu^3}{2} + \frac{\mu^3}{3}\right]\right\}$$

$$= \frac{1}{c^2}\left\{\frac{1}{2}(c\mu^2 - \mu^3 - c^3 + 3c^2\mu - 3c\mu^2 + \mu^3) + \frac{1}{3}(\mu^3 - \mu^3 + 3\mu^2 c - 3\mu c^2 + c^3)\right\}$$

$$+ \frac{1}{c^2}\left\{\frac{1}{2}(c^3 + 3c^2\mu + 3c\mu^2 + \mu^3 - c\mu^2 - \mu^3) + \frac{1}{3}(-c^3 - 3c\mu^2 - 3c\mu - \mu^2 + \mu^3)\right\}$$

$$= \frac{1}{c^2}\left\{\frac{1}{2}(-c^3 + 3c^2\mu - 2c\mu^2) + \frac{1}{3}(3\mu^2 c - 3\mu c^2 + c^3)\right\}$$

$$+ \frac{1}{c^2}\left\{\frac{1}{2}(c^3 + 3c^2\mu + 2c\mu^2) + \frac{1}{3}(-c^3 - 3c^2\mu - 3c\mu^2)\right\}$$

$$= \frac{1}{c^2}\left\{\frac{1}{2}(6c^2\mu) + \frac{1}{3}(-6c^2\mu)\right\} = 3\mu - 2\mu = \mu$$

c. $E(y^2) = \int_{\mu-c}^{\mu} \frac{(c - \mu)y^2 + y^3}{c^2}dy + \int_{\mu}^{\mu+c} \frac{(c + \mu)y^2 - y^3}{c^2}dy$

$$= \frac{1}{c^2}\left[\frac{(c - \mu)y^3}{3} + \frac{y^4}{4}\right]\Bigg]_{\mu-c}^{\mu} + \frac{1}{c^2}\left[\frac{(c + \mu)y^3}{3} - \frac{y^4}{4}\right]\Bigg]_{\mu}^{\mu+c}$$

$$= \frac{1}{c^2}\left\{\frac{(c - \mu)\mu^3}{3} + \frac{\mu^4}{4} - \left[\frac{(c - \mu)(\mu - c)^3}{3} + \frac{(\mu - c)^4}{4}\right]\right\}$$

$$+ \frac{1}{c^2}\left\{\frac{(c + \mu)(\mu + c)^3}{3} - \frac{(\mu + c)^4}{4} - \left[\frac{(c + \mu)\mu^3}{3} - \frac{\mu^4}{4}\right]\right\}$$

$$= \frac{1}{c^2}\left\{\frac{1}{3}(c\mu^3 - \mu^4 + \mu^4 - 4\mu^3c + 6\mu^2c^2 - 4\mu c^3 + c^4)\right.$$

$$+ \frac{1}{4}(\mu^4 - \mu^4 + 4\mu^3c - 6\mu^2c^2 + 4\mu c^3 - c^4)\Big\}$$

$$+ \frac{1}{c^2}\left\{\frac{1}{3}(\mu^4 + 4\mu^3c + 6\mu^2c^2 + 4\mu c^3 + c^4 - c\mu^3 - \mu^4)\right.$$

$$+ \frac{1}{4}(-\mu^4 - 4\mu^3c - 6\mu^2c^2 - 4\mu c^3 - c^4 + \mu^4)\Big\}$$

$$= \frac{1}{c^2}\left\{\frac{1}{3}(-3c\mu^3 + 6\mu^2c^2 - 4\mu c^3 + c^4) + \frac{1}{4}(4\mu^3c - 6\mu^2c^2 + 4\mu c^3 - c^4)\right\}$$

$$+ \frac{1}{c^2}\left\{\frac{1}{3}(3\mu^3c + 6\mu^2c^2 + 4\mu c^3 + c^4) + \frac{1}{4}(-4\mu^3c - 6\mu^2c^2 - 4\mu c^3 - c^4)\right\}$$

$$= \frac{1}{c^2}\left\{\frac{1}{3}(12\mu^2c^2 + 2c^4) + \frac{1}{4}(-12\mu^2c^2 - 2c^4)\right\}$$

$$= \frac{1}{c^2}\left[4\mu^2c^2 + \frac{2}{3}c^4 - 3\mu^2c^2 - \frac{1}{2}c^4\right] = \mu^2 + \frac{c^2}{6}$$

Thus, $\sigma^2 = E(y^2) - \mu^2 = \mu^2 + \frac{c^2}{6} - \mu^2 = \frac{c^2}{6}$

5.15 $\sigma^2 = E[(y - \mu^2)] = E[y^2 - 2y\mu + \mu^2]$
$= E[y^2] - E[2y\mu] + E[\mu^2]$
$= E(y^2) - 2\mu E(y) + \mu^2$
$= E(y)^2 - 2\mu(\mu) + \mu^2$
$= E(y^2) - \mu^2$

5.17   a.   When y is uniformly distributed with $a = .5$ and $b = 2.25$,

$$\mu = \frac{a + b}{2} = \frac{.5 + 2.25}{2} = 1.375$$

$$\sigma = \frac{b - a}{\sqrt{12}} = \frac{2.25 - .5}{\sqrt{12}} = .5052$$

$$\sigma^2 = (.5052)^2 = .2552$$

b. $\mu \pm 2\sigma \Rightarrow 1.375 \pm 2(.51) \Rightarrow 1.375 \pm 1.02 \Rightarrow (.355, 2.395)$

The height of the rectangle is $\dfrac{1}{b - a} = \dfrac{1}{2.25 - .5} = \dfrac{1}{1.75} = .57$

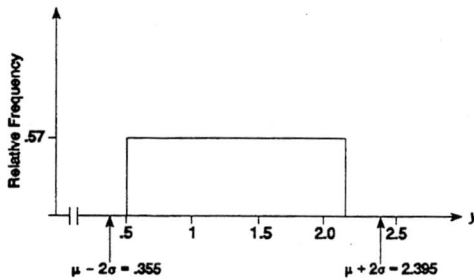

$P(\mu - 2\sigma < y < \mu + 2\sigma) = P(.355 < y < 2.395) = 1$

According to the Empirical Rule, we would expect the probability to be approximately .95. This probability is greater than what we would expect.

c. $P(y < 1) = \dfrac{c - a}{b - a} = \dfrac{1 - .5}{2.25 - .5} = \dfrac{.5}{1.75} = .286$

5.19   a. $\mu = \dfrac{a + b}{2} = \dfrac{6.5 + 7.5}{2} = \dfrac{14}{2} = 7$

$\sigma^2 = \dfrac{(b - a)^2}{12} = \dfrac{(7.5 - 6.5)^2}{12} = \dfrac{1^2}{12} = .0833$

$\sigma = \sqrt{.0833} = .289$

$$\mu \pm \sigma \Rightarrow 7 \pm .289$$
$$\Rightarrow (6.711, 7.289)$$
$$\mu \pm 2\sigma \Rightarrow 7 \pm 2(.289)$$
$$\Rightarrow (6.422, 7.578)$$

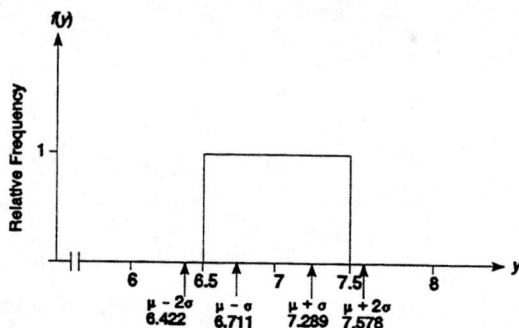

b. $P(y > 7.2) = \int_{7.2}^{7.5} 1 \, dy = y \Big]_{7.2}^{7.5} = 7.5 - 7.2 = .3$

5.21 a. $f(y) = \dfrac{1}{b - a}$

$$\mu = E(y) = \int_a^b y \left[\frac{1}{b - a} \right] dy = \frac{y^2}{2(b - a)} \Bigg]_a^b = \frac{b^2}{2(b - a)} - \frac{a^2}{2(b - a)}$$

$$= \frac{b^2 - a^2}{2(b - a)} = \frac{(b - a)(b + a)}{2(b - a)} = \frac{a + b}{2}$$

$$\sigma^2 = E(y^2) - \mu^2$$

$$E(y^2) = \int_a^b y^2 \left[\frac{1}{b - a} \right] dy = \frac{y^3}{3(b - a)} \Bigg]_a^b = \frac{b^3}{3(b - a)} - \frac{a^3}{3(b - a)} = \frac{b^3 - a^3}{3(b - a)}$$

$$\sigma^2 = E(y^2) - \mu^2 = \frac{b^3 - a^3}{3(b - a)} - \frac{(a + b)^2}{4} = \frac{4(b^3 - a^3) - 3(b - a)(b^2 + 2ab + a^2)}{12(b - a)}$$

$$= \frac{4b^3 - 4a^3 - 3b^3 - 6ab^2 - 3a^2b + 3ab^2 + 6a^2b + 3a^3}{12(b - a)}$$

$$= \frac{b^3 - 3ab^2 + 3a^2b - a^3}{12(b - a)} = \frac{(b - a)^3}{12(b - a)} = \frac{(b - a)^2}{12}$$

5.23   For $a \geq 0$, $b \geq 0$, and $(a + b) \leq 1$,

$$P(a < y < a + b) = \int_a^{a+b} 1 \, dy = y \Big]_a^{a+b} = a + b - a = b$$

5.25   a. $P(z > z_0) = .05$

$A_1 = .5 - .05 = .4500$

Look up .4500 in body of Table 4. Since .4500 is not in the Table, the two closest values are .4495 and .4505. The z-scores that correspond to .4495 and .4505 are 1.64 and 1.65. We average the two z-scores to get

$$z_0 = \frac{1.64 + 1.65}{2} = 1.645$$

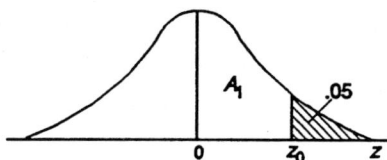

b. $P(z > z_0) = .025$

$A_1 = .5 - .025 = .4750$

Look up .4750 in body of Table 4. $z_0 = 1.96$.

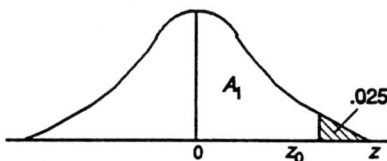

c. $P(z > z_0) = .80$

$A_1 + A_2 = .80$

$A_2 = .5$. Thus,

$A_1 = .80 - .5 = .3000$

Look up .3000 in body of Table 4. The closest value to .3000 is .2995 with a z-score of .84. Since $z_0 < 0$, $z_0 = -.84$.

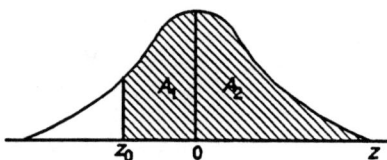

d. $P(z < z_0) = .0013$

$A_1 = .5 - .0013 = .4987$

Look up .4987 in body of Table 4. The z-score corresponding to .4987 are 3.00, 3.01, and 3.02. Since $z_0 < 0$, we will use $z_0 = -3.01$.

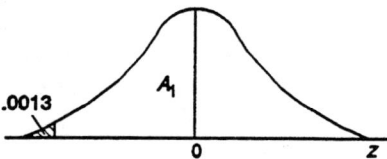

e. $P(z < z_0) = .97$

$A_1 + A_2 = .97$

$\Rightarrow A_1 = .97 - .5 = .4700$

Look up .4700 in body of Table 4. The closest value to .4700 is .4699 with a corresponding z-score of 1.88. Thus, $z_0 = 1.88$.

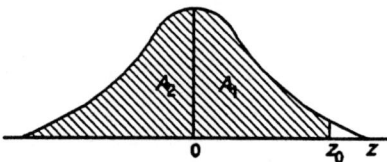

f. $P(z < z_0) = .5596$

$A_1 + A_2 = .5596$

$\Rightarrow A_1 = .5596 - .5 = .0596$

Look up .0596 in body of Table 4. The z-score corresponding to .0596 is .15. Thus, $z_0 = .15$.

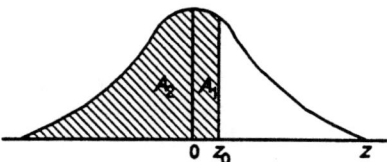

5.27   a.   Let y = ingestion time $z = \dfrac{y - \mu}{\sigma} = \dfrac{4 - 2.83}{.79} = 1.48$

$P(y \geq 4) = P(z \geq 1.48) = .5 - P(0 \leq z \leq 1.48) = .5 - .4306 = .0694$

   b. $z = \dfrac{y - \mu}{\sigma} = \dfrac{2 - 2.83}{.79} = -1.05$ $z = \dfrac{y - \mu}{\sigma} = \dfrac{3 - 2.83}{.79} = .22$

$P(2 \leq y \leq 3) = P(-1.05 \leq z \leq .22) = .3531 + .0871 = .4402$

5.29   a.   To be out-of-spec, $y < .304$ or $y > .322$

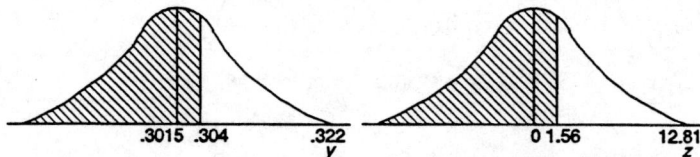

The probability we are interested in is the shaded portion of the graphs. To find this
probability, we need to find the z-values for $y = .304$ and $y = .322$.

$z = \dfrac{y - \mu}{\sigma} = \dfrac{.304 - .3015}{.0016} = 1.56$ $z = \dfrac{y - \mu}{\sigma} = \dfrac{.332 - .3015}{.0016} = 12.81$

Thus, $P(y < .304) + P(y > .322)$
$= P(z < 1.56) + P(z > 12.81)$
$= .5 + P(0 < z < 1.56) + .5 - P(0 < z < 12.81)$
$= .5 + .4406 + .5 - .5 = .9406$ from Table 4

   b.   For this part, the only things that change are μ from .3015 to .3146 and σ from .0016 to
.003. We again find the z-values for $y = .304$ and $y = .322$.

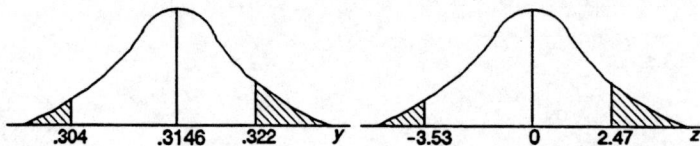

$z = \dfrac{y - \mu}{\sigma} = \dfrac{.304 - .3146}{.003} = -3.53$ $z = \dfrac{y - \mu}{\sigma} = \dfrac{.322 = .3146}{.003} = 2.47$

Thus, $P(y < .304) + P(y > .322)$
$= P(z < -3.53) + P(z > 2.47)$
$= .5 - P(-3.53 < z < 0) + .5 - P(0 < z < 2.47)$
$= .5 - .5 + .5 - .4932 = .0068$ from Table 4

5.31   a.   Let y = lifelength of standard fluorescent tube
          x = lifelength of new compact tube

$z = \dfrac{y - \mu}{\sigma} = \dfrac{9000 - 7000}{1000} = 2$

. . . .

$$P(y > 9000) = P(z > 2) = .5 - P(0 < z \le 2) = .5 - .4772 = .0228$$

$$z = \frac{x - \mu}{\sigma} = \frac{9000 - 7500}{1200} = 1.25$$

$$P(x > 9000) = P(z > 1.25) = .5 - P(0 < z \le 1.25) = .5 - .3944 = .1056$$

Thus, the new compact tube is more likely to have a lifelength greater than 9000 hours.

b. $$z = \frac{y - \mu}{\sigma} = \frac{5000 - 7000}{1000} = -2$$

$$P(y < 5000) = P(z < -2) = .5 - P(-2 < z < 0) = .5 - .4772 = .0228$$

$$z = \frac{x - \mu}{\sigma} = \frac{5000 - 7000}{1200} = -2.08$$

$$P(x < 5000) = P(z < -2.08) = .5 - P(-2.08 < z < 0) = .5 - .4812 = .0188$$

Thus, the standard fluorescent tube is more likely to have a lifelength less than 5000 hours.

5.33 a. $$z = \frac{y - \mu_T}{\sigma_T} = \frac{11 - 13.6}{\sqrt{2}} = -1.84$$

$$P(y \ge 11) = P(z \ge -1.84) = P(-1.84 \le z \le 0) + .5 = .4671 + .5 = .9671$$

b. $$z = \frac{y - \mu_N}{\sigma_N} = \frac{11 - 10.1}{\sqrt{2}} = .64$$

$$P(y \ge 11) = P(z \ge .64) = .5 - P(0 \le z \le .64) = .5 - .2389 = .2611$$

c. No. To make the probability in part a larger, C would have to be less than 11. To make the probability in part b smaller, C would have to be greater than 11.

5.35 Using SAS Procedure UNIVARIATE, the following was obtained:

Univariate Procedure

Variable=SIZE

Moments

N	28	Sum Wgts	28
Mean	17.79643	Sum	498.3
Std Dev	4.469774	Variance	19.97888
Skewness	0.777576	Kurtosis	-0.24567
USS	9407.39	CSS	539.4296
CV	25.11613	Std Mean	0.844708
T:Mean=0	21.06815	Pr>¦T¦	0.0001
Num ^=0	28	Num > 0	28
M(Sign)	14	Pr>=¦M¦	0.0001
Sgn Rank	203	Pr>=¦S¦	0.0001

100% Max	27.3	99%	27.3
75% Q3	19.65	95%	26.9
50% Med	16.75	90%	25.7
25% Q1	14.2	10%	12.7
0% Min	12.4	5%	12.4
		1%	12.4

Range	14.9
Q3-Q1	5.45
Mode	12.4

Extremes

Lowest	Obs	Highest	Obs
12.4(	11)	23.2(	24)
12.4(	1)	24.8(	4)
12.7(	28)	25.7(	18)
12.8(	27)	26.9(	8)
12.9(	3)	27.3(	9)

Univariate Procedure

Variable=SIZE

```
Stem Leaf                    #      Boxplot
29 93                        2         |
24 87                        2         |
22 2                         1         |
20 04                        2         |
18 6813                      4      +-----+
16 2336939                   7      *--+--*
14 909                       3      +-----+
12 4478925                   7         |
   ----+----+----+----+
```

Normal Probability Plot

```
27+                                *  +*++
   |                            * *+++++
   |                            *+++
   |                         ++**+
   |                       ++****
   |                     +**+***
   |                  ++**+***
   |               +*****
   |             ++****
13+    *    *  *+**+**
   +----+----+----+----+----+----+----+----+
      -2       -1        0       +1       +2
```

From the stem-and-leaf display, the data appear to be somewhat skewed to the right.

The interquartile range is 5.45 and $s = 4.47$. The ratio of IQR/s = 5.45/4.47 = 1.21. This is close to 1.3 which indicates that the data are normal.

The normal probability plot is a fairly straight line, which indicates that the data are normal.

5.37 From the histogram, the data do not appear to follow a normal distribution. The histogram is not mound-shaped, but rather looks to have two mounds.

5.39 The variable SCORE does not appear to have a normal distribution. From the stem-and-leaf display, the data appear to be skewed to the right.

The variable ENROLL does appear to have a normal distribution. From the stem-and-leaf display, the data appear to be fairly mound-shaped.

The variable RESEARCH does appear to have a normal distribution. From the stem-and-leaf display, the data appear to be fairly mound-shaped.

The variable RATIO does appear to have a normal distribution. From the stem-and-leaf display, the data appear to be fairly mound-shaped.

The variable ACCRATE does appear to have a normal distribution. From the stem-and-leaf display, the data appear to be fairly mound-shaped.

5.41 We know $\int_0^\infty cy^2 e^{-y/2} dy = 1$

We know for the gamma-type random variable $\int_0^\infty \frac{y^{\alpha-1} e^{-y/\beta}}{\beta^\alpha \Gamma(\alpha)} = 1$

Thus, $\beta = 2$, $\alpha - 1 = 2 \Rightarrow \alpha = 3$

Therefore, $c = \dfrac{1}{2^3 \Gamma(3)} = \dfrac{1}{2^3(3-1)!} = \dfrac{1}{8(2)} = \dfrac{1}{16}$

5.43 a. For the exponential distribution,

$$\mu = \beta = 1000$$
$$\sigma^2 = \beta^2 = 1000^2 = 1{,}000{,}000$$

b. $P(y \geq 2000) = \displaystyle\int_{2000}^\infty \frac{e^{-y/1000}}{1000} dy = -e^{-y/1000} \Big|_{2000}^\infty$

$= -e^{-\infty/1000} - \left(-e^{-2000/1000}\right) = 0 + e^{-2} = .135$

c. $P(y \leq 1500) = \displaystyle\int_0^{1500} \frac{e^{-y/1000}}{1000} dy = -e^{-y/1000} \Big|_0^{1500}$

$= -e^{-1500/1000} - \left(-e^{-0/1000}\right) = -e^{-1.5} + 1$

$= -.233 + 1 = .777$

5.45 a. Let y = time until next programming error.

$$P(y \geq 60) = \int_{60}^\infty \frac{1}{24^1 \Gamma(1)} y^{1-1} e^{-y/24} dy = \int_{60}^\infty \frac{e^{-y/24}}{24} dy = -e^{-y/24} \Big|_{60}^\infty$$

$= -e^{-\infty/24} - \left(-e^{-60/24}\right) = 0 + e^{-2.5} = .082$

b. If the mean time is 24 days, it is very unlikely (probability .082) that it will take at least 60 days to find the next error. We would infer the mean time is probably greater than 24.

5.47 a. Let y = number of industrial accidents per hour.

$$P(y \geq 1) = \int_1^\infty \frac{e^{-y/.5}}{.5} dy = -e^{-y/.5} \Big]_1^\infty = -e^{-\infty/.5} - \left(-e^{-1/.5}\right) = 0 + e^{-2} = .1353$$

$$P(y < 2) = \int_0^2 \frac{e^{-y/.5}}{.5} dy = -e^{-y/.5} \Big]_0^2 = -e^{-2/.5} - \left(-e^{-0/.5}\right) = -e^{-4} + 1$$
$$= -.0183 + 1 = .9817$$

5.49 $F(a) = \int_0^a \frac{e^{-y/\beta}}{\beta} dy = -e^{-y/\beta} \Big]_0^a = -e^{-a/\beta} - \left(-e^{-0/\beta}\right) = -e^{-a/\beta} + 1$ or $1 - e^{-a/\beta}$

$P(y > a) = 1 - P(y \leq a) = 1 - F(a) = 1 - \left(1 - e^{-a/\beta}\right) = e^{-a/\beta}$

5.51 Show $\Gamma(\alpha) = (\alpha - 1)\Gamma(\alpha - 1)$

From the definition, $\Gamma(\alpha) = (\alpha - 1)!$

$\Gamma(\alpha) = (\alpha - 1)! = (\alpha - 1)(\alpha - 2)! = (\alpha - 1)\Gamma(\alpha - 1)$

5.53 $f(y) = \frac{\alpha}{\beta} y^{\alpha-1} e^{-y^\alpha/\beta}$, $y \geq 0$

For $\alpha = 2$, $\beta = 100$, $f(y) = \frac{2}{100} y\, e^{-y^2/100}$ $y \geq 0$

$f(2) = \frac{2}{50} e^{-2^2/100} = \frac{2}{50} e^{-.04} = .03843$

$f(5) = \frac{5}{50} e^{-5^2/100} = \frac{5}{50} e^{-.25} = .07788$

$f(8) = \frac{8}{50} e^{-8^2/100} = \frac{8}{50} e^{-.64} = .08437$

$f(11) = \frac{11}{50} e^{-11^2/100} = \frac{11}{50} e^{-1.21} = .06560$

$f(14) = \frac{14}{50} e^{-14^2/100} = \frac{11}{50} e^{-1.96} = .03944$

$f(17) = \frac{17}{50} e^{-17^2/100} = \frac{11}{50} e^{-2.89} = .018896$

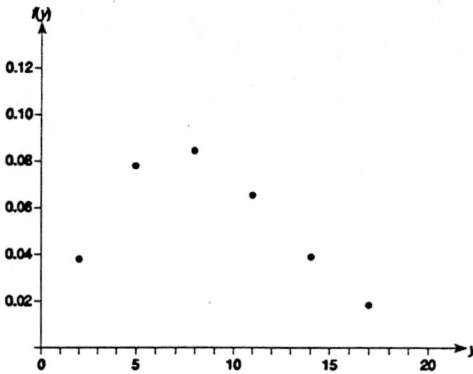

5.55 For the Weibull distribution, $F(y) = 1 - e^{-y^{\alpha}/\beta}$ (see Example 5.13)

For $\alpha = 4$, $\beta = 100$

$$F(y) = 1 - e^{-y^4/100}$$

a. $F(5) = 1 - e^{-5^4/100} = 1 - e^{-6.25} = 1 - .0019 = .9981$

b. $P(y \geq 3) = 1 - P(y < 3) = 1 - F(3) = 1 - \left[1 - e^{-3^4/100}\right] = 1 - 1 + e^{-.81}$
$$= .4449$$

c. $\mu = \beta^{1/2}\Gamma\left[\dfrac{\alpha + 1}{\alpha}\right] = 100^{1/4}\Gamma\left[\dfrac{4 + 1}{4}\right] = 3.1623\Gamma(1.25) = 3.1623(.9064) = 2.866$

$\sigma^2 = \beta^{2/\alpha}\left[\Gamma\left[\dfrac{\alpha + 2}{\alpha}\right] - \Gamma^2\left[\dfrac{\alpha + 1}{\alpha}\right]\right] = 100^{2/4}\left[\Gamma\left[\dfrac{4 + 2}{4}\right] - \Gamma^2\left[\dfrac{4 + 1}{4}\right]\right]$
$\quad = 10[\Gamma(1.5) - \Gamma^2(1.25)] = 10[.88623 - .9064^2]$
$\quad = 10[.06467] = .6467$

$\sigma = \sqrt{.6467} = .8042$

d. $P(\mu - 2\sigma \leq y \leq \mu + 2\sigma)$
$\quad = P(2.866 - 2(.8042) \leq y \leq 2.866 + 2(.8042))$
$\quad = P(1.2576 \leq y \leq 4.4744) = F(4.4744) - F(1.2576)$
$\quad = 1 - e^{-4.4744^4/100} - \left[1 - e^{-1.2576^4/100}\right]$
$\quad = 1 - e^{-4.008} - [1 - e^{-.0250}] = 1 - .0182 - [1 - .9753] = .9571$

5.57 a. $\mu = \beta^{1/\alpha}\Gamma\left[\dfrac{\alpha + 1}{\alpha}\right] = \left[\dfrac{v}{2}\right]^{1/1}\Gamma\left[\dfrac{1 + 1}{1}\right] = \dfrac{v}{2}\Gamma(1) = \dfrac{v}{2}$

For $\nu = 11.3$, $\mu = \dfrac{11.3}{2} = 5.65$

$$\sigma^2 = \beta^{2/\alpha}\left[\Gamma\left[\dfrac{\alpha + 2}{\alpha}\right] - \Gamma^2\left[\dfrac{\alpha + 1}{\alpha}\right]\right] = \left[\dfrac{\nu}{2}\right]^{2/1}\left[\Gamma\left[\dfrac{1 + 2}{1}\right] - \Gamma^2\left[\dfrac{1 + 1}{1}\right]\right]$$

$$= \left[\dfrac{\nu}{2}\right]^2\left[\Gamma(3) - \Gamma^2(2)\right] = \left[\dfrac{\nu}{2}\right]^2\left[2! - 1^2\right] = \dfrac{\nu^2}{4}$$

For $\nu = 11.3$, $\sigma^2 = \dfrac{11.3^2}{4} = 31.9225$

b. $P(y < 6) = F(6)$. From Example 5.14, $F(y) = 1 - e^{-y^\alpha/\beta}$

$\quad\quad \alpha = 1,\ \beta = \dfrac{\nu}{2} = \dfrac{11.3}{2} = 5.65$

$\quad F(6) = 1 - e^{-6^1/5.65} = 1 - e^{-1.062} = 1 - .3458 = .6542$

c. $P(y > 10) = 1 - P(y \le 10) = 1 - F(10) = 1 - \left[1 - e^{-10/5.65}\right] = e^{-1.7699} = .1703$

5.59   a. $\alpha = 2$

$\quad\quad \mu = \beta^{1/\alpha}\Gamma\left[\dfrac{\alpha + 1}{\alpha}\right] = \beta^{1/2}\Gamma\left[\dfrac{2 + 1}{2}\right] = \beta^{1/2}\Gamma(1.5) = .88623\sqrt{\beta}$

b. $\sigma^2 = \beta^{2/\alpha}\left[\Gamma\left[\dfrac{\alpha + 2}{\alpha}\right] - \Gamma^2\left[\dfrac{\alpha + 1}{\alpha}\right]\right] = \beta^{2/2}\left[\Gamma\left[\dfrac{2 + 2}{2}\right] - \Gamma^2\left[\dfrac{2 + 1}{2}\right]\right]$

$\quad\quad = \beta[\Gamma(2) - \Gamma^2(1.5)] = \beta[1 - .88623^2] = .2146\beta$

c. $P(y > C) = 1 - P(y \le C) = 1 - F(C)$

$\quad$ From Example 5.14, $F(y) = 1 - e^{-y^\alpha/\beta}$

$\quad P(y > C) = 1 - F(C) = 1 - \left[1 - e^{-C^2/\beta}\right] = e^{-C^2/\beta}$

5.61   If y has a Weibull distribution, then

$$f(y) = \begin{cases} \dfrac{\alpha}{\beta}y^{\alpha-1}e^{-y^\alpha/\beta} & 0 \le y < \infty \\ 0 & \text{elsewhere} \end{cases}$$

$$E(y^2) = \int_0^\infty \dfrac{\alpha}{\beta}y^{\alpha+1}e^{-y^\alpha/\beta}dy = \dfrac{1}{\beta}\int_0^\infty \alpha y^{\alpha-1}y^2e^{-y^\alpha/\beta}dy$$

Let $z = y^\alpha$. Then $dz = \alpha y^{\alpha-1}dy$. $y = 0 \Rightarrow z = 0^\alpha = 0$

$\quad\quad\quad\quad\quad\quad\quad\quad\quad\quad\quad\quad\quad\quad y = \infty \Rightarrow z = \infty^\alpha = \infty$

Thus, $E(y^2) = \frac{1}{\beta}\int_0^\infty z^{2/\alpha}e^{-z/\beta}dz$

We know from the Gamma distribution that $\int_0^\infty y^{\alpha-1}e^{-y/\beta} = \Gamma(\alpha)\beta^\alpha$

So, $E(y^2) = \frac{1}{\beta}\int_0^\infty z^{2/\alpha}e^{-z/\beta}dz = \frac{1}{\beta}\Gamma\left[\frac{2}{\alpha}+1\right]\beta^{(2/\alpha)+1} = \Gamma\left[\frac{2+\alpha}{\alpha}\right]\beta^{2/\alpha}$

From Exercise 5.60, $\mu = \beta^{1/\alpha}\Gamma\left[\frac{\alpha+1}{\alpha}\right]$

Thus, $\sigma^2 = E(y^2) - \mu^2 = \Gamma\left[\frac{2+\alpha}{\alpha}\right]\beta^{2/\alpha} - \beta^{2/\alpha}\Gamma^2\left[\frac{\alpha+1}{\alpha}\right]$

$$= \beta^{2/\alpha}\left[\Gamma\left[\frac{\alpha+2}{\alpha}\right] - \Gamma^2\left[\frac{\alpha+1}{\alpha}\right]\right]$$

5.63 We know $\int_0^1 \dfrac{y^{\alpha-1}(1-y)^{\beta-1}}{\dfrac{\Gamma(\alpha)\Gamma(\beta)}{\Gamma(\alpha+\beta)}} = 1$

$\Rightarrow 5 = \alpha - 1 \Rightarrow \alpha = 6$
$\Rightarrow 2 = \beta - 1 \Rightarrow \beta = 3$

Thus, $c = \dfrac{\Gamma(\alpha+\beta)}{\Gamma(\alpha)\Gamma(\beta)} = \dfrac{\Gamma(6+3)}{\Gamma(6)\Gamma(3)} = \dfrac{\Gamma(9)}{\Gamma(6)\Gamma(3)} = \dfrac{8!}{5!2!} = 168$

5.65   a.   For the beta distribution,

$$\mu = \frac{\alpha}{\alpha+\beta} = \frac{1}{1+25} = .0385$$
$$\sigma^2 = \frac{\alpha\beta}{(\alpha+\beta)^2(\alpha+\beta+1)} = \frac{1(25)}{(1+25)^2(1+25+1)} = \frac{25}{26^2(27)} = .00137$$

b.   From Section 5.9, for cases where α and β are integers,

$$P(y \le p) = F(p) = \sum_{y=\alpha}^n p(y) \qquad \text{where } p(y) \text{ is a binomial probability distribution} \\ \text{with } n = \alpha + \beta - 1 \text{ and } p.$$

Thus, $n = \alpha + \beta - 1 = 1 + 25 - 1 = 25$, $p = .01$, and

$$P(y > .01) = 1 - P(y \le .01) = 1 - F(.01)$$

$$= 1 - \sum_{y=1}^{25} p(y) = \sum_{y=0}^{0} p(y) = p(0) = .7778$$

(from Table 1, Appendix II)

5.67 **a.** Let y = proportion of small data processing firms that make a profit during their first year of operation.

From Section 5.9, for cases where α and β are integers,

$$P(y \leq p) = F(p) = \sum_{y=\alpha}^{n} p(y) \quad \text{where } p(y) \text{ is a binomial probability distribution}$$

with $n = \alpha + \beta - 1 = 5 + 6 - 1 = 10$ and p.

Thus, $P(y \leq .60) = F(.6) = \sum_{y=5}^{10} p(y) = 1 - \sum_{y=0}^{4} p(y)$

From Table 1, Appendix II, with $n = 10$, $p = .6$,

$$1 - \sum_{y=0}^{4} p(y) = 1 - .1662 = .8338$$

b. $P(y \geq .8) = 1 - F(.8) = 1 - \sum_{y=5}^{10} p(y) = \sum_{y=0}^{4} p(y)$

From Table 1, Appendix II, with $n = 10$, $p = .8$,

$$\sum_{y=0}^{4} p(y) = .0064$$

5.69 If y has a beta distribution, then

$$f(y) = \begin{cases} \dfrac{y^{\alpha-1}(1-y)^{\beta-1}}{B(\alpha, \beta)} & 0 \leq y \leq 1 \\ 0 & \text{elsewhere} \end{cases}$$

where $B(\alpha, \beta) = \dfrac{\Gamma(\alpha)\Gamma(\beta)}{\Gamma(\alpha + \beta)}$

For $\alpha = \beta = 1$, $B(\alpha, \beta) = \dfrac{\Gamma(1)\Gamma(1)}{\Gamma(1 + 1)} = \dfrac{0!0!}{1!} = 1$

Thus, $f(y) = \dfrac{y^{1-1}(1-y)^{1-1}}{1} = \begin{cases} 1 & 0 \leq y \leq 1 \\ 0 & \text{elsewhere} \end{cases}$

Therefore, y has a uniform distribution over the interval $0 \leq y \leq 1$.

5.71 From Table 5.1, the moment generating function for the normal random variable is

$$m(t) = e^{ut + t^2\sigma^2/2}$$

$$\mu_1' = \frac{dm(t)}{dt}\bigg]_{t=0} = \frac{d[e^{ut+t^2\sigma^2/2}]}{dt}\bigg]_{t=0} = e^{ut+t^2\sigma^2/2}\left[u + \frac{2t\sigma^2}{2}\right]\bigg]_{t=0} = e^0\left[\mu + \frac{2(0)\sigma^2}{2}\right] = \mu$$

$$\mu_2' = \frac{d^2m(t)}{dt^2}\bigg]_{t=0} = \frac{d\left[e^{ut+t^2\sigma^2/2}\left(\mu + \frac{2t\sigma^2}{2}\right)\right]}{dt}\bigg]_{t=0}$$

$$= e^{ut+t^2\sigma^2/2}\left[\frac{2\sigma^2}{2}\right] + \left[\mu + \frac{2t\sigma^2}{2}\right]\left[e^{ut+t^2\sigma^2/2}\left(u + \frac{2t\sigma^2}{2}\right)\right]\bigg]_{t=0}$$

$$= e^0\left[\frac{2\sigma^2}{2}\right] + \left[\mu + \frac{2(0)\sigma^2}{2}\right]\left[e^0\left(\mu + \frac{2(0)\sigma^2}{2}\right)\right] = \sigma^2 + \mu^2$$

$$\mu = \mu_1' = \mu$$
$$\sigma^2 = \mu_2' - (\mu_1')^2 = \sigma^2 + \mu^2 - \mu^2 = \sigma^2$$

5.73 $m(t) = E(e^{ty}) = \int_a^b \frac{e^{ty}}{(b-a)}dy = \frac{1}{b-a} \cdot \frac{e^{ty}}{t}\bigg]_a^b = \frac{1}{b-a}\left[\frac{e^{bt} - e^{at}}{t}\right] = \frac{e^{bt} - e^{at}}{t(b-a)}$

5.75 **a.** We know $\int_{-\infty}^{\infty} f(y)dy = 1$

Thus, $\int_0^{\infty} cye^{-y^2} = 1$

$$\Rightarrow \frac{-c}{2}e^{-y^2}\bigg]_0^{\infty} = 1 \Rightarrow \frac{-c}{2}e^{-(\infty)^2} - \left[\frac{-c}{2}e^{-0^2}\right] = 0 + \frac{c}{2} = 1 \Rightarrow c = 2$$

b. $f(y) = P(Y < y) = \int_0^y f(t)dt = \int_0^y 2te^{-t^2}dt$

Let $z = t^2 \Rightarrow z^{1/2} = t \qquad 0 \leq t \leq y \Rightarrow 0 \leq z \leq y^2$

Then $dz = 2t\,dt$ or $dt = \frac{1}{2t}dz = \frac{1}{2}z^{-1/2}dz$

$$F(y) = \int_0^{y^2} 2z^{1/2}e^{-z}\left[\frac{1}{2}z^{-1/2}\right]dz = \int_0^{y^2} e^{-z}dz = -e^{-z}\bigg]_0^{y^2} = -e^{-y^2} - (-e^{-0}) = 1 - e^{-y^2}$$

c. $P(y > 2.5) = 1 - P(y \leq 2.5) = 1 - F(2.5) = 1 - \left[1 - e^{-2.5^2}\right] = e^{-6.25} = .00193$

5.77 **a.** Let y = interrarival time

$$P(y < 10) = \int_0^{10} \frac{e^{-y/20}}{20} dy = -e^{-y/20}\Big]_0^{10} = -e^{-10/20} - \left(-e^{-0/20}\right) = -e^{-.5} + e^0$$

$$= -.6065 + 1 = .3935$$

b. Let y = number of interarrival times that are less than 10 seconds. Then y has a binomial probability distribution with $n = 4$ and $p = .3935$. (We assume the interarrival times are independent of each other.)

$$P(y = 4) = \begin{bmatrix} 4 \\ 4 \end{bmatrix} .3935^4 (.6065)^{4-4} = .3935^4 = .0240$$

c. $P(y > 60) = \int_{60}^{\infty} \frac{e^{-y/20}}{20} dy = -e^{-y/20}\Big]_{60}^{\infty} = e^{-\infty/2} - \left(-e^{-60/20}\right) = 0 + e^{-3} = .0498$

5.79 **a.** Let y = interarrival time. Then y has an exponential distribution with $\mu = 1.25$.

$$P(y < 1) = \int_0^1 \frac{1}{1.25} e^{-y/1.25} dy = -e^{-y/1.25}\Big]_0^1 = -e^{-1/1.25} - \left(-e^{-0/1.25}\right)$$

$$= -.4493 + 1 = .5507$$

b. Let y = amount of time the machine operates before breaking down. Then y has an exponential distribution with $\mu = 540$ minutes.

$$P(y > 720) = \int_{720}^{\infty} \frac{1}{540} e^{-y/540} dy = -e^{-y/540}\Big]_{720}^{\infty} = -e^{-\infty/540} - \left(-e^{-720/540}\right)$$

$$= 0 + e^{-1.3333} = .2636$$

c. Let y = repair time for the machine. Then y has a gamma distribution with $\alpha = 2$ and $\beta = 30$.

$$\mu = \alpha\beta = 2(30) = 60$$
$$\sigma^2 = \alpha\beta^2 = 2(30^2) = 1,800$$

The average repair time for the machine is 60 minutes. The standard deviation is $\sqrt{1800}$ = 42.43 minutes.

d. Define y as in part c.

$$P(y > 120) = \int_{120}^{\infty} \frac{1}{\Gamma(2)30^2} y^{2-1} e^{-y/30} dy = \int_{120}^{\infty} \frac{1}{900} y e^{-y/30} dy$$

Integration by parts:

$$\mu = y \qquad dv = \frac{1}{900}e^{-y/30}dy$$

$$du = dy \qquad v = \frac{-1}{30}e^{-y/30}$$

$$= \left. \frac{-y}{30}e^{-y/30}\right]_{120}^{\infty} + \int_{120}^{\infty}\frac{1}{30}e^{-y/30}dy$$

Now, $\lim\limits_{y\to\infty} \dfrac{-y}{30}e^{-y/30} = 0$

$$= 0 + \left. \frac{120}{30}e^{-120/30} - e^{-y/30}\right]_{120}^{\infty}$$

$$= 4e^{-4} - 0 + e^{-120/30} = 5e^{-4} = .0916$$

5.81 **a.** From Example 5.14,

$$P(y \le y_0) = F(y_0) = 1 - e^{-y_0^{\alpha}/\beta}$$

Let y = service life

$$P(y < 12.2) = F(12.2) = 1 - e^{-12.2^{1.5}/110} = 1 - e^{-.3874} = 1 - .679 = .321$$

b. $P(y < 12.2) = \displaystyle\int_0^{12.2}\frac{e^{-y/110}}{110}dy = \left. -e^{-y/110}\right]_0^{12.2} = -e^{-12.2/110} - \left(-e^{-0/110}\right)$

$$= -e^{-.1109} + 1 = -.8950 + 1 = .1050$$

This probability is much smaller than that found in part a.

5.83 Let y = time of occurrence. Then y has a uniform distribution on the interval 0 to 30.

$$P(y > 25) = \int_{25}^{30}\frac{1}{30}dy = \left.\frac{y}{30}\right]_{25}^{30} = \frac{30}{30} - \frac{25}{30} = \frac{5}{30} = \frac{1}{6}$$

5.85 **a.** $8 = \alpha - 1 \Rightarrow \alpha = 9$
$1 = \beta - 1 \Rightarrow \beta = 2$

b. $\mu = \dfrac{\alpha}{\alpha + \beta} = \dfrac{9}{9 + 2} = \dfrac{9}{11} = .818$

$$\sigma^2 = \frac{\alpha\beta}{(\alpha + \beta)^2(\alpha + \beta + 1)} = \frac{9(2)}{(9 + 2)^2(9 + 2 + 1)} = \frac{18}{121(12)} = .0124$$

c.   Let y = proportion of impurities

$$P(y > .8) = 1 - P(y \le .8) = 1 - F(.8) = 1 - \sum_{y=9}^{10} p(y) \quad \text{(from Section 5.8)}$$

where $p(y)$ is a binomial probability distribution with $n = \alpha + \beta - 1 = 9 + 2 - 1 = 10$ and $p = .8$.

From Table 1, Appendix B,

$$1 - \sum_{y=9}^{10} p(y) = 1 - \left[1 - \sum_{y=0}^{8} p(y) \right] = \sum_{y=0}^{\infty} p(y) = .624$$

5.87   From Example 5.14,

$$P(y < y_0) = F(y_0) = 1 - e^{-y_0^{\alpha}/\beta} = 1 - e^{-y_0^2/2}$$

$$P(y < 1) = 1 - e^{-1^2/2} = 1 - e^{-.5} = 1 - .6065 = .3935$$

5.89 $\int_0^1 f(y)dy = \int_0^1 \left[wg(y, \lambda_1) + (1 - w)g(1 - y, \lambda_2) \right] dy$

$$= \int_0^1 \left[w \frac{(1 - y)e^{\lambda_1 y}}{\int_0^1 (1 - y)e^{\lambda_1 y}dy} + (1 - w)\frac{ye^{\lambda_2 y}}{\int_0^1 ye^{\lambda_2 y}dy} \right] dy$$

Now, $\int_0^1 (1 - y)e^{\lambda_1 y}dy$ is constant with respect to y once it has been evaluated.

Similarly, $\int_0^1 ye^{\lambda_2 y}dy$ is constant with respect to y.

Thus, $\int_0^1 \left[w \frac{(1 - y)e^{\lambda_1 y}}{\int_0^1 (1 - y)e^{\lambda_1 y}dy} + (1 - w)\frac{ye^{\lambda_2 y}}{\int_0^1 ye^{\lambda_2 y}dy} \right] dy$

$$= \frac{w\int_0^1 (1 - y)e^{\lambda_1 y}dy}{\int_0^1 (1 - y)e^{\lambda_1 y}dy} + \frac{(1 - w)\int_0^1 ye^{\lambda_2 y}dy}{\int_0^1 ye^{\lambda_2 y}dy} = w + (1 - w) = 1$$

CHAPTER SIX

....

Bivariate Probability Distributions

6.1   a.   The properties for a discrete bivariate probability distribution are:

       1. $0 \leq p(x, y) \leq 1$ for all values of x and y

       2. $\sum_y \sum_x p(x, y) = 1$

From the table of probabilities, it is clear that $0 \leq p(x, y) \leq 1$

$$\sum_y \sum_x p(x, y) = 0 + .05 + .025 + 0 + .025 + 0 + .2 + .05 + 0 + .3 + 0$$

$$+ 0 + .1 + 0 + 0 + 0 + .1 + .15 = 1$$

   b.   To find the marginal probability distribution for x, we need to find $P(x = 0)$, $P(x = 1)$, $P(x = 2)$, $P(x = 3)$, $P(x = 4)$, and $P(x = 5)$. Since $x = 0$ can occur when $y = 0$, 1 or 2 occurs, then $P(x = 0) = p_1(0)$ is calculated by summing the probabilities of 3 mutually exclusive events:

$$P(x = 0) = p_1(0) = p(0, 0) + p(0, 1) + p(0, 2) = 0 + .200 + .100 = .300$$

Similarly,

$$P(x = 1) = p_1(1) = p(1, 0) + p(1, 1) + p(1, 2) = .050 + .050 + 0 = .100$$
$$P(x = 2) = p_1(2) = p(2, 0) + p(2, 1) + p(2, 2) = .025 + 0 + 0 = .025$$
$$P(x = 3) = p_1(3) = p(3, 0) + p(3, 1) + p(3, 2) = 0 + .300 + 0 = .300$$
$$P(x = 4) = p_1(4) = p(4, 0) + p(4, 1) + p(4, 2) = .025 + 0 + .100 = .125$$
$$P(x = 5) = p_1(5) = p(5, 0) + p(5, 1) + p(5, 2) = 0 + 0 + .150 = .150$$

The marginal probability distribution $p_1(x)$ is given as:

x	0	1	2	3	4	5
$p_1(x)$	.300	.100	.025	.300	.125	.150

   c.   The marginal probability distribution for y is found as in part a.

$$P(y = 0) = p_2(0) = p(0, 0) + p(1, 0) + p(2, 0) + p(3, 0) + p(4, 0) + p(5, 0)$$
$$= 0 + .050 + .025 + 0 + .025 + 0 = .100$$
$$P(y = 1) = p_2(1) = p(0, 1) + p(1, 1) + p(2, 1) + p(3, 1) + p(4, 1) + p(5, 1)$$
$$= .200 + .050 + 0 + .300 + 0 + 0 = .550$$
$$P(y = 2) = p_2(2) = p(0, 2) + p(1, 2) + p(2, 2) + p(3, 2) + p(4, 2) + p(5, 2)$$
$$= .100 + 0 + 0 + 0 + .100 + .150 = .350$$

The marginal probability distribution $p_1(x)$ is given as:

y	0	1	2
$p_2(y)$	.100	.550	.350

d. The conditional probability of x given y is $p_1(x \mid y) = \dfrac{p(x, y)}{p_2(y)}$

Since there are 3 levels of y, there are 3 conditional probability distributions for x.

When $y = 0$, $p_1(x \mid 0) = \dfrac{p(x, 0)}{p_2(0)}$

$$p_1(0 \mid 0) = \frac{p(0, 0)}{p_2(0)} = \frac{0}{.100} = 0$$

$$p_1(1 \mid 0) = \frac{p(1, 0)}{p_2(0)} = \frac{.050}{.100} = .500$$

$$p_1(2 \mid 0) = \frac{p(2, 0)}{p_2(0)} = \frac{.025}{.100} = .250$$

$$p_1(3 \mid 0) = \frac{p(3, 0)}{p_2(0)} = \frac{0}{.100} = 0$$

$$p_1(4 \mid 0) = \frac{p(4, 0)}{p_2(0)} = \frac{.025}{.100} = .250$$

$$p_1(5 \mid 0) = \frac{p(5, 0)}{p_2(0)} = \frac{0}{.100} = 0$$

The conditional probability distribution of x given $y = 0$ is given in the table:

x	0	1	2	3	4	5
$p_1(x \mid 0)$	0	.500	.250	0	.250	0

When $y = 1$, $p_1(x \mid 1) = \dfrac{p(x, 1)}{p_2(1)}$

$$p_1(0 \mid 1) = \frac{p(0, 1)}{p_2(1)} = \frac{.200}{.550} = .364$$

$$p_1(1 \mid 1) = \frac{p(1, 1)}{p_2(1)} = \frac{.050}{.550} = .091$$

$$p_1(2 \mid 1) = \frac{p(2, 1)}{p_2(1)} = \frac{0}{.550} = 0$$

$$p_1(3 \mid 1) = \frac{p(3, 1)}{p_2(1)} = \frac{.300}{.550} = .545$$

$$p_1(4 \mid 1) = \frac{p(4, 1)}{p_2(1)} = \frac{0}{.550} = 0$$

$$p_1(5 \mid 1) = \frac{p(5, 1)}{p_2(1)} = \frac{0}{.550} = 0$$

The conditional probability distribution of x given $y = 1$ is given in the table:

x	0	1	2	3	4	5
$p_1(x \mid 1)$	.364	.091	0	.545	0	0

When $y = 2$, $p_1(x \mid 2) = \dfrac{p(x, 2)}{p_2(2)}$

$$p_1(0 \mid 2) = \frac{p(0, 2)}{p_2(2)} = \frac{.100}{.350} = .286$$

$$p_1(1 \mid 2) = \frac{p(1, 2)}{p_2(2)} = \frac{0}{.350} = 0$$

$$p_1(2 \mid 2) = \frac{p(2, 2)}{p_2(2)} = \frac{0}{.350} = 0$$

$$p_1(3 \mid 2) = \frac{p(3, 2)}{p_2(2)} = \frac{0}{.350} = 0$$

$$p_1(4 \mid 2) = \frac{p(4, 2)}{p_2(2)} = \frac{.100}{.350} = .286$$

$$p_1(5 \mid 2) = \frac{p(5, 2)}{p_2(2)} = \frac{.150}{.350} = .429$$

The conditional probability distribution of x given $y = 2$ is given in the table:

x	0	1	2	3	4	5
$p_1(x \mid 2)$	.286	0	0	0	.286	.429

e. Similar to part d, the conditional probability of y given x is $p_2(y \mid x) = \dfrac{p(x, y)}{p_1(x)}$. Since there are 6 levels of x, there are 6 conditional probability distributions of y.

When $x = 0$, $p_2(y \mid 0) = \dfrac{p(0, y)}{p_1(0)}$

$$p_2(0 \mid 0) = \frac{p(0, 0)}{p_1(0)} = \frac{0}{.300} = 0$$

$$p_2(1 \mid 0) = \frac{p(0, 1)}{p_1(0)} = \frac{.200}{.300} = .667$$

$$p_2(2 \mid 0) = \frac{p(0, 2)}{p_1(0)} = \frac{.100}{.300} = .333$$

The conditional probability distribution for y given $x = 0$ is given in the table:

y	0	1	2
$p_2(y \mid 0)$	0	.667	.333

When $x = 1$, $p_2(y \mid 1) = \dfrac{p(1, y)}{p_1(1)}$

$$p(0 \mid 1) = \frac{p(1, 0)}{p_1(1)} = \frac{.050}{.100} = .500$$

$$p(1 \mid 1) = \frac{p(1, 1)}{p_1(1)} = \frac{.050}{.100} = .500$$

$$p(2 \mid 1) = \frac{p(1, 2)}{p_1(1)} = \frac{0}{.100} = 0$$

The conditional probability distribution for y given $x = 1$ is given in the table:

y	0	1	2
$p_2(y \mid 1)$	.500	.500	0

When $x = 2$, $p_2(y \mid 2) = \dfrac{p(2, y)}{p_1(2)}$

$$p(0 \mid 2) = \frac{p(2, 0)}{p_1(2)} = \frac{.025}{.025} = 1$$

$$p(1 \mid 2) = \frac{p(2, 1)}{p_1(2)} = \frac{0}{.025} = 0$$

$$p(2 \mid 2) = \frac{p(2, 2)}{p_1(2)} = \frac{0}{.025} = 0$$

The conditional probability distribution for y given $x = 2$ is given in the table:

y	0	1	2
$p_2(y \mid 2)$	1	0	0

When $x = 3$, $p_2(y \mid 3) = \dfrac{p(3, y)}{p_1(3)}$

$$p(0 \mid 3) = \frac{p(3, 0)}{p_1(3)} = \frac{0}{.300} = 0$$

$$p(1 \mid 3) = \frac{p(3, 1)}{p_1(3)} = \frac{.300}{.300} = 1$$

$$p(2 \mid 3) = \frac{p(3, 2)}{p_1(3)} = \frac{0}{.300} = 0$$

The conditional probability distribution for y given $x = 3$ is given in the table:

y	0	1	2
$p_2(y \mid 3)$	0	1	0

When $x = 4$, $p_2(y \mid 4) = \dfrac{p(4, y)}{p_1(4)}$

$$p(0 \mid 4) = \frac{p(4, 0)}{p_1(4)} = \frac{.025}{.125} = .200$$

$$p(1 \mid 4) = \frac{p(4, 1)}{p_1(4)} = \frac{0}{.125} = 0$$

$$p(2 \mid 4) = \frac{p(4, 2)}{p_1(4)} = \frac{.100}{.125} = .800$$

The conditional probability distribution for y given $x = 4$ is given in the table:

y	0	1	2
$p_2(y \mid 4)$	.200	0	.800

When $x = 5$, $p_2(y \mid 5) = \dfrac{p(5, y)}{p_1(5)}$

$$p(0 \mid 5) = \frac{p(5, 0)}{p_1(5)} = \frac{0}{.150} = 0$$

$$p(1 \mid 5) = \frac{p(5, 1)}{p_1(5)} = \frac{0}{.150} = 0$$

$$p(2 \mid 5) = \frac{p(5, 2)}{p_1(5)} = \frac{.150}{.150} = 1$$

The conditional probability distribution for y given $x = 5$ is given in the table:

y	0	1	2
$p_2(y \mid 5)$	0	0	1

6.3 a. The total number of ways to get two conditions is a combination of 8 conditions taken 2 at a time, or $\begin{pmatrix} 8 \\ 2 \end{pmatrix} = 28$.

The 28 events in this sample space are (where the numbers correspond to the experiment number):

1, 2						
1, 3	2, 3					
1, 4	2, 4	3, 4				
1, 5	2, 5	3, 5	4, 5			
1, 6	2, 6	3, 6	4, 6	5, 6		
1, 7	2, 7	3, 7	4, 7	5, 7	6, 7	
1, 8	2, 8	3, 8	4, 8	5, 8	6, 8	7, 8

Now, x = number of these two conditions with steel material and y = number of these two conditions with .25 inch drill size. Then $x = 0$, 1, or 2 and $y = 0$, 1, or 2.

$P(x = 0, y = 0) = P$(neither of the 2 conditions has steel and neither of the 2 conditions has .25 inch drill size)

The only pair above that has neither of the conditions steel is (1, 2), but this pair has 2 conditions with .25 drill size. Thus, $P(x = 0, y = 0) = 0$.

$P(x = 1, y = 0) = P$(one of the 2 conditions has steel and neither of the 2 conditions has .25 inch drill size)

The pairs above that have one of the conditions steel are: (1, 3); (1, 4); (1, 5); (1, 6); (1, 7); (1, 8); (2, 3); (2, 4); (2, 5); (2, 6); (2, 7); (2, 8). Of all of these pairs, none has neither of the 2 conditions with .25 drill size. Thus, $P(x = 1, y = 0) = 0$.

$P(x = 2, y = 0) = P$(both of the 2 conditions have steel and neither of the 2 conditions has .25 inch drill size)

The pairs above that have both of the conditions steel are: (3, 4); (3, 5); (3, 6); (3, 7); (3, 8); (4, 5); (4, 6); (4, 7); (4, 8); (5, 6); (5, 7); (5, 8); (6, 7); (6, 8); and (7, 8). Of all of these pairs, only the following have neither of the 2 conditions with .25 drill size: (6, 7); (6, 8); and (7, 8). Thus, $P(x = 2, y = 0) = 3/28$.

$P(x = 0, y = 1) = P$(neither of the 2 conditions has steel and one of the 2 conditions has .25 inch drill size)

None of the pairs above have neither of the 2 conditions steel and one of the conditions .25 drill size. Thus, $P(x = 0, y = 1) = 0$.

$P(x = 1, y = 1) = P$(one of the 2 conditions has steel and one of the 2 conditions has .25 inch drill size)

The pairs above that have one of the conditions steel and one of the 2 conditions .25 drill size are: (1, 6); (1, 7); (1, 8); (2, 6); (2, 7); and (2, 8). Thus, $P(x = 1, y = 1) = 6/28$.

$P(x = 2, y = 1) = P$(both of the 2 conditions have steel and one of the 2 conditions has .25 inch drill size)

The pairs above that have both of the conditions steel and one of the conditions .25 drill size are: (3, 6); (3, 7); (3, 8); (4, 6); (4, 7); (4, 8); (5, 6); (5, 7); and (5, 8). Thus, $P(x = 2, y = 1) = 9/28$.

$P(x = 0, y = 2) = P$(neither of the 2 conditions has steel and both of the 2 conditions have .25 inch drill size)

The pair above that has neither of the conditions steel and both of the conditions .25 drill size is: (1, 2). Thus, $P(x = 0, y = 2) = 1/28$.

$P(x = 1, y = 2) = P(\text{one of the 2 conditions has steel and both of the 2 conditions}$
$\qquad\qquad\qquad\text{have .25 inch drill size})$

The pairs above that have one of the conditions steel and both of the conditions .25 drill size are: (1, 3); (1, 4); (1, 5); (2, 3); (2, 4); and (2, 5). Thus, $P(x = 1, y = 2) = 6/28$.

$P(x = 2, y = 2) = P(\text{both of the 2 conditions have steel and both of the 2 conditions}$
$\qquad\qquad\qquad\text{have .25 inch drill size})$

The pairs above that have both of the conditions steel and both of the conditions .25 drill size are: (3, 4); (3, 5); and (4, 5). Thus, $P(x = 2, y = 2) = 3/28$.

In table form, the bivariate probability distribution of (x, y) is

			x	
		0	1	2
	0	0	0	3/28
y	1	0	6/28	9/28
	2	1/28	6/28	3/28

b. $P_2(y = 0) = 0 + 0 + 3/28 = 3/28$
$P_2(y = 1) = 0 + 6/28 + 9/28 = 15/28$
$P_2(y = 2) = 1/28 + 6/28 + 3/28 = 10/28$

In table form, the marginal probability distribution of y is

y	0	1	2
$P_2(y)$	3/28	15/28	10/28

c. By definition, $p_1(x \mid y) = \dfrac{p(x, y)}{p_2(y)}$

For $y = 0$, the conditional probability of x is:

$$p_1(0 \mid 0) = \frac{p(0, 0)}{p_2(0)} = \frac{0}{3/28} = 0$$

$$p_1(1 \mid 0) = \frac{p(1, 0)}{p_2(0)} = \frac{0}{3/28} = 0$$

$$p_1(2 \mid 0) = \frac{p(2, 0)}{p_2(0)} = \frac{3/28}{3/28} = 1$$

In table form, the conditional probability distribution of x given $y = 0$ is:

x	0	1	2
$p_1(x \mid 0)$	0	0	1

For $y = 1$, the conditional probability of x is:

$$p_1(0 \mid 1) = \frac{p(0, 1)}{p_2(1)} = \frac{0}{15/28} = 0$$

$$p_1(1 \mid 1) = \frac{p(1, 1)}{p_2(1)} = \frac{6/28}{15/28} = \frac{6}{15} = \frac{2}{5}$$

$$p_1(2 \mid 1) = \frac{p(2, 1)}{p_2(1)} = \frac{9/28}{15/28} = \frac{9}{15} = \frac{3}{5}$$

In table form, the conditional probability distribution of x given $y = 1$ is:

x	0	1	2
$p_1(x \mid 1)$	0	2/5	3/5

For $y = 2$, the conditional probability of x is:

$$p_1(0 \mid 2) = \frac{p(0, 2)}{p_2(2)} = \frac{1/28}{10/28} = \frac{1}{10}$$

$$p_1(1 \mid 2) = \frac{p(1, 2)}{p_2(2)} = \frac{6/28}{10/28} = \frac{6}{10}$$

$$p_1(2 \mid 2) = \frac{p(2, 2)}{p_2(2)} = \frac{3/28}{10/28} = \frac{3}{10}$$

In table form, the conditional probability distribution of x given $y = 2$ is:

x	0	1	2
$p_1(x \mid 2)$	1/10	6/10	3/10

6.5   a.   The marginal probability of y is:

$$P(y = 0) = p_2(0) = p(0, 0) + p(1, 0) + p(2, 0) + p(3, 0)$$
$$= .01 + .02 + .07 + .01 = .11$$
$$P(y = 1) = p_2(1) = p(0, 1) + p(1, 1) + p(2, 1) + p(3, 1)$$
$$= .03 + .06 + .10 + .06 = .25$$
$$P(y = 2) = p_2(2) = p(0, 2) + p(1, 2) + p(2, 2) + p(3, 2)$$
$$= .05 + .12 + .15 + .08 = .40$$
$$P(y = 3) = p_2(3) = p(0, 3) + p(1, 3) + p(2, 3) + p(3, 3)$$
$$= .02 + .09 + .08 + .05 = .24$$

The marginal probability distribution $p_2(y)$ is given as:

y	0	1	2	3
$p_2(y)$	.11	.25	.40	.24

b. The conditional probability of y, given x is 2 is $p_2(y \mid 2) = \dfrac{p(2, y)}{p_1(2)}$

$$p(x = 2) = p_1(2) = p(2, 0) + p(2, 1) + p(2, 2) + p(2, 3)$$
$$= .07 + .10 + .15 + .08 = .40$$

$$p_2(0 \mid 2) = \frac{p(2, 0)}{p_1(2)} = \frac{.07}{.40} = .175$$

$$p_2(1 \mid 2) = \frac{p(2, 1)}{p_1(2)} = \frac{.10}{.40} = .250$$

$$p_2(2 \mid 2) = \frac{p(2, 2)}{p_1(2)} = \frac{.15}{.40} = .375$$

$$p_2(3 \mid 2) = \frac{p(2, 3)}{p_1(2)} = \frac{.08}{.40} = .200$$

The conditional distribution of y given $x = 2$ is given as:

y	0	1	2	3
$p_2(y \mid 2)$	.175	.250	.375	.200

6.7 a. Verify $F_1(a) = \displaystyle\sum_{x \le a} \sum_{y} p(x, y)$

By definition, $\displaystyle\sum_{y} p(x, y) = p_1(x)$

Thus, $F_1(a) = \displaystyle\sum_{x \le a} \sum_{y} p(x, y) = \sum_{x \le a} p_1(x) = P(x \le a) = F_1(a)$

b. Verify $F_1(a \mid y) = \dfrac{\displaystyle\sum_{x \le a} p(x, y)}{p_2(y)}$

By definition, $p_1(x \mid y) = \dfrac{p(x, y)}{p_2(y)}$

Given $F_1(a \mid y) = P(x \le a \mid y)$

Thus, $F_1(a \mid y) = P(x \le a \mid y) = \dfrac{\displaystyle\sum_{x \le a} p(x, y)}{p_2(y)} = F_1(a \mid y)$

6.9   a.   Note $\int_{-\infty}^{+\infty}\int_{-\infty}^{+\infty} f(x,\,y)dx\,\,dy\,=\,1$

Therefore, $\int_{-\infty}^{+\infty}\int_{-\infty}^{+\infty} cxy\,\,dx\,\,dy\,=\,\int_0^1\int_0^1 cxy\,\,dx\,\,dy\,=\,\int_0^1 cy\dfrac{x^2}{2}\bigg]_0^1\,dy$

$$=\,\int_0^1 cy\left[\dfrac{1^2}{2}\,-\,\dfrac{0^2}{2}\right]dy\,=\,\int_0^1\dfrac{cy}{2}dy\,=\,\dfrac{cy^2}{2(2)}\bigg]_0^1\,=\,\dfrac{c}{4}\,=\,1$$

Since $\dfrac{c}{4}\,=\,1\Rightarrow c\,=\,4$

b. $f_1(x)\,=\,\int_{-\infty}^{+\infty} f(x,\,y)dy\,=\,\int_0^1 4xy\,\,dy\,=\,\dfrac{4xy^2}{2}\bigg]_0^1\,=\,\dfrac{4x}{2}\,=\,2x$

$f_2(y)\,=\,\int_{-\infty}^{+\infty} f(x,\,y)dx\,=\,\int_0^1 4xy\,\,dx\,=\,\dfrac{4yx^2}{2}\bigg]_0^1\,=\,\dfrac{4y}{2}\,=\,2y$

c. $f_1(x\mid y)\,=\,\dfrac{f(x,\,y)}{f_2(y)}$

From a, $f(x,\,y)\,=\,4xy$; and from b, $f_1(x)\,=\,2x,\,f_2(y)\,=\,2y$

Therefore, $f_1(x\mid y)\,=\,\dfrac{f(x,\,y)}{f_2(y)}\,=\,\dfrac{4xy}{2y}\,=\,2x$

$f_2(y\mid x)\,=\,\dfrac{f(x,\,y)}{f_1(x)}\,=\,\dfrac{4xy}{2x}\,=\,2y$

6.11   a.   To verify that $f(x,\,y)$ is a bivariate joint probability distribution function, we must show:

1. $f(x,\,y)\,\geq\,0$ for all $x,\,y$
2. $\int_{-\infty}^{\infty}\int_{-\infty}^{\infty} f(x,\,y)dx\,\,dy\,=\,1$

Show $f(x,\,y)\,\geq\,0$

For $x\,<\,0,\,-\infty\,<\,y\,<\,\infty$ or $x\,>\,2,\,-\infty\,<\,y\,<\,\infty,\,f(x,\,y)\,=\,0$
For $y\,<\,0,\,-\infty\,<\,x\,<\,\infty$ or $y\,>\,2,\,-\infty\,<\,x\,<\,\infty,\,f(x,\,y)\,=\,0$
For $0\,\leq\,x\,\leq\,1,\,0\,\leq\,y\,\leq\,1,\,f(x,\,y)\,=\,xy\,\geq\,0$
For $1\,\leq\,x\,\leq\,2,\,0\,\leq\,y\,\leq\,1,\,f(x,\,y)\,=\,(2\,-\,x)y\,\geq\,0$
For $0\,\leq\,x\,\leq\,1,\,1\,\leq\,y\,\leq\,2,\,f(x,\,y)\,=\,x(2\,-\,y)\,\geq\,0$
For $1\,\leq\,x\,\leq\,2,\,1\,\leq\,y\,\leq\,2,\,f(x,\,y)\,=\,(2\,-\,x)(2\,-\,y)\,\geq\,0$

Thus, $f(x,\,y)\,\geq\,0$

Show $\int_{-\infty}^{\infty}\int_{-\infty}^{\infty} f(x, y)dx\, dy = 1$

$\int_{-\infty}^{\infty}\int_{-\infty}^{\infty} f(x, y)dx\, dy$

$$= \int_0^1\int_0^1 xy\, dx\, dy + \int_0^1\int_1^2 (2 - x)y\, dx\, dy + \int_1^2\int_0^1 x(2 - y)dx\, dy$$

$$+ \int_1^2\int_1^2 (2 - x)(2 - y)dx\, dy$$

$$= \int_0^1 y\frac{x^2}{2}\Big]_0^1 dy + \int_0^1 2yx - y\frac{x^2}{2}\Big]_1^2 dy + \int_1^2 x^2 - y\frac{x^2}{2}\Big]_0^1 dy + \int_1^2 (4 - 2y)x - x^2 + y\frac{x^2}{2}\Big]_1^2 dy$$

$$= \int_0^1 y\left[\frac{1^2}{2}\right] dy + \int_0^1 2y(2 - 1) - y\left[\frac{2^2 - 1^2}{2}\right] dy + \int_1^2 1^2 - y\left[\frac{1^2}{2}\right] dy$$

$$+ \int_1^2 (4 - 2y)(2 - 1) - (2^2 - 1^2) + y\left[\frac{2^2 - 1^2}{2}\right] dy$$

$$= \int_0^1 \frac{y}{2}dy + \int_0^1 \frac{y}{2}dy + \int_1^2 1 - \frac{y}{2}dy + \int_1^2 1 - \frac{y}{2}dy$$

$$= 2\int_0^1 \frac{y}{2}dy + 2\int_1^2 1 - \frac{y}{2}dy = \int_0^1 y\, dy + \int_1^2 2 - y\, dy$$

$$= \frac{y^2}{2}\Big]_0^1 + 2y - \frac{y^2}{2}\Big]_1^2 = \frac{1^2}{2} + 2(2 - 1) - \left[\frac{2^2}{2} - \frac{1^2}{2}\right] = \frac{1}{2} + 2 - \frac{3}{2} = 1$$

Thus, $f(x, y)$ is a bivariate joint probability distribution function.

b. $P(x > .8), y > .8) = \int_{.8}^1\int_{.8}^1 xy\, dx\, dy + \int_{.8}^1\int_1^2 (2 - x)y\, dx\, dy + \int_1^2\int_{.8}^1 x(2 - y)dx\, dy$

$$+ \int_1^2\int_1^2 (2 - x)(2 - y)dx\, dy$$

$$= \int_{.8}^1 y\frac{x^2}{2}\int_{.8}^1 dy + \int_{.8}^1 2yx - y\frac{x^2}{2}\Big]_1^2 dy + \int_1^2 (2 - y)\frac{x^2}{2}\Big]_{.8}^1 dy$$

$$+ \int_1^2 (2 - y)2x - (2 - y)\frac{x^2}{2}\Big]_1^2 dy$$

$$= \int_{.8}^{1} y \left[\frac{1^2}{2} - \frac{.8^2}{2} \right] + \int_{.8}^{1} \left[2y(2-1) - y \left(\frac{2^2}{2} - \frac{1^2}{2} \right) \right] dy$$

$$+ \int_{1}^{2} (2-y) \left[\frac{1^2}{2} - \frac{.8^2}{2} \right] dy$$

$$+ \int_{1}^{2} \left[(2-y)(2(2) - 2(1)) - (2-y) \left(\frac{2^2}{2} - \frac{1^2}{2} \right) \right] dy$$

$$= \int_{.8}^{1} .18y \, dy + \int_{.8}^{1} .5y \, dy + \int_{1}^{2} (2-y).18dy + \int_{1}^{2} (2-y).5dy$$

$$= \int_{.8}^{1} .68y \, dy + \int_{1}^{2} (2-y).68dy$$

$$= .34y^2 \Big|_{.8}^{1} + 1.36y - .34y^2 \Big|_{1}^{2}$$

$$= .34(1^2 - .8^2) + 1.36(2-1) - .34(2^2 - 1^2)$$

$$= .1224 + 1.36 - 1.02 = .4624$$

6.13 a. $\int_{0}^{\infty} \int_{0}^{\infty} ce^{-(x+y)} dy \, dx$

$$= \int_{0}^{\infty} \int_{0}^{\infty} ce^{-x} e^{-y} dy \, dx = \int_{0}^{\infty} ce^{-x} dx(-e^{-y}) \Big|_{0}^{\infty}$$

$$= \int_{0}^{\infty} ce^{-x} dx[0 - (-1)] = \int_{0}^{\infty} ce^{-x} dx = -ce^{-x} \Big|_{0}^{\infty} = 0 - (-c) = c$$

$$\Rightarrow c = 1$$
$$f(x, y) = e^{-(x+y)}$$

b. $f_1(x) = \int_{0}^{\infty} e^{-(x+y)} dy = e^{-x}(-e^{-y}) \Big|_{0}^{\infty} = e^{-x}[0 - (-1)] = e^{-x}$

c. $f_2(y) = \int_{0}^{\infty} e^{-(x+y)} dx = e^{-y}(-e^{-x}) \Big|_{0}^{\infty} = e^{-y}[0 - (-1)] = e^{-y}$

d. $f_1(x|y) = \dfrac{f(x, y)}{f_2(y)} = \dfrac{e^{-x}e^{-y}}{e^{-y}} = e^{-x}$

e. $f_2(y|x) = \dfrac{f(x, y)}{f_1(x)} = \dfrac{e^{-x}e^{-y}}{e^{-x}} = e^{-y}$

f. $P(x \le 1, y \le 1) = \int_0^1 \int_0^1 e^{-x} e^{-y} dy \, dx = \int_0^1 e^{-x} dx \left[-e^{-y} \Big|_0^1 \right]$

$$= \int_0^1 e^{-x} dx [-.36788 - (-1)] = .63212 \int_0^1 e^{-x} dx$$

$$= .63212 \left(-e^{-x} \right) \Big|_0^1 = .63212[.36788 - (-1)] = .63212^2 = .39958$$

6.15 a. $E(x) = \displaystyle\sum_{x=0}^{3} x p_1(x) = 0(.11) + 1(.29) + 2(.4) + 3(.2) = 1.69$ red lights

b. $E(x + y) = E(x) + E(y)$

Need to find $E(y)$.

$$E(y) = \sum_{y=0}^{3} y p_2(y) = 0(.11) + 1(.25) + 2(.4) + 3(.24) = 1.77 \text{ red lights}$$

Therefore, $E(x + y) = E(x) + E(y) = 1.69 + 1.77 = 3.46$ red lights

6.17 a. From Exercise 6.8, $c = -1$.

Therefore, $f(x, y) = \begin{cases} x - y & \text{if } 1 \le x \le 2;\ 0 \le y \le 1 \\ 0 & \text{elsewhere} \end{cases}$

$$E(x) = \int_{-\infty}^{+\infty} x f_1(x) dx$$

We need to find the marginal distribution of x.

$$f_1(x) = \int_0^1 f(x, y) dy = \int_0^1 (x - y) dy = xy - \frac{y^2}{2} \Bigg|_0^1 = x - \frac{1}{2}$$

$$E(x) = \int_1^2 x \left(x - \frac{1}{2} \right) dx = \int_1^2 \left(x^2 - \frac{1}{2}x \right) dx = \frac{x^3}{3} - \frac{1}{2(2)} x^2 \Bigg|_1^2$$

$$= \frac{2^3}{3} - \frac{2^2}{2(2)} - \left[\frac{1^3}{3} - \frac{1^2}{2(2)} \right] = \frac{8}{3} - \frac{4}{4} - \frac{1}{3} + \frac{1}{4} = \frac{7}{3} - \frac{3}{4} = \frac{28}{12} - \frac{9}{12}$$

$$= \frac{19}{12}$$

b. We found the marginal of y which is $f_2(y) = \dfrac{3}{2} - y \quad 0 \le y \le 1$

$$\text{Therefore, } E(y) = \int_0^1 y\left[\frac{3}{2} - y\right] dy = \int_0^1 \left[\frac{3}{2}y - y^2\right] dy = \frac{3}{2(2)}y^2 - \frac{y^3}{3}\Bigg]_0^1$$

$$= \frac{3(1^2)}{4} - \frac{1^3}{3} = \frac{9}{12} - \frac{4}{12} = \frac{5}{12}$$

c. $E(x + y) = E(x) + E(y) = \dfrac{19}{12} + \dfrac{5}{12} = \dfrac{24}{12} = 2$

d.
$$E(xy) = \int_{-\infty}^{+\infty}\int_{-\infty}^{+\infty} xy\, f(x, y)dy\, dx = \int_1^2 \int_0^1 xy(x - y)dy\, dx$$

$$= \int_1^2 \int_0^1 (x^2 y - xy^2)dy\, dx = \int_1^2 \frac{2x^2 y^2}{2} - \frac{xy^3}{3}\Bigg]_0^1 dx$$

$$= \int_1^2 \frac{2x^2}{2} - \frac{x}{3}dx = \frac{x^3}{2(3)} - \frac{x^2}{3(2)}\Bigg]_1^2$$

$$= \frac{2^3}{6} - \frac{2^2}{6} - \left[\frac{1^3}{6} - \frac{1^2}{6}\right] = \frac{4}{6} = \frac{2}{3}$$

6.21 (1) $E(c) = \displaystyle\int_{-\infty}^{+\infty}\int_{-\infty}^{+\infty} cf(x, y)dy\, dx = c\int_{-\infty}^{+\infty}\int_{-\infty}^{+\infty} f(x, y)dy\, dx = c \cdot 1 = c$

(2) $E(cg(x, y)) = \displaystyle\int_{-\infty}^{+\infty}\int_{-\infty}^{+\infty} c \cdot g(x, y)f(x, y)dy\, dx$

$$= c\int_{-\infty}^{+\infty}\int_{-\infty}^{+\infty} g(x, y)f(x, y)dy\, dx = cE[g(x, y)]$$

(3) $E[g_1(x, y) + g_2(x, y) + \cdots + g_k(x, y)]$

$$= \int_{-\infty}^{+\infty}\int_{-\infty}^{+\infty} [g_1(x, y) + g_2(x, y) + \cdots + g_k(x, y)]f(x, y)dy\, dx$$

$$= \int_{-\infty}^{+\infty}\int_{-\infty}^{+\infty} [g_1(x, y)f(x, y) + g_2(x, y)f(x, y) + \cdots + g_k(x, y)f(x, y)]dy\, dx$$

$$= \int_{-\infty}^{+\infty}\int_{-\infty}^{+\infty} g_1(x, y)f(x, y)dy\, dx + \int_{-\infty}^{+\infty}\int_{-\infty}^{+\infty} g_2(x, y)f(x, y)dy\, dx$$

$$+ \int_{-\infty}^{+\infty}\int_{-\infty}^{+\infty} g_k(x, y)f(x, y)dy\, dx$$

$$= E[g_1(x, y)] + E[g_2(x, y)] + \cdots + E[g_k(x, y)]$$

6.23 From Exercise 6.2, $p(x, y) = \dfrac{1}{36}$

$$p_1(x) = \frac{1}{6}, p_2(y) = \frac{1}{6} \quad \text{for } x, y = 1, 2, \dots, 6$$

To show independence,

$$p(x, y) = p_1(x)p_2(y)$$
$$\frac{1}{36} = \frac{1}{6} \cdot \frac{1}{6}$$

Thus, x and y are independent.

6.25   From Exercise 6.4, $p(x, y) = p^{x+y}q^{2-(x+y)}$, $x = 0, 1; y = 0, 1; 0 \leq p \leq 1, q = 1 - p$

To determine if x and y are independent, we must see if $p(x, y) = p_1(x)p_2(y)$

$$p_1(x) = \sum_{y=0}^{1} p(x, y) = \sum_{y=0}^{1} p^{x+y}q^{2-(x+y)} = p^{x+0}q^{2-(x+0)} + p^{x+1}q^{2-(x+1)}$$

$$= p^x q^{2-x} + p^{x+1}q^{1-x} = p^x q^{1-x}(q + p) = p^x q^{1-x}$$

$$p_2(x) = \sum_{x=0}^{1} p(x, y) = \sum_{x=0}^{1} p^{x+y}q^{2-(x+y)} = p^{0+y}q^{2-(0+y)} + p^{1+y}q^{2-(1+y)}$$

$$= p^y q^{2-y} + p^{y+1}q^{1-y} = p^y q^{1-y}(q + p) = p^y q^{1-y}$$

Now, $p_1(x)p_2(y) = p^x q^{1-x}p^y q^{1-y} = p^{x+y}q^{2-(x+y)} = p(x, y)$. Thus, x and y are independent.

6.27   From Exercise 6.9, $f(x, y) = 4xy$ $0 \leq x \leq 1; 0 \leq y \leq 1$

$$f_1(x) = 2x; f_2(y) = 2y$$

To determine if x and y are independent, we must see if $f(x, y) = f_1(x)f_2(y)$.

$$f_1(x)f_2(y) = 2x(2y) = 4xy = f(x, y)$$

Thus, x and y are independent.

6.29 $E(xy) = \int_{\infty}^{\infty} \int_{\infty}^{\infty} xyf(x, y)dx \ dy$

$$= \int_{-\infty}^{\infty} \int_{-\infty}^{\infty} xyf_1(x)f_2(y)dx \ dy \quad \text{by independence}$$

$$= \int_{-\infty}^{\infty} \int_{-\infty}^{\infty} xf_1(x)yf_2(y)dx \ dy$$

$$= \int_{-\infty}^{\infty} xf_1(x)dx \int_{-\infty}^{\infty} yf_2(y)dy = E(x)E(y)$$

6.31   From Exercise 6.11, $f(x, y) = $

xy	$0 \leq x \leq 1; 0 \leq y \leq 1$
$(2 - x)y$	$1 \leq x \leq 2; 0 \leq y \leq 1$
$x(2 - y)$	$0 \leq x \leq 1; 1 \leq y \leq 2$
$(2 - x)(2 - y)$	$1 \leq x \leq 2; 1 \leq y \leq 2$

From Exercise 6.30, the theorem indicates that x and y are independent if we can write $f(x, y)$ $= g(x)h(y)$ where $g(x)$ is a nonnegative function of x only and $h(y)$ is a nonnegative function of y only.

For each region above, $f(x, y)$ can be written as $g(x)h(y)$.

For $xy = g(x)h(y)$ where $g(x) = x$ and $h(y) = y$. Both $g(x)$ and $h(y)$ are nonnegative functions of x and y respectively.

For $(2 - x)y = g(x)h(y)$ where $g(x) = 2 - x$ and $h(y) = y$. Both $g(x)$ and $h(y)$ are nonnegative functions of x and y respectively.

For $x(2 - y) = g(x)h(y)$ where $g(x) = x$ and $h(y) = 2 - y$. Both $g(x)$ and $h(y)$ are nonnegative functions of x and y respectively.

For $(2 - x)(2 - y) = g(x)h(y)$ where $g(x) = 2 - x$ and $h(y) = 2 - y$. Both $g(x)$ and $h(y)$ are nonnegative functions of x and y respectively.

Thus, x and y are independent.

6.33 $\text{Cov}(x, y) = E(xy) - \mu_1\mu_2$

$$E(xy) = \sum_{x=1}^{6} \sum_{y=1}^{6} xy\, p(x, y)$$

$$= 1(1)\left[\frac{1}{36}\right] + 1(2)\left[\frac{1}{36}\right] + 1(3)\left[\frac{1}{36}\right] + \cdots + 6(6)\left[\frac{1}{36}\right] = \frac{441}{36}$$

$$\mu_1 = E(x) = \sum_{x=1}^{6} xp_1(x) = 1\left[\frac{1}{6}\right] + 2\left[\frac{1}{6}\right] + \cdots + 6\left[\frac{1}{6}\right] = \frac{21}{6}$$

$$\mu_2 = E(y) = \sum_{y=1}^{6} yp_2(y) = 1\left[\frac{1}{6}\right] + 2\left[\frac{1}{6}\right] + \cdots + 6\left[\frac{1}{6}\right] = \frac{21}{6}$$

$$\text{Cov}(x, y) = \frac{441}{36} - \frac{21}{6}\left[\frac{21}{6}\right] = 0$$

6.35   a.   From Exercise 6.15, $E(x) = 1.69$ and $E(y) = 1.77$

$$\text{Cov}(x, y) = E(xy) - E(x)E(y) = \sum_{x=0}^{3} \sum_{y=0}^{3} xyp(x, y) - 1.69(1.77)$$

$= 0(0)(.01) + 0(1)(.03) + 0(2)(.05) + 0(3)(.02) + 1(0)(.02) + 1(1)(.06)$
$\quad + 1(2)(.12) + 1(3)(.09) + 2(0)(.07) + 2(1)(.10) + 2(2)(.15)$
$\quad + 2(3)(.08) + 3(0)(.01) + 3(1)(.06) + 3(2)(.08) + 3(3)(.05)$
$\quad - 1.69(1.77)$
$= 2.96 - 2.9913 = -.0313$

b. $\sigma_1^2 = E(x^2) - \mu_1^2 = \sum_{x=0}^{3} x^2 p_1(x) - 1.69^2$

$$= 0^2(.11) + 1^2(.29) + 2^2(.4) + 3^2(.2) - 2.8561 = .8339$$

$\sigma_1 = \sqrt{.8339} = .9132$

$\sigma_2^2 = E(y^2) - \mu_2^2 = \sum_{y=0}^{3} y^2 p_2(y) - 1.77^2$

$$= 0^2(.11) + 1^2(.25) + 2^2(.4) + 3^2(.24) - 3.1329 = .8771$$

$\sigma_2 = \sqrt{.8771} = .9365$

$$\rho = \frac{\text{Cov}(x, y)}{\sigma_1 \sigma_2} = \frac{-.0313}{(.9132)(.9365)} = -.0366$$

6.37 a. From Exercise 6.9,

$$f(x, y) = 4xy, f_1(x) = 2x, \text{ and } f_2(y) = 2y$$

$$E(x) = \int_0^1 x\, 2x\, dx = \left. \frac{2x^3}{3} \right|_0^1 = \frac{2(1^3)}{3} = \frac{2}{3}$$

$$E(y) = \int_0^1 y\, 2y\, dy = \left. \frac{2y^3}{3} \right|_0^1 = \frac{2(1^3)}{3} = \frac{2}{3}$$

$E(xy) = E(x)E(y)$ because x and y are independent (see Exercise 6.27)

$$= \frac{2}{3}\left[\frac{2}{3}\right] = \frac{4}{9}$$

$$\text{Cov}(x, y) = E(xy) - E(x)E(y)) = \frac{4}{9} - \frac{2}{3}\left[\frac{2}{3}\right] = \frac{4}{9} - \frac{4}{9} = 0$$

b. Thus, $\rho = \dfrac{\text{Cov}(x, y)}{\sigma_1 \sigma_2} = \dfrac{0}{\sigma_1 \sigma_2} = 0$

6.39 $\text{Cov}(x, y) = E(xy) - E(x)E(y)$

$$= \sum_{x=0}^{a_1} \sum_{y=0}^{a_2} xy\, p(x, y) - \sum_{x=0}^{a_1} x p_1(x) \sum_{y=0}^{a_2} y p_2(y)$$

If x and y are independent, then $p(x, y) = p_1(x)p_2(y)$

Thus, $\text{Cov}(x, y) = \sum\limits_{x=0}^{a_1} \sum\limits_{y=0}^{a_2} xy\, p_1(x)p_2(y) - \sum\limits_{x=0}^{a_1} xp_1(x) \sum\limits_{y=0}^{a_2} yp_2(y)$

$$= \sum\limits_{x=0}^{a_1} xp_1(x) \sum\limits_{y=0}^{a_2} yp_2(y) - \sum\limits_{x=0}^{a_1} xp_1(x) \sum\limits_{y=0}^{a_2} yp_2(y) = 0$$

6.41 $\text{Cov}(x, y) = E(xy) - E(x)E(y)$

$E(xy) = \sum\limits_{x=-1}^{1} \sum\limits_{y=-1}^{1} xy\, p(x, y)$

$= (-1)(-1)\left[\dfrac{1}{12}\right] + (-1)(0)\left[\dfrac{2}{12}\right] + (-1)(1)\left[\dfrac{1}{12}\right] + 0(-1)\left[\dfrac{2}{12}\right] + 0(0)(0)$

$+ \ 0(1)\left[\dfrac{2}{12}\right] + 1(-1)\left[\dfrac{1}{12}\right] + 1(0)\left[\dfrac{2}{12}\right] + 1(1)\left[\dfrac{1}{12}\right] = 0$

To find $E(x)$ and $E(y)$, we must find the marginal distribution of x and y.

$P(x = -1) = p_1(-1) = p(-1, -1) + p(-1, 0) + p(-1, 1)$

$$= \frac{1}{12} + \frac{2}{12} + \frac{1}{12} = \frac{4}{12} = \frac{1}{3}$$

$P(x = 0) = p_1(0) = p(0, -1) + p(0, 0) + p(0, 1)$

$$= \frac{2}{12} + 0 + \frac{2}{12} = \frac{4}{12} = \frac{1}{3}$$

$P(x = 1) = p_1(1) = p(1, -1) + p(1, 0) + p(1, 1)$

$$= \frac{1}{12} + \frac{2}{12} + \frac{1}{12} = \frac{4}{12} = \frac{1}{3}$$

$E(x) = \sum\limits_{x=-1}^{1} xp_1(x) = -1\left[\dfrac{1}{3}\right] + 0\left[\dfrac{1}{3}\right] + 1\left[\dfrac{1}{3}\right] = 0$

Similarly, $p_2(-1) = \dfrac{1}{3}$, $p_2(0) = \dfrac{1}{3}$, and $p_2(1) = \dfrac{1}{3}$

$E(y) = \sum\limits_{y=-1}^{1} yp_2(y) = -1\left[\dfrac{1}{3}\right] + 0\left[\dfrac{1}{3}\right] + 1\left[\dfrac{1}{3}\right] = 0$

$\text{Cov}(x, y) = E(xy) - E(x)E(y) = 0 - 0(0) = 0$

To show x and y are not independent, we must show $p(x, y) \neq p_1(x)p_2(y)$ for at least one pair (x, y). Let $x = -1$ and $y = -1$.

$$p(-1, -1) = \frac{1}{12}, \ p_1(-1) = \frac{1}{3}, \text{ and } p_2(-1) = \frac{1}{3}$$

$$\frac{1}{12} \neq \frac{1}{3}\left[\frac{1}{3}\right] = \frac{1}{9}$$

Thus, x and y are not independent, but $\text{Cov}(x, y) = 0$

6.43 $E(\ell) = E\left[\frac{1}{2}y_1 - y_2 + 2y_3\right] = \frac{1}{2}E(y_1) - E(y_2) + 2E(y_3) = \frac{1}{2}(0) - (-1) + 2(5) = 11$

$V(\ell) = E\left[\frac{1}{2}y_1 - y_2 + 2y_3\right]$

$\quad = \left[\frac{1}{2}\right]^2\sigma_1^2 + (-1)^2\sigma_2^2 + 2^2\sigma_3^2 + 2\left[\frac{1}{2}\right](-1)\text{Cov}(y_1, y_2) + 2\left[\frac{1}{2}\right](2)\text{Cov}(y_1, y_3)$

$\qquad + 2(-1)(2)\text{Cov}(y_2, y_3)$

$\quad = \frac{1}{4}(2) + 3 + 4(9) - 1 + 2(4) - 4(-2)$

$\quad = 54.5$

6.45 From Exercise 6.2, $p(x, y) = \frac{1}{36}$,

$\quad p_1(x) = \frac{1}{6}$ for $x = 1, 2, \ldots, 6$

$\quad$ and $p_2(y) = \frac{1}{6}$ for $y = 1, 2, 3, \ldots, 6$

$E(x) = \sum_{x=1}^{6} x p_1(x) = 1\left[\frac{1}{6}\right] + 2\left[\frac{1}{6}\right] + 3\left[\frac{1}{6}\right] + 4\left[\frac{1}{6}\right] + 5\left[\frac{1}{6}\right] + 6\left[\frac{1}{6}\right] = 3.5$

$E(y) = \sum_{y=1}^{6} y p_2(y) = 1\left[\frac{1}{6}\right] + 2\left[\frac{1}{6}\right] + 3\left[\frac{1}{6}\right] + 4\left[\frac{1}{6}\right] + 5\left[\frac{1}{6}\right] + 6\left[\frac{1}{6}\right] = 3.5$

$E(x + y) = E(x) + E(y) = 3.5 + 3.5 = 7$

$V(x + y) = \sigma_1^2 + \sigma_2^2 + 2\text{ Cov}(x, y)$

First, find σ_1^2, σ_2^2, Cov(x, y)

$E(x^2) = \sum_{x=1}^{6} x^2 p_1(x) = 1^2\left[\frac{1}{6}\right] + 2^2\left[\frac{1}{6}\right] + 3^2\left[\frac{1}{6}\right] + 4^2\left[\frac{1}{6}\right] + 5^2\left[\frac{1}{6}\right] + 6^2\left[\frac{1}{6}\right]$

$\qquad = \frac{1}{6} + \frac{4}{6} + \frac{9}{6} + \frac{16}{6} + \frac{25}{6} + \frac{36}{6} = \frac{91}{6} = 15.1667$

$\sigma_1^2 = E(x^2) - [E(x)]^2 = 15.1667 - (3.5)^2 = 2.9167$

$E(y^2) = \sum_{y=1}^{6} y^2 p_2(y) = 1^2\left[\frac{1}{6}\right] + 2^2\left[\frac{1}{6}\right] + 3^2\left[\frac{1}{6}\right] + 4^2\left[\frac{1}{6}\right] + 5^2\left[\frac{1}{6}\right] + 6^2\left[\frac{1}{6}\right]$

$\qquad = \frac{1}{6} + \frac{4}{6} + \frac{9}{6} + \frac{16}{6} + \frac{25}{6} + \frac{36}{6} = \frac{91}{6} = 15.1667$

$\sigma_2^2 = E(y^2) - [E(y)]^2 = 15.1667 - (3.5)^2 = 2.9167$

$\text{Cov}(x, y) = E(xy) - E(x)E(y)$

$$E(xy) = \sum_{x=1}^{6} \sum_{y=1}^{6} xy\, p(x, y) = 1(1)\left[\frac{1}{36}\right] + 1(2)\left[\frac{1}{36}\right] + \cdots + 6(6)\left[\frac{1}{36}\right] = \frac{441}{36} = 12.25$$

$\text{Cov}(x, y) = E(xy) - E(x)E(y) = 12.25 - 3.5(3.5) = 0$

Thus, $V(x + y) = \sigma_1^2 + \sigma_2^2 + 2\,\text{Cov}(x, y) = 2.9167 + 2.9167 + 2(0) = 5.8334$

6.47   From Exercise 6.27, we know x and y are independent. Thus, $\text{Cov}(x, y) = 0$. This implies $V(x - y) = \sigma_1^2 + (-1)^2\sigma_2^2$

From Exercise 6.9, $f_1(x) = 2x$ and $f_2(y) = 2y$

$$E(x) = \int_0^1 x\, 2x\, dx = \frac{2x^3}{3}\Bigg]_0^1 = \frac{2(1^3)}{3} = \frac{2}{3}$$

Similarly, $E(y) = \frac{2}{3}$

To find the variance of x, we first find $E(x^2)$.

$$E(x^2) = \int_0^1 x^2\, 2x\, dx = \frac{2x^4}{4}\Bigg]_0^1 = \frac{2(1^4)}{4} = \frac{2}{4} = \frac{1}{2}$$

$$\sigma_1^2 = E(x^2) - [E(x)]^2 = \frac{1}{2} - \left[\frac{2}{3}\right]^2 = \frac{1}{2} - \frac{4}{9} = \frac{1}{18}$$

Similarly, $\sigma_2^2 = \frac{1}{18}$

$$V(x - y) = \sigma_1^2 + \sigma_2^2 = \frac{1}{18} + \frac{1}{18} = \frac{2}{18} = \frac{1}{9}$$

6.49 $y_1, y_2, \ldots, y_n$ are independent random variables from a gamma distribution with $\alpha = 1$ and $\beta = 2$.

We know $\mu_i = E(y_i) = \alpha\beta = 1(2) = 2$
    and $\sigma_1^2 = V(y_i) = \alpha\beta^2 = 1(2^2) = 4$ (Section 5.7)

$$\bar{y} = \frac{\sum_{i=1}^{n} y_i}{n} = \frac{y_1}{n} + \frac{y_2}{n} + \cdots + \frac{y_n}{n}$$

From Theorem 6.7,

$$E(\bar{y}) = \frac{1}{n}E(y_1) + \frac{1}{n}E(y_2) + \cdots + \frac{1}{n}E(y_n) = \frac{1}{n}\mu_1 + \frac{1}{n}\mu_2 + \cdots + \frac{1}{n}\mu_n$$

$$= \frac{1}{n}(2) + \frac{1}{n}(2) + \cdots + \frac{1}{n}(2) = \frac{n}{n}(2) = 2$$

and

$$\sigma_{\bar{y}}^2 = V(\bar{y}) = \left[\frac{1}{n}\right]^2\sigma_1^2 + \left[\frac{1}{n}\right]^2\sigma_2^2 + \cdots + \frac{1}{n^2}\sigma_n^2 \qquad \text{(since } y_i\text{'s are independent)}$$

$$= \frac{1}{n^2}(4) + \frac{1}{n^2}(4) + \cdots + \frac{1}{n^2}(4)$$

$$= \frac{n}{n^2}(4) = \frac{4}{n}$$

A gamma distribution with $\alpha = n$ and $\beta = \dfrac{2}{n}$ has

$$\mu = \alpha\beta = n\left[\frac{2}{n}\right] = 2 \text{ and}$$

$$\sigma^2 = \alpha\beta^2 = n\left[\frac{2}{n}\right]^2 = \frac{4n}{n^2} = \frac{4}{n}$$

6.51   a. $f_1(x) = \displaystyle\int_0^1 (x + y)dy = xy + \left.\dfrac{y^2}{2}\right]_0^1 = x + \dfrac{1^2}{2} = x + \dfrac{1}{2}$

       $f_2(y) = \displaystyle\int_0^1 (x + y)dx = \left.\dfrac{x^2}{2} + xy\right]_0^1 = \dfrac{1^2}{2} + y = y + \dfrac{1}{2}$

   b. $\displaystyle\int_{-\infty}^{\infty} f_1(x)dx = \int_0^1\left[x + \dfrac{1}{2}\right]dx = \left.\dfrac{x^2}{2} + \dfrac{x}{2}\right]_0^1 = \dfrac{1^2}{2} + \dfrac{1}{2} = 1$

       $\displaystyle\int_{-\infty}^{\infty} f_2(y)dy = \int_0^1\left[\dfrac{1}{2} + y\right]dy = \left.\dfrac{y}{2} + \dfrac{y^2}{2}\right]_0^1 = \dfrac{1}{2} + \dfrac{1^2}{2} = 1$

   c. $f_1(x\mid y) = \dfrac{f(x, y)}{f_2(y)} = \dfrac{x + y}{y + \dfrac{1}{2}}$

       $f_2(y\mid x) = \dfrac{f(x, y)}{f_1(x)} = \dfrac{x + y}{x + \dfrac{1}{2}}$

d. $\int_{-\infty}^{\infty} f_1(x|y)dx = \int_0^1 \dfrac{x+y}{y+\dfrac{1}{2}}dx = \dfrac{1}{y+\dfrac{1}{2}}\int_0^1 (x+y)dx = \dfrac{1}{y+\dfrac{1}{2}}\left[\dfrac{x^2}{2}+xy\right]_0^1$

$$= \dfrac{1}{y+\dfrac{1}{2}}\left[\dfrac{1^2}{2}+y\right] = 1$$

$\int_{-\infty}^{\infty} f_2(y|x)dy = \int_0^1 \dfrac{x+y}{x+\dfrac{1}{2}}dy = \dfrac{1}{x+\dfrac{1}{2}}\int_0^1 (x+y)dy = \dfrac{1}{x+\dfrac{1}{2}}\left[xy+\dfrac{y^2}{2}\right]_0^1$

$$= \dfrac{1}{x+\dfrac{1}{2}}\left[x+\dfrac{1^2}{2}\right] = 1$$

e. If $\rho \neq 0$, then x and y are correlated. We know

$$\rho = \dfrac{\text{Cov}(x,y)}{\sigma_1\sigma_2}. \text{ Thus, if } \text{Cov}(x,y) \neq 0, \text{ then } \rho \neq 0.$$

$\text{Cov}(x,y) = E(xy) - E(x)E(y)$

$$E(x) = \int_0^1 x\left(x+\dfrac{1}{2}\right)dx = \int_0^1 \left[x^2 - \dfrac{x}{2}\right]dx = \dfrac{x^3}{3}+\dfrac{x^2}{2(2)}\Big|_0^1 = \dfrac{1^3}{3}+\dfrac{1^2}{4} = \dfrac{7}{12}$$

Similarly, $E(y) = \dfrac{7}{12}$

$E(xy) = \int_0^1\int_0^1 xy(x+y)dx\,dy = \int_0^1\int_0^1 (x^2y + xy^2)dx\,dy = \int_0^1\left[\dfrac{x^3y}{3}+\dfrac{x^2y^2}{2}\right]_0^1 dy$

$$= \int_0^1\left[\dfrac{1^3y}{3}+\dfrac{1^2y^2}{2}\right]dy = \int_0^1\left[\dfrac{y}{3}+\dfrac{y^2}{2}\right]dy = \dfrac{y^2}{2(3)}+\dfrac{y^3}{3(2)}\Big|_0^1$$

$$= \dfrac{1^2}{6}+\dfrac{1^3}{6} = \dfrac{2}{6} = \dfrac{1}{3}$$

$$\text{Cov}(x,y) = E(xy) = E(x)E(y) = \dfrac{1}{3} - \dfrac{7}{12}\left(\dfrac{7}{12}\right) = \dfrac{1}{3}-\dfrac{49}{144} = -\dfrac{1}{144}$$

Since the Cov$(x, y) \neq 0$, x and y are correlated.

To determine if x and y are independent, we must see if $f(x, y) = f_1(x)f_2(y)$

From a, $f_1(x)f_2(y) = \left[x + \dfrac{1}{2}\right]\left[y + \dfrac{1}{2}\right] = xy + \dfrac{x}{2} + \dfrac{y}{2} + \dfrac{1}{4}$

$f_1(x)f_2(y) = x + y$

Since $f(x, y) \neq f_1(x)f_2(y)$, x and y are not independent.

f. $E(d) = E\left[1 - \dfrac{(x + y)}{2}\right] = 1 - \dfrac{1}{2}[E(x) + E(y)] = 1 - \dfrac{1}{2}\left[\dfrac{7}{12} + \dfrac{7}{12}\right]$

$$= 1 - \dfrac{1}{2}\left[\dfrac{14}{12}\right] = \dfrac{10}{24} = \dfrac{5}{12}$$

$V(d) = V\left[1 - \dfrac{x + y}{2}\right] = V\left[\dfrac{-1}{2}x + \dfrac{-1}{2}y\right]$

$$= \left[\dfrac{-1}{2}\right]^2 V(x) + \left[\dfrac{-1}{2}\right]^2 V(y) + 2\left[\dfrac{-1}{2}\right]\left[\dfrac{-1}{2}\right]\text{Cov}(x, y)$$

$V(x) = E(x^2) - [E(x)]^2$

$E(x^2) = \int_0^1 x^2\left[x + \dfrac{1}{2}\right]dx = \int_0^1\left[x^3 + \dfrac{x^2}{2}\right]dx = \dfrac{x^4}{4} + \dfrac{x^3}{3(2)}\Bigg]_0^1 = \dfrac{1^4}{4} + \dfrac{1^3}{6} = \dfrac{5}{12}$

$V(x) = E(x^2) - [E(x)]^2 = \dfrac{5}{12} - \left[\dfrac{7}{12}\right]^2 = \dfrac{5}{12} - \dfrac{49}{144} = \dfrac{11}{144}$

Similarly, $V(y) = \dfrac{11}{144}$

$V(d) = \dfrac{1}{4}\left[\dfrac{11}{144}\right] + \dfrac{1}{4}\left[\dfrac{11}{144}\right] + \dfrac{2}{4}\left[-\dfrac{1}{144}\right] = \dfrac{20}{576} = \dfrac{5}{144}$

We would expect d to fall within 3 standard deviations of the mean.

$$\mu \pm 3\sigma \Rightarrow \dfrac{5}{12} \pm 3\sqrt{\dfrac{5}{144}} \Rightarrow .42 \pm .559 \Rightarrow (-.139, .979)$$

6.53 $f_2(y \mid x)f_1(x) = \dfrac{f(x, y)}{f_1(x)} \cdot f_1(x) = f(x, y)$

$f_1(x \mid y)f_2(y) = \dfrac{f(x, y)}{f_2(y)} \cdot f_2(y) = f(x, y)$

Thus, $f_2(y \mid x)f_1(x) = f_1(x \mid y)f_2(y)$

6.55   **a.**   To find c,

$$\int_0^1 \int_0^2 \int_0^\infty c(y_1 + y_2)e^{-y_3}dy_3dy_2dy_1$$

$$= \int_0^1 \int_0^2 c(y_1 + y_2)\left(-e^{-y_3}\right)\Big]_0^\infty dy_2dy_1 = \int_0^1 \int_0^2 c(y_1 + y_2)(1)dy_2dy_1$$

$$= c\int_0^1 \left. y_1y_2 + \frac{y_2^2}{2}\right]_0^2 dy_1 = c\int_0^1 \left[2y_1 + \frac{2^2}{2}\right]dy_1 = c\int_0^1 (2y_1 + 2)dy_1$$

$$= c\left(y_1^2 + 2y_1\right]_0^1\right) = c(1^2 + 2(1)) = c(3)$$

Since $3c = 1 \Rightarrow c = \dfrac{1}{3}$

b. $f_1(y_1) = \displaystyle\int_0^2 \int_0^\infty \frac{1}{3}(y_1 + y_2)e^{-y_3}dy_3dy_2 = \frac{1}{3}\int_0^2 (y_1 + y_2)\left(-e^{-y_3}\right)\Big]_0^\infty dy_2$

$$= \frac{1}{3}\int_0^2 (y_1 + y_2)(1)dy_2 = \frac{1}{3}\left[\left. y_1y_2 + \frac{y_2^2}{2}\right]\right]_0^2 = \frac{1}{3}\left[2y_1 + \frac{2^2}{2}\right] = \frac{1}{3}(2y_1 + 2)$$

$f_2(y_2) = \displaystyle\int_0^1 \int_0^\infty \frac{1}{3}(y_1 + y_2)e^{-y_3}dy_3dy_1 = \frac{1}{3}\int_0^1 (y_1 + y_2)\left(-e^{-y_3}\right)\Big]_0^\infty dy_1$

$$= \frac{1}{3}\int_0^1 (y_1 + y_2)(1)dy_1 = \frac{1}{3}\left[\left.\frac{y_1^2}{2} + y_1y_2\right]\right]_0^1 = \frac{1}{3}\left[\frac{1^2}{2} + y_2\right] = \frac{1}{3}\left[y_2 + \frac{1}{2}\right]$$

$f_3(y_3) = \displaystyle\int_0^1 \int_0^2 \frac{1}{3}(y_1 + y_2)e^{-y_3}dy_2dy_1 = \frac{1}{3}e^{-y_3}\int_0^1 \left[\left. y_1y_2 + \frac{y_2^2}{2}\right]\right]_0^2 dy_1$

$$= \frac{1}{3}e^{-y_3}\int_0^1 \left[2y_1 + \frac{2^2}{2}\right]dy_1 = \frac{1}{3}e^{-y_3}\int_0^1 (2y_1 + 2)dy_1$$

$$= \frac{1}{3}e^{-y_3}\left(y_1^2 + 2y_1\right)\Big]_0^1 = \frac{1}{3}e^{-y_3}(1^2 + 2) = e^{-y_3}$$

$$f_1(y_1)f_2(y_2)f_3(y_3) = \frac{1}{3}(2y_1 + 2)\frac{1}{3}\left[y_2 + \frac{1}{2}\right]e^{-y_3} \neq f(y_1, y_2, y_3)$$

Thus, the 3 variables are not independent.

CHAPTER SEVEN

. . . .

Sampling Distributions

7.1   The data points are numbered from 1 to 144. Since there are 144 observations in the data set, we will have to use 3 digits from the Random Number Table. We select a starting point in the Random Number Table (Table 6) at random. Suppose we start at row 55, column 6, use the middle three digits, and go down the column. Any three digits between 1 and 144 can be used. If the random number is between 145 and 999, 144 is repeatedly subtracted from the random number until a number between 1 and 144 is obtained. This procedure is followed until 10 unique numbers are obtained.

The first random number selected is 579. Since this number is greater than 144, we subtract 144 from it and get $579 - 144 = 435$. This number is still greater than 144, so 144 is subtracted from 435 to obtain 291. Again 144 is subtracted from 291 to get 147. Finally, 144 is subtracted from 147 to get 3. Since 3 is between 1 and 144, it is the first observation selected. The rest of the sample is selected in a similar manner and is:

Random Number	Observation Number	DDT Value
579	3	23.00
239	95	5.70
216	72	15.00
413	125	12.00
448	16	9.10
258	114	0.43
736	16*	
566	134	3.90
545	113	44.00
450	18	4.10
564	132	2.40

*Observation 16 was previous selected.

7.3   The data points are numbered 1 to 390. Since there are 390 observations in the data set, we will have to use 3 digits from the Random Number Table. We select a starting point in the Random Number Table (Table 6) at random. Suppose we start at row 37, column 13, use the last three digits, and go down the column. Any three digits between 1 and 390 can be used. If the random number is between 391 and 999, 390 is repeatedly subtracted from the random number until a number between 1 and 390 is obtained. This procedure is followed until 5 unique numbers are obtained.

. . . .

The first random number selected is 103. The second random number selected is 562. Since this number is greater than 390, we subtract 390 from it and get $562 - 390 = 172$. This number is between 1 and 390; it is the second observation selected. The rest of the sample is selected in a similar manner and is:

Random Number	Observation Number	% Iron
103	103	64.96
562	172	66.34
509	119	65.75
490	100	65.19
880	100*	
775	385	66.54

*Observation 100 was previous selected.

7.7   a.   Let $w = y^2$. The range for y is $0 \leq y \leq 1$, which implies

$$0^2 \leq w \leq 1^2 \text{ or } 0 \leq w \leq 1$$

$P(w \leq w_0) = G(w) = F(y_0) = P(y \leq y_0)$. This implies $y_0 = \sqrt{w_0}$

$$F(y_0) = F\left(\sqrt{w_0}\right) = \int_0^{\sqrt{w_0}} 2y \, dy = y^2 \Big]_0^{\sqrt{w_0}} = \left(\sqrt{w_0}\right)^2 - 0^2 = w_0$$

The cumulative distribution for w is $F(w) = w$

The density function for w is $\dfrac{dF(w)}{dw} = \dfrac{d(w)}{dw} = 1 = f(w)$

Thus, $f(w) = \begin{cases} 1 & 0 \leq w \leq 1 \\ 0 & \text{elsewhere} \end{cases}$

b.   Let $w = 2y - 1$. The range for y is $0 \leq y \leq 1$, which implies

$$2(0) - 1 \leq w \leq 2(1) - 1 \text{ or } -1 \leq w \leq 1$$

$P(w \leq w_0) = G(w) = F(y_0) = P(y \leq y_0)$. This implies $y_0 = \dfrac{w + 1}{2}$

$$F(y_0) = F\left[\frac{w_0 + 1}{2}\right] = \int_0^{\frac{w_0+1}{2}} 2y \, dy = y^2 \Big]_0^{\frac{w_0+1}{2}} = \left[\frac{w_0 + 1}{2}\right]^2 - 0^2$$

$$= \frac{1}{4}\left(w_0^2 + 2w_0 + 1\right)$$

The cumulative distribution for w is $F(w) = \dfrac{1}{4}(w^2 + 2w + 1)$

The density function for w is

$$\frac{dF(w)}{dw} = \frac{d\left[\frac{1}{4}[w^2 + 2w + 1]\right]}{dw} = \frac{1}{4}(2w + 2) = \frac{1}{2}(w + 1)$$

Thus, $f(w) = \begin{cases} \frac{1}{2}(w + 1) & -1 \le w \le 1 \\ 0 & \text{elsewhere} \end{cases}$

c.   Let $w = \frac{1}{y}$. The range for y is $0 \le y \le 1$, which implies

$$\frac{1}{0} \ge w \ge \frac{1}{1} \text{ or } 1 \le w \le \infty$$

$$P(w \le w_0) = G(w_0) = P(y \ge y_0) = 1 - F(y_0)$$

Let $y_0 = \frac{1}{w_0}$

$$F(y_0) = F\left[\frac{1}{w_0}\right] = \int_0^{\frac{1}{w_0}} 2y \, dy = y^2 \Big]_0^{\frac{1}{w_0}} = \left[\frac{1}{w_0}\right]^2 - 0^2 = \frac{1}{w_0^2}$$

The cumulative distribution function for w is $F(w) = 1 - F(y) = 1 - \frac{1}{w^2}$

The density function for w is $\dfrac{dF(w)}{dw} = \dfrac{d\left[1 - \dfrac{1}{w^2}\right]}{dw} = \dfrac{2w}{w^4} = \dfrac{2}{w^3}$

Thus, $f(w) = \begin{cases} \dfrac{2}{w^3} & 1 \le w \le \infty \\ 0 & \text{elsewhere} \end{cases}$

7.9 $f(y) = \begin{cases} \dfrac{1}{5}e^{-y/5} & 0 \le y \le \infty \\ 0 & \text{elsewhere} \end{cases}$

$c = 3y + 2$

The range of y is $0 \le y < \infty$ which implies $3(0) + 2 \le c < 3(\infty) + 2$ or $2 \le c < \infty$

$$P(c \le c_0) = G(c_0) = F(y_0) = P(y \le y_0)$$

This implies $y_0 = \dfrac{c_0 - 2}{3}$

$$F(y_0) = F\left[\frac{c_0 - 2}{3}\right] = \int_0^{\frac{c_0-2}{3}} \frac{1}{5}e^{-y/5}dy = -e^{-y/5}\Big]_0^{\frac{c_0-2}{3}}$$

$$= -e^{\left(\frac{c_0-2}{3(5)}\right)} - \left(-e^{-0/5}\right) = 1 - e^{-\frac{c_0-2}{15}}$$

The cumulative distribution for c is $F(c) = 1 - e^{-\frac{c-2}{15}}$

The density function for c is $\dfrac{dF(c)}{dc} = \dfrac{d\left[1 - e^{-\frac{c-2}{15}}\right]}{dc} = -e^{-\frac{c-2}{15}}\left(-\frac{1}{15}\right) = \frac{1}{15}e^{-\frac{c-2}{15}}$

Thus, $f(c) = \begin{cases} \dfrac{1}{15}e^{-\frac{c-2}{15}} & 2 \le c < \infty \\ 0 & \text{elsewhere} \end{cases}$

7.11 $f(y) = \begin{cases} \dfrac{y}{\mu}e^{-y^2/2\mu} & y > 0 \\ 0 & \text{elsewhere} \end{cases}$

Let $w = y^2$. Then $y_0 = \sqrt{w_0}$

$$F(y_0) = F\left(\sqrt{w_0}\right) = \int_0^{\sqrt{w_0}} \frac{y}{\mu}e^{-y^2/2\mu}dy = -e^{-y^2/2\mu}\Big]_0^{\sqrt{w_0}} = -e^{-\frac{\left(\sqrt{w_0}\right)^2}{2\mu}} - \left(-e^{-0/2\mu}\right)$$

$$= -e^{-w_0/2\mu} + 1 = 1 - e^{-w_0/2\mu}$$

The cumulative distribution function for w is $F(w) = 1 - e^{-w/2\mu}$

The density function for w is $\dfrac{dF(w)}{dw} = f(w) = \begin{cases} \dfrac{1}{2\mu}e^{-w/2\mu} & w > 0 \\ 0 & \text{otherwise} \end{cases}$

w has an exponential distribution with a mean of 2μ.

7.13   The density function for the beta distribution with $\alpha = 2$ and $\beta = 1$ is

$$f(y) = \begin{cases} \dfrac{\Gamma(3)}{\Gamma(2)\Gamma(1)}y & 0 \le y \le 1 \\ 0 & \text{elsewhere} \end{cases}$$

where $\Gamma(3) = 2! = 2$, $\Gamma(2) = 1$, and $\Gamma(1) = 1$

Thus, $f(y) = \begin{cases} 2y & 0 \le y \le 1 \\ 0 & \text{elsewhere} \end{cases}$

The cumulative distribution function is

$$F(y) = \int_0^y 2t \, dt = t^2 \Big|_0^y = y^2$$

If we let $w = F(y) = y^2$, then Theorem 7.1 tells us that w has a uniform density function over the interval $0 \le w \le 1$.

To draw a random number y from the beta distribution, we first randomly draw a value w from the uniform distribution. This can be done by drawing a random number from Table 6 of Appendix II. Suppose we draw the random number 91646 (1st number in column 6). This corresponds to the random selection of the value $w_1 = .91646$ from a uniform distribution over the interval $0 \le w_1 \le 1$. Substituting this value of w_1 into the formula for $w = F(y)$ and solving for y, we obtain

$$w_1 = F(y) = y^2$$

or $.91646 = y^2$

or $y = .95732 \approx .957$

The next 4 random numbers selected are 89198, 64809, 16376 and 91782. Thus, $w_2 = .89198$, $w_3 = .64809$, $w_4 = .16376$ and $w_5 = .91782$.

Substituting these values into the formula $w = y^2$ yields

$$y_2 = \sqrt{.89198} = .94445 \approx .944$$
$$y_3 = \sqrt{.64809} = .80504 \approx .805$$
$$y_4 = \sqrt{.16376} = .40467 \approx .405$$
$$y_5 = \sqrt{.91782} = .95803 \approx .958$$

7.17   a. $\mu_{\bar{y}} = \mu = 293$

$$\sigma_{\bar{y}} = \frac{\sigma}{\sqrt{n}} = \frac{847}{\sqrt{50}} = 119.8$$

b. By the Central Limit Theorem, the distribution of $\bar{y}$ is approximately normal.

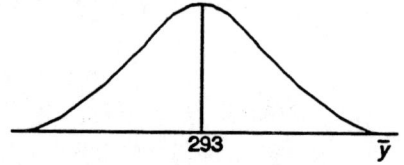

c. $P(\bar{y} > 550) = P\left(z > \dfrac{550 - 293}{119.8}\right) = P(z > 2.15) = .5 - .4842 = .0158$

7.19 **a.** By the Theorem 7.2, the sampling distribution of $\bar{y}$ is approximately normal with

$\mu_{\bar{y}} = \mu = -7.02$ and $\sigma_{\bar{y}} = \dfrac{\sigma}{\sqrt{n}} = \dfrac{24.66}{\sqrt{50}} = 3.487$

b. $P(\bar{y} < 0) = P\left(z < \dfrac{0 - (-7.02)}{3.487}\right) = P(z < 2.01) = .5 + .4778 = .9778$

c. $P(\bar{y} < 17.83) = P\left(z < \dfrac{-17.83 - (-7.02)}{3.487}\right) = P(z < -3.10) = .5 - .4987$

$\qquad\qquad\qquad\qquad\qquad\qquad\qquad\qquad\qquad\qquad\qquad\qquad = .0013$

Since this probability is so small, it is unlikely that all these contracts are likely to have five bidders.

7.21 **a.** We know $\mu = 51$ and $\sigma = 14$. Therefore, $\sigma^2 = 14^2 = 196$. By Theorem 7.2, the sampling distribution of $\bar{y}$ is approximately normal with mean $\mu_{\bar{y}} = \mu = 51$ and standard deviation

$\sigma_{\bar{y}} = \dfrac{\sigma}{\sqrt{n}} = \dfrac{14}{\sqrt{45}} = 2.09$

b. $P(\bar{y} > 52) = P\left(z > \dfrac{52 - 51}{2.09}\right) = P(z > .48) = .5 - P(0 \le z \le .48)$

$\qquad\qquad\qquad\qquad\qquad\qquad\qquad\qquad\quad = .5 - .1844 = .3156$

c. $P(49.5 \le \bar{y} \le 50.5) = P\left(\dfrac{49.5 - 51}{2.09} \le z \le \dfrac{50.5 - 51}{2.09}\right) = P(-.72 \le z \le -.24)$

$\qquad\qquad\qquad\qquad\qquad = P(0 \le z \le .72) - P(0 \le z \le .24)$

$\qquad\qquad\qquad\qquad\qquad = .2642 - .0948 = .1694$

7.23 **a.** If y has a gamma distribution with $\alpha = 1$ and $\beta = 60$, then

$\mu = E(y) = \alpha\beta = 1(60) = 60$
and $\quad \sigma^2 = V(y) = \alpha\beta^2 = 1(60)^2 = 3600$

$\mu_{\bar{y}} = E(\bar{y}) = \mu = 60, \quad \sigma_{\bar{y}}^2 = V(\bar{y}) = \dfrac{\sigma^2}{n} = \dfrac{3600}{100} = 36$

b. By the Central Limit Theorem, the sampling distribution of $\bar{y}$ is approximately normal.

c. $P(\bar{y} \le 30) = P\left(z \le \dfrac{30 - 60}{6}\right) = P(z \le -5) = .5 - P(-5 \le z \le 0) \approx .5 - .5$
$$= 0$$

7.25 Let $\hat{p}_1 - \hat{p}_2 = a_1 y_1 + a_2 y_2$ where $a_1 = \dfrac{1}{n_1}$ and $a_2 = \dfrac{1}{n_2}$

$E(\hat{p}_1 - \hat{p}_2) = \dfrac{1}{n_1} E(y_1) - \dfrac{1}{n_2} E(y_2) = \dfrac{1}{n_1}(n_1 p_1) - \dfrac{1}{n_2}(n_2 p_2) = p_1 - p_2$

$V(\hat{p}_1 - \hat{p}_2) = \left[\dfrac{1}{n_1}\right]^2 V(y_1) + \left[\dfrac{1}{n_2}\right]^2 V(y_2) + 2\left[\dfrac{1}{n_1}\right]\left[\dfrac{-1}{n_2}\right] Cov(y_1, y_2)$

$\qquad = \left[\dfrac{1}{n_1}\right]^2 (n_1 p_1 q_1) + \left[\dfrac{1}{n_2}\right]^2 (n_2 p_2 q_2) + 2\left[\dfrac{1}{n_1}\right]\left[\dfrac{-1}{n_2}\right](0)$

$\qquad = \dfrac{p_1 q_1}{n_1} + \dfrac{p_2 q_2}{n_2}$

$\Rightarrow$ The sampling distribution of $\hat{p}_1 - \hat{p}_2$ has an approximate normal distribution for large values of n_1 and n_2 with mean $\mu_{\hat{p}_1 - \hat{p}_2} = p_1 - p_2$ and standard deviation

$$\sigma_{\hat{p}_1 - \hat{p}_2} = \sqrt{\dfrac{p_1 q_1}{n_1} + \dfrac{p_2 q_2}{n_2}}$$

$\Rightarrow$ $z = \dfrac{(\hat{p}_1 - \hat{p}_2) - (p_1 - p_2)}{\sqrt{\dfrac{p_1 q_1}{n_1} + \dfrac{p_2 q_2}{n_2}}}$ has a standard normal distribution.

7.27 a. Using Table 1 of Appendix II, $P(y \le 8) = .9848$

b. $P(y \le 8) \approx P\left(z \le \dfrac{(8 + .5) - 15(.3)}{\sqrt{15(.3)(.7)}}\right) = P(z \le 2.25) = .5 + P(0 < z < 2.25)$
$$= .5 + .4878 = .9879$$

7.29 a. Using the normal approximation to the binomial, the distribution of y is approximately normal with mean $\mu = np = 330(.54) = 178.2$ and standard deviation
$\sigma = \sqrt{npq} = \sqrt{330(.54)(.46)} = 9.0538.$

$\qquad P(y < 100) \approx P\left(z < \dfrac{99.5 - 178.2}{9.0538}\right) = P(z < -8.69) = .5 - P(0 \le z \le 8.69)$
$$\approx .5 - .5 = 0$$

b. $P(y \geq 200) \approx P\left[z \geq \dfrac{199.5 - 178.2}{9.0538}\right] = P(z \geq 2.35) = .5 - P(0 \leq z \leq 2.35)$

$$\approx .5 - .4906 = .0094$$

7.31 Using the normal approximation to the binomial, the distribution of y is approximately normal with mean $\mu = np = 150(.6) = 90$ and standard deviation $\sigma = \sqrt{npq} = \sqrt{150(.6)(.4)} = 6$.

$$P(y \geq 75) \approx P\left[z \geq \dfrac{75 - .5 - 90}{6}\right] = P(z \geq -2.58) = .5 + P(0 \leq z \leq 2.58)$$

$$= .5 + .4951 = .9951$$

7.33 **a.** Using the normal approximation to the binomial, the distribution of y is approximately normal with mean $\mu = np = 50(.34) = 17$ and standard deviation

$\sigma = \sqrt{npq} = \sqrt{50(.34)(.66)} = 3.3496$.

$$P(y \leq 10) \approx P\left[z \leq \dfrac{10.5 - 17}{3.3496}\right] = P(z \leq -1.94) = .5 - P(0 \leq z \leq 1.94)$$

$$= .5 - .4738 = .0262$$

b. $P(y \geq 25) \approx P\left[z \geq \dfrac{24.5 - 17}{3.3496}\right] = P(z \geq 2.24) = .5 - P(0 \leq z \leq 2.24)$

$$= .5 - .4875 = .0125$$

c. $P(20 \leq y \leq 30) \approx P\left[\dfrac{19.5 - 17}{3.3496} \leq z \leq \dfrac{30.5 - 17}{3.3496}\right] = P(.75 \leq z \leq 4.03)$

$$= P(0 \leq z \leq 4.03) - P(0 \leq z \leq .75)$$

$$\approx .5 - .2734 = .2266$$

7.35 **a.** The sampling distribution of $\dfrac{(n-1)s^2}{\sigma^2}$ is chi-square with $\nu = (n-1) = (30-1) = 29$ degrees of freedom.

b. $P(s^2 > 3.30) = P\left[\chi^2 > \dfrac{(30-1)3.30}{1.5^2}\right] = P(\chi^2 > 42.533) \approx .05$

(from Table 7 Appendix II)

7.37 From Theorem 7.4,

$$\dfrac{(n_1 - 1)s_1^2}{\sigma^2} \sim \chi^2 \text{ with } \nu = n_1 - 1 \text{ and } \dfrac{(n_2 - 1)s_2^2}{\sigma^2} \sim \chi^2 \text{ with } \nu = n_2 - 1$$

From Theorem 7.5, $\dfrac{(n_1 - 1)s_1^2}{\sigma^2} + \dfrac{(n_2 - 1)s_2^2}{\sigma^2} \sim \chi^2$ with $\nu = n_1 - 1 + n_2 - 1$

$$= n_1 + n_2 - 2$$

Now, $\dfrac{(n_1 + n_2 - 2)s^2}{\sigma^2} = \dfrac{n_1 + n_2 - 2}{\sigma^2}\left[\dfrac{(n_1 - 1)s_1^2 + (n_2 - 1)s_2^2}{n_1 + n_2 - 2}\right]$

$$= \dfrac{(n_1 - 1)s_1^2 + (n_2 - 1)s_2^2}{\sigma^2}$$

$$= \dfrac{(n_1 - 1)s_1^2}{\sigma^2} + \dfrac{(n_2 - 1)s_2^2}{\sigma^2} \sim \chi^2 \text{ with } \nu = n_1 + n_2 - 2$$

7.39 $\quad f(y) = \begin{cases} \dfrac{1}{\sigma y \sqrt{2\pi}} e^{-\frac{(\ln y - \mu)^2}{2\sigma^2}} & y > 0 \\[4mm] 0 & \text{otherwise} \end{cases}$

$x = \ln y \Rightarrow dx = \dfrac{1}{y}dy, \ -\infty < x < \infty$

Thus, by substitution, $f(x) = \dfrac{1}{\sqrt{2\pi}\,\sigma} e^{-\frac{(x-\mu)^2}{2\sigma^2}} \quad -\infty < x < \infty$

Then x has a normal distribution with mean μ and variance σ^2.

7.41 $\quad$ Let $w = 10y - 2$ and $y = \dfrac{w + 2}{10}$

For $0 \le y \le 1$, the range of w is $-2 \le w \le 8$

$$F(y_0) = F\left[\dfrac{w_0 + 2}{10}\right] = \int_0^{\frac{w_0+2}{10}} \dfrac{y}{2}\,dy = \dfrac{y^2}{4}\Bigg]_0^{\frac{w_0+2}{10}} = \dfrac{(w_0 + 2)^2}{400} \text{ if } -2 \le w_0 \le 8$$

For $1 \le y \le 2.5$, the range of w is $8 \le w \le 23$

$$F(y_0) = F\left[\dfrac{w_0 + 2}{10}\right] = \int_0^1 \dfrac{y}{2}\,dy + \int_0^{\frac{w_0+2}{10}} \dfrac{1}{2}\,dy = \dfrac{y^2}{4}\Bigg]_0^1 + \dfrac{y}{2}\Bigg]_1^{\frac{w_0+2}{10}}$$

$$= \dfrac{1^2}{4} + \dfrac{w_0 + 2}{20} - \dfrac{1}{2} = \dfrac{w_0 - 3}{20} \text{ if } 8 \le w_0 \le 23$$

The cumulative distribution of w is

$$F(w) = \frac{(w + 2)^2}{400} \quad \text{if } -2 \le w \le 8$$

$$\text{and} \quad F(w) = \frac{(w - 3)}{20} \quad \text{if } 8 \le w \le 23$$

The density function of w is

$$\frac{dF(w)}{dw} = \frac{d\left[\dfrac{(w + 2)^2}{400}\right]}{dw} = \frac{2(w + 2)}{400} = \frac{w + 2}{200} \quad \text{if } -2 \le w \le 8$$

$$\text{and} \quad \frac{dF(w)}{dw} = \frac{d\left[\dfrac{(w - 3)^2}{20}\right]}{dw} = \frac{1}{20} \quad \text{if } 8 \le w \le 23$$

$$\text{Thus, } f(w) = \begin{cases} \dfrac{w + 2}{200} & \text{if } -2 \le w \le 8 \\[2mm] \dfrac{1}{20} & \text{if } 8 \le w \le 23 \\[2mm] 0 & \text{elsewhere} \end{cases}$$

7.43   a.   By Theorem 7.2, the sampling distribution of $\bar{y}$ is approximately normal with mean $\mu_{\bar{y}} = \mu = 121.74$ and standard deviation $\sigma_{\bar{y}} = \dfrac{\sigma}{\sqrt{n}} = \dfrac{27.52}{\sqrt{32}} = 4.8649$.

b. $P(118 \le \bar{y} \le 130) \approx P\left[\dfrac{118 - 121.74}{4.8649} \le z \le \dfrac{130 - 121.74}{4.8649}\right]$

$$= P(-.77 \le z \le 1.70)$$
$$= P(0 \le z \le 1.70) + P(0 \le z \le .77)$$
$$= .4554 + .2794 = .7348$$

7.45   To use Theorem 7.1, we must first find the cumulative distribution function of y.

$$F(y) = \int_0^y 2te^{-t^2}dt = -e^{-t^2}\Big]_0^y = -e^{-y^2} - \left(-e^{-0^2}\right) = 1 - e^{-y^2}$$

Let $w = F(y) = 1 - e^{-y^2}$. By Theorem 7.1, w has a uniform density function over the interval $0 \le w \le 1$.

To draw a random number y from this distribution, we first randomly draw a value w from the uniform distribution. This can be done by drawing a random number from Table 6, Appendix II. Suppose we draw the random number 20969 (1st number in column 11). This corresponds to the random selection of the value $w_1 = .20969$ from a uniform distribution over the interval $0 \leq w_1 \leq 1$. Substituting this value of w_1 into the formula for $w = F(y)$ and solving for y, we get

$$w = 1 - e^{-y^2} \Rightarrow e^{-y^2} = 1 - w \Rightarrow -y^2 = \ln(1 - w) \Rightarrow y = \sqrt{-\ln(1 - w)}$$

For $w_1 = .20969$, $y_1 = \sqrt{-\ln(1 - .20969)} = .4851$

The next 4 random numbers selected are 52666, 30680, 00849, and 14110. Thus,

$$w_2 = .52666 \Rightarrow y_2 = \sqrt{-\ln(1 - .52666)} = .8648$$
$$w_3 = .30680 \Rightarrow y_3 = \sqrt{-\ln(1 - .30680)} = .6053$$
$$w_4 = .00849 \Rightarrow y_4 = \sqrt{-\ln(1 - .00849)} = .0923$$
$$w_5 = .14110 \Rightarrow y_5 = \sqrt{-\ln(1 - .14110)} = .3900$$

7.47   a.   By Theorem 7.2, the sampling distribution of $\bar{y}$ is approximately normal with mean $\mu_{\bar{y}} = \mu = 9.2$ and standard deviation

$$\sigma_{\bar{y}} = \frac{\sigma}{\sqrt{n}} = \frac{\sqrt{3.24}}{\sqrt{80}} = .201$$

$$P(\bar{y} < 8.80) = P\left(z < \frac{8.80 - 9.2}{.201}\right) = P(z < -1.99) = .5 - P(0 \leq z \leq 1.99)$$
$$= .5 - .4767 = .0233$$

   b.   Since the probability of observing a $\bar{y}$ of 8.8 or less is quite small (.0233), we have either seen a very rare event, or the true mean is less than 9.2.

7.49   Using the normal approximation to the binomial, the distribution of y (number of reflected particles out of 2000) is approximately normal with mean $\mu = np = 2000(.16) = 320$ and standard deviation $\sigma = \sqrt{npq} = \sqrt{2000(.16)(.84)} = 16.3951$.

$$P(y \leq 280) \approx P\left(z \leq \frac{280.5 - 320}{16.3951}\right) = P(z \leq -2.41) = .5 - P(0 \leq z \leq 2.41)$$
$$= .5 - .4920 = .0080$$

7.51   a.   The sampling distribution of $\frac{(n - 1)s^2}{\sigma^2}$ is chi-square with $\nu = (n - 1) = (100 - 1) = 99$ degrees of freedom.

b. $P(s^2 > 500^2) = P\left[\chi^2 > \dfrac{(100-1)250{,}000}{422^2}\right] = P(\chi^2 > 138.979)$

Using Table 7, Appendix II, this probability is between .005 and .01. Since $.005 < P(s^2 > 500^2) < .01$, then $.005 < P(s > 500) < .01$.

7.53   Since y_1 and y_2 are independent,

$$f(y_1, y_2) = f(y_1)f(y_2) = \left[\frac{1}{\beta}e^{-y_1/\beta}\right]\left[\frac{1}{\beta}e^{-y_2/\beta}\right]$$

$$= \begin{cases} \dfrac{1}{\beta^2}e^{-(y_1+y_2)/\beta} & \text{for } y_1 > 0 \text{ and } y_2 > 0 \\ 0 & \text{elsewhere} \end{cases}$$

For $w = y_1 + y_2$,

$$P(w \le w_0) = \int_0^{w_0}\int_0^{w_0-y_1} \frac{1}{\beta^2}e^{-(y_1+y_2)/\beta}\,dy_2\,dy_1$$

$$= \int_0^{w_0}\frac{1}{\beta}e^{-y_1/\beta}\left[-e^{-y_2/\beta}\right]_0^{w_0-y_1}dy_1 = \int_0^{w_0}\frac{1}{\beta}e^{-y_1/\beta}\left(1 - e^{-(w_0-y_1)/\beta}\right)dy_1$$

$$= \int_0^{w_0}\left[\frac{1}{\beta}e^{-y_1/\beta} - \frac{1}{\beta}e^{-w_0/\beta}\right]dy_1 = \left[-e^{-y_1/\beta}\Big|_0^{w_0} - \frac{1}{\beta}e^{-w_0/\beta}y_1\right]_0^{w_0}$$

$$= 1 - e^{-w_0/\beta} - \frac{1}{\beta}e^{-w_0/\beta}(w_0) = 1 - e^{-w_0/\beta} - \frac{w_0}{\beta}e^{-w_0/\beta}$$

To find the density function of w, we take the derivative of the cumulative distribution function with respect to w.

$$F(w) = 1 - e^{-w/\beta} - \frac{w}{\beta}e^{-w/\beta}$$

$$\frac{dF(w)}{dw} = \frac{d\left[1 - e^{-w/\beta} - \frac{w}{\beta}e^{-w/\beta}\right]}{dw}$$

$$= -\left[-\frac{1}{\beta}e^{-w/\beta}\right] - \frac{1}{\beta}\left[w\left(-\frac{1}{\beta}e^{-w/\beta}\right) + e^{-w/\beta}\right] = \frac{1}{\beta}e^{-w/\beta} + \frac{w}{\beta^2}e^{-w/\beta} - \frac{1}{\beta}e^{-w/\beta}$$

$$= \frac{w}{\beta^2}e^{-w/\beta} \quad \text{for } w > 0$$

Thus, the distribution of w is gamma with $\alpha = 2$ and unknown β.

7.55   a. $E(y_1 - y_2) = E(y_1) - E(y_2) = \mu - \mu = 0$

$V(y_1 - y_2) = V(y_1) + V(y_2) + 2(1)(-1) \, \text{Cov}(y_1, y_2)$
$$= \sigma^2 + \sigma^2 + 2(1)(-1)(0) = 2\sigma^2$$

$$z = \frac{y_1 - y_2}{\sqrt{2\sigma^2}} = \frac{y_1 - y_2}{\sqrt{2}\,\sigma} \text{ has a standard normal distribution.}$$

(The sum or difference of two independent normal random variables is normal.)

b. $s^2 = \dfrac{\sum y^2 - \dfrac{(\sum y)^2}{n}}{n-1} = \dfrac{y_1^2 + y_2^2 - \dfrac{(y_1 + y_2)^2}{2}}{1} = y_1^2 + y_2^2 - \dfrac{(y_1 + y_2)^2}{2}$

$$= \frac{2y_1^2 + 2y_2^2 - y_1^2 - y_2^2 - 2y_1y_2}{2} = \frac{y_1^2 + y_2^2 - 2y_1y_2}{2} = \frac{(y_1 - y_2)^2}{2}$$

$$\chi^2 = \frac{(n-1)s^2}{\sigma^2} = \frac{s^2}{\sigma^2} = \frac{(y_1 - y_2)^2}{2\sigma^2} = z^2$$

has a χ^2 distribution with $\nu = n - 1 = 1$ degrees of freedom (by applying Theorem 7.4).

CHAPTER EIGHT

···

Estimation

8.1   **a.** $E(\hat{\theta}_1) = E(\bar{y}) = E\left[\dfrac{y_1 + y_2 + y_3}{3}\right] = \dfrac{1}{3}E(y_1 + y_2 + y_3) = \dfrac{1}{3}[E(y_1) + E(y_2) + E(y_3)]$

$$= \dfrac{1}{3}(\theta + \theta + \theta) = \theta$$

$E(\hat{\theta}_2) = E(y_1) = \theta$

$E(\hat{\theta}_3) = E\left[\dfrac{y_1 + y_2}{2}\right] = \dfrac{1}{2}E(y_1 + y_2) = \dfrac{1}{2}[E(y_1) + E(y_2)] = \dfrac{1}{2}(\theta + \theta) = \theta$

b. $V(\hat{\theta}_1) = V(\bar{y}) = V\left[\dfrac{y_1 + y_2 + y_3}{3}\right] = \dfrac{1}{3^2}(V(y_1) + V(y_2) + V(y_3))$

$\quad = \dfrac{1}{9}(\theta^2 + \theta^2 + \theta^2) = \dfrac{\theta^2}{3}$ (Since y_i's are independent, the covariances are 0.)

$V(\hat{\theta}_2) = V(y_1) = \theta^2$

$V(\hat{\theta}_3) = V\left[\dfrac{y_1 + y_2}{2}\right] = \dfrac{1}{2^2}(V(y_1) + V(y_2)) = \dfrac{1}{4}(\theta^2 + \theta^2) = \dfrac{\theta^2}{2}$

Thus, $\hat{\theta}_1$ has the smallest variance.

8.3   **a.** The mean of the binomial distribution is $\mu = E(y) = np$.

$$E(\hat{p}) = E\left[\dfrac{y}{n}\right] = \dfrac{1}{n}E(y) = \dfrac{1}{n}(np) = p$$

b. The variance of the binomial distribution is $\sigma^2 = npq$.

$$V(\hat{p}) = V\left[\dfrac{y}{n}\right] = \dfrac{1}{n^2}V(y) = \dfrac{1}{n^2}(npq) = \dfrac{pq}{n}$$

8.5
$$E[(\hat{\theta} - \theta)^2] = E\left\{[\hat{\theta} - E(\hat{\theta})] + [E(\hat{\theta}) - \theta]\right\}^2$$
$$= E\left\{[\hat{\theta} - E(\hat{\theta})]^2 + 2[\hat{\theta} - E(\hat{\theta})][E(\hat{\theta}) - \theta] + [E(\hat{\theta}) - \theta]^2\right\}$$
$$= V(\hat{\theta}) + 2[E(\hat{\theta}) - \theta]E[\hat{\theta} - E(\hat{\theta})] + E[E(\hat{\theta}) - \theta]^2$$
$$= V(\hat{\theta}) + 2[E(\hat{\theta}) - \theta]0 + [E(\hat{\theta}) - \theta]^2$$
$$= V(\hat{\theta}) + B^2$$

8.7 From Theorem 7.4, we know that when sampling from a normal distribution,

$$\frac{(n - 1)s^2}{\sigma^2} = \chi^2$$

where χ^2 is a chi-square random variable with $\nu = (n - 1)$ degrees of freedom. Rearranging terms yields

$$s^2 = \frac{\sigma^2}{n - 1}\chi^2$$

We know from Section 5.6 that $E(\chi^2) = \nu = n - 1$ and $V(\chi^2) = 2\nu = 2(n - 1)$.

Thus, $V(s^2) = V\left[\frac{\sigma^2}{n - 1}\chi^2\right] = \frac{\sigma^4}{(n - 1)^2}V(\chi^2) = \frac{\sigma^4}{(n - 1)^2}2(n - 1) = \frac{2\sigma^4}{(n - 1)}$

8.9 **a.** If $y_1, y_2, \ldots, y_n$ is a random sample of n observations from a Poisson distribution, then the likelihood function is

$$L = p(y_1)p(y_2) \cdots p(y_n) = \left[\frac{e^{-\lambda}\lambda^{y_1}}{y_1!}\right]\left[\frac{e^{-\lambda}\lambda^{y_2}}{y_2!}\right] \cdots \left[\frac{e^{-\lambda}\lambda^{y_n}}{y_n!}\right] = \frac{e^{-n\lambda}\lambda^{\sum y_i}}{\prod\limits_{i=1}^{n} y_i!}$$

Then, $\ln L = -n\lambda + \sum y_i \ln \lambda - \ln \prod\limits_{i-1}^{n} y_i!$

The derivative of $\ln L$ with respect to λ is

$$\frac{d \ln L}{d\lambda} = \frac{d\left[-n\lambda + \sum y_i \ln \lambda - \ln \prod\limits_{i=1}^{n} y_i!\right]}{d\lambda} = -n + \sum y_i\left(\frac{1}{\lambda}\right)$$

Setting this equal to 0 and solving, we get

$$-n + \sum y_i \left[\frac{1}{\hat{\lambda}} \right] = 0$$

$$\frac{\sum y_i}{\hat{\lambda}} = n$$

$$\hat{\lambda} = \frac{\sum y_i}{n} = \bar{y}$$

b. To determine if the maximum likelihood estimator is unbiased, we must find its expected value. We know $E(y) = \lambda$ if y has a Poisson distribution with parameter λ.

$$E(\hat{\lambda}) = E(\bar{y}) = E\left[\frac{y_1 + y_2 + \cdots + y_n}{n} \right] = \frac{1}{n} E(y_1 + y_2 + \cdots + y_n)$$

$$= \frac{1}{n}(\lambda + \lambda + \cdots + \lambda) = \frac{1}{n}(n\lambda) = \lambda$$

Therefore $\hat{\lambda}$ is an unbiased estimator of λ.

8.11 a. Since we only have one unknown parameter, β, to estimate, the moment estimator is found by setting the first population moment, $E(y)$, equal to the first sample moment, $\bar{y}$. For the gamma distribution, $E(y) = \alpha\beta = 2\beta$ ($\alpha = 2$).

Thus, the moment estimator is:

$$2\hat{\beta} = \bar{y} \Rightarrow \hat{\beta} = \frac{\bar{y}}{2}$$

b. $$E(\hat{\beta}) = E\left[\frac{\bar{y}}{2} \right] = \frac{1}{2} E\left[\frac{y_1 + y_2 + \cdots + y_n}{n} \right] = \frac{1}{2n} E(y_1 + y_2 + \cdots + y_n)$$

$$= \frac{1}{2n}(2\beta + 2\beta + \cdots + 2\beta) = \frac{1}{2n}(2n\beta) = \beta$$

The variance of a gamma distribution is $V(y) = \alpha\beta^2 = 2\beta^2$ when $\alpha = 2$.

$$V(\hat{\beta}) = V\left[\frac{\bar{y}}{2} \right] = \frac{1}{2^2} V\left[\frac{y_1 + y_2 + \cdots + y_n}{n} \right] = \frac{1}{2^2 n^2}(V(y_1) + V(y_2) + \cdots + V(y_n))$$

$$= \frac{1}{4n^2}(2\beta^2 + 2\beta^2 + \cdots + 2\beta^2) = \frac{1}{4n^2}(2n\beta^2) = \frac{\beta^2}{2n}$$

(Note: Since the y_i's are a random sample, they are independent of each other. Therefore, all the covariances are equal to zero.)

8.13   a.   Since we only have one parameter, β, to estimate, the moment estimator is found by setting the first population moment, $E(y)$, equal to the first sample moment, $\bar{y}$. For the exponential distribution, $E(y) = \beta$.

Thus, the moment estimator is $\hat{\beta} = \bar{y}$.

b. $E(\hat{\beta}) = E(\bar{y}) = E\left[\dfrac{y_1 + y_2 + \cdots + y_n}{n}\right] = \dfrac{1}{n}E(y_1 + y_2 + \cdots + y_n)$

$$= \dfrac{1}{n}(\beta + \beta + \cdots + \beta) = \dfrac{1}{n}(n\beta) = \beta$$

Thus, $\hat{\beta} = \bar{y}$ is an unbiased estimator of β.

c. $V(\hat{\beta}) = V(\bar{y}) = V\left[\dfrac{y_1 + y_2 + \cdots + y_n}{n}\right] = \dfrac{1}{n^2}(V(y_1) + V(y_2) + \cdots + V(y_n))$

$$= \dfrac{1}{n^2}(\beta^2 + \beta^2 + \cdots + \beta^2) = \dfrac{1}{n^2}(n\beta^2) = \dfrac{\beta^2}{n}$$

(The variance of the exponential distribution is β^2. Also, because $y_1, y_2, \ldots, y_n$ is a random sample, the y_i's are independent of each other. Therefore, all the covariances are zero.)

8.15   Using degrees of freedom $\nu = \infty$, we find from Table 7:

$t_{.05} = 1.645 = z_{.05}$
$t_{.025} = 1.96 = z_{.025}$
$t_{.01} = 2.326 = z_{.01}$

8.17   By Theorem 7.2, the sampling distribution of $\bar{y}$ is approximately normal with mean $\mu_{\bar{y}} = \mu = \lambda$ and standard deviation $\sigma_{\bar{y}} = \sigma/\sqrt{n} = \sqrt{\lambda/n}$.

Thus $z = \dfrac{\bar{y} - \lambda}{\sqrt{\lambda/n}}$ has an approximate standard normal distribution.

Using z as the pivotal statistic, the confidence interval for λ is:

$$P\left(-z_{\alpha/2} \le z \le z_{\alpha/2}\right) = P\left[-z_{\alpha/2} \le \dfrac{\bar{y} - \lambda}{\sqrt{\lambda/n}} \le z_{\alpha/2}\right] = 1 - \alpha$$

Now substitute $\bar{y}$ for λ in the denominator. (We know the maximum likelihood estimate of λ is $\bar{y}$ from Exercise 8.9.)

$$P\left[-z_{\alpha/2} \leq \frac{\bar{y} - \lambda}{\sqrt{\bar{y}/n}} \leq z_{\alpha/2}\right] = P\left(-z_{\alpha/2}\sqrt{\bar{y}/n} \leq \bar{y} - \lambda \leq z_{\alpha/2}\sqrt{\bar{y}/\sqrt{n}}\right)$$

$$= P\left(-\bar{y} - z_{\alpha/2}\sqrt{\bar{y}/n} \leq -\lambda \leq -\bar{y} + z_{\alpha/2}\sqrt{\bar{y}/n}\right)$$

$$= P\left(\bar{y} - z_{\alpha/2}\sqrt{\bar{y}/n} \leq \lambda \leq \bar{y} + z_{\alpha/2}\sqrt{\bar{y}/n}\right) = 1 - \alpha$$

The $1 - \alpha$ confidence interval for λ is $\bar{y} \pm z_{\alpha/2}\sqrt{\bar{y}/n}$.

8.19 By Theorem 7.2, the sampling normal distribution of $\bar{y}_1 - \bar{y}_2$ has an approximate normal distribution with mean $\mu_{\bar{y}_1 - \bar{y}_2} = \mu_1 - \mu_2$ and standard deviation $\sigma_{\bar{x}_1 - \bar{x}_2} = \sqrt{\dfrac{\sigma_1^2}{n_1} + \dfrac{\sigma_2^2}{n_2}}$.

Thus, $z = \dfrac{(\bar{y}_1 - \bar{y}_2) - (\mu_1 - \mu_2)}{\sqrt{\dfrac{\sigma_1^2}{n_1} + \dfrac{\sigma_2^2}{n}}}$ has a standard normal distribution.

Using z as the pivotal statistic, the confidence interval for $\mu_1 - \mu_2$ is:

$$P\left(-z_{\alpha/z} \leq z \leq z_{\alpha/2}\right) = P\left[-z_{\alpha/2} \leq \frac{(\bar{y}_1 - \bar{y}_2) - (\mu_1 - \mu_2)}{\sqrt{\dfrac{\sigma_1^2}{n_1} + \dfrac{\sigma_2^2}{n_2}}} \leq z_{\alpha/2}\right] = 1 - \alpha$$

Now, substitute s_1^2 and s_2^2 (the maximum likelihood estimates) for σ_1^2 and σ_2^2 in the denominator.

$$P\left[-z_{\alpha/2} \leq \frac{(\bar{y}_1 - \bar{y}_2) - (\mu_1 - \mu_2)}{\sqrt{\dfrac{s_1^2}{n_1} + \dfrac{s_2^2}{n_2}}} \leq z_{\alpha/2}\right]$$

$$= P\left[-z_{\alpha/2}\sqrt{\frac{s_1^2}{n_1} + \frac{s_2^2}{n_2}} \leq (\bar{y}_1 - \bar{y}_2) - (\mu_1 - \mu_2) \leq z_{\alpha/2}\sqrt{\frac{s_1^2}{n_2} + \frac{s_2^2}{n_2}}\right]$$

$$= P\left[-(\bar{y}_1 - \bar{y}_2) - z_{\alpha/2}\sqrt{\frac{s_1^2}{n_1} + \frac{s_2^2}{n_2}} \leq -(\mu_1 - \mu_2) \leq -(\bar{y}_1 - \bar{y}_2) + z_{\alpha/2}\sqrt{\frac{s_1^2}{n_1} + \frac{s_2^2}{n_2}}\right]$$

$$= P\left[(\bar{y}_1 - \bar{y}_2) - z_{\alpha/2}\sqrt{\frac{s_1^2}{n_1} + \frac{s_2^2}{n_2}} \leq \mu_1 - \mu_2 \leq (\bar{y}_1 - \bar{y}_2) + z_{\alpha/2}\sqrt{\frac{s_1^2}{n_1} + \frac{s_2^2}{n_2}}\right]$$

$$= 1 - \alpha$$

Thus, the $1 - \alpha$ confidence interval for $\mu_1 - \mu_2$ is $(\bar{y}_1 - \bar{y}_2) \pm z_{\alpha/2}\sqrt{\dfrac{s_1^2}{n_1} + \dfrac{s_2^2}{n_2}}$

8.21   If χ_1^2 and χ_2^2 are two chi-square random variables with $\nu_1 = n_1 - 1$ and $\nu_2 = n_2 - 1$ degrees of freedom, respectively, then, by Theorem 7.5,

$$\chi^2 = \chi_1^2 + \chi_2^2 = \frac{(n_1 - 1)s_1^2}{\sigma^2} + \frac{(n_2 - 1)s_2^2}{\sigma^2} = \frac{(n_1 - 1)s_1^2 + (n_2 - 1)s_2^2}{\sigma^2}$$

is a chi-square random variable with $(\nu_1 + \nu_2) = (n_1 - 1) + (n_2 - 1) = n_1 + n_2 - 2$ degrees of freedom.

8.23 Using the pivotal statistic $t = \dfrac{(\bar{y}_1 - \bar{y}_2) - (\mu_1 - \mu_2)}{s_p\sqrt{\dfrac{1}{n_1} + \dfrac{1}{n_2}}}$, the confidence interval for $\mu_1 - \mu_2$ is:

$$P(-t_{\alpha/2} \leq t \leq t_{\alpha/2}) = P\left[-t_{\alpha/2} \leq \frac{(\bar{y}_1 - \bar{y}_2) - (\mu_1 - \mu_2)}{s_p\sqrt{\dfrac{1}{n_2} + \dfrac{1}{n_2}}} \leq t_{\alpha/2}\right]$$

$$= P\left[-t_{\alpha/2}s_p\sqrt{\frac{1}{n_1} + \frac{1}{n_2}} \leq (\bar{y}_1 - \bar{y}_2) - (\mu_1 - \mu_2) \leq t_{\alpha/2}s_p\sqrt{\frac{1}{n_1} + \frac{1}{n_2}}\right]$$

$$= P\left[-(\bar{y}_1 - \bar{y}_2) - t_{\alpha/2}s_p\sqrt{\frac{1}{n_1} + \frac{1}{n_2}} \leq -(\mu_1 - \mu_2) \leq -(\bar{y}_1 - \bar{y}_2) + t_{\alpha/2}s_p\sqrt{\frac{1}{n_1} + \frac{1}{n_2}}\right]$$

$$= P\left[(\bar{y}_1 - \bar{y}_2) - t_{\alpha/2}s_p\sqrt{\frac{1}{n_1} + \frac{1}{n_2}} \leq (\mu_1 - \mu_2) \leq (\bar{y}_1 - \bar{y}_2) + t_{\alpha/2}s_p\sqrt{\frac{1}{n_1} + \frac{1}{n_2}}\right]$$

$$= 1 - \alpha$$

The $1 - \alpha$ confidence interval for $\mu_1 - \mu_2$ is $(\bar{y}_1 - \bar{y}_2) \pm t_{\alpha/2}s_p\sqrt{\dfrac{1}{n_1} + \dfrac{1}{n_2}}$ where $t_{\alpha/2}$ is based of $n_1 + n_2 - 2$ degrees of freedom.

8.25 **a.** We are 95% confident the true mean thermal relaxation time of sand is between 13.6 and 26.4 seconds.

b. If we were to repeatedly collect a sample of size n from the population of non-homogeneous materials and construct a 95% confidence interval for each sample, then we expect 95% of the intervals to enclose the true mean thermal relaxation time.

8.27 **a.** The small sample confidence interval for μ is

$$\bar{y} \pm t_{\alpha/2}\left(s/\sqrt{n}\right)$$

For confidence coefficient .99, $\alpha = 1 - .99 = .01$ and $\alpha/2 = .01/2 = .005$. From Table 7, Appendix II, $t_{.005} = 2.807$ with df $= n - 1 = 24 - 1 = 23$. The 99% confidence interval for μ is

$$9.9 \pm 2.807\left(8.4/\sqrt{24}\right) \Rightarrow 9.9 \pm 4.813 \Rightarrow (5.087, 14.713)$$

We are 99% confident the mean lead concentration in water samples is between 5.087 and 14.713.

b. The 99% confidence interval is

$$6.7 \pm 2.087\left(10.8/\sqrt{24}\,\right) \Rightarrow 6.7 \pm 6.188 \Rightarrow (.512,\ 12.888)$$

We are 99% confident the mean aluminum concentration in water samples is between .512 and 12.888.

c. We must assume the lead concentrations have a normal distribution and the aluminum concentrations have a normal distribution.

8.29 a. We are 95% confident that the true mean number of cigarettes smoked per day by all smokers is between 19.7 and 20.3

b. We must assume that the sample of smokers is random and representative of the population of all smokers.

c. Since the mean number of cigarettes smoked per day claimed by the tobacco industry researcher, 15, is not in the 95% confidence interval, there is strong evidence that the claim is false.

8.31 Since $n = 52$ is sufficiently large, we can use the large sample confidence interval for μ.

$$\bar{y} \pm z_{\alpha/2}\frac{\sigma}{\sqrt{n}}$$

For confidence coefficient .95, $\alpha = 1 - .95 = .05$ and $\alpha/2 = .05/2 = .025$. From Table 4, Appendix II, $z_{.025} = 1.96$. The 95% confidence interval is

$$.812 \pm 1.96\frac{1.50476}{\sqrt{52}} \Rightarrow .812 \pm .409 \Rightarrow (.403,\ 1.221)$$

We are 95% confident that the true mean solution time for the hybrid algorithm is between .403 and 1.221 seconds.

8.33 Let μ_1 = mean protein uptake of fast muscles and μ_2 = mean protein uptake of slow muscles. Since both sample sizes are less than 30, we must use the small sample confidence interval for $\mu_1 - \mu_2$.

$$\bar{y}_1 - \bar{y}_2 \pm t_{\alpha/2}s_p\sqrt{\frac{1}{n_1} + \frac{1}{n_2}}$$

$$s_p^2 = \frac{(n_1 - 1)s_1^2 + (n_2 - 1)s_2^2}{n_1 + n_2 - 2} = \frac{(12 - 1).104^2 + (12 - 1).035^2}{12 + 12 - 2} = .006$$

$$s_p = \sqrt{.006} = .0776$$

For confidence coefficient .95, $\alpha = 1 - .95 = .05$ and $\alpha/2 = .05/2 = .025$. From Table 7, Appendix II, with df $= n_1 + n_2 = 12 + 12 - 2 = 22$, $t_{.025} = 2.074$. The 95% confidence interval is

$$(.57 - .37) \pm 2074(.0776)\sqrt{\frac{1}{12} + \frac{1}{12}} \Rightarrow .2 \pm .066 \Rightarrow (.134, .266)$$

We are 95% confident that the difference in mean protein uptake of fast and slow muscles is between .134 and .266. Since zero is not in the interval, we are confident that the mean protein uptake of fast muscles is greater than the mean protein uptake of slow muscles.

8.35   Let μ_1 = mean shear stress for Southern Pine and μ_2 = mean shear stress for Ponderosa Pine. The large sample confidence interval for $\mu_1 - \mu_1$ is

$$(\bar{y}_1 - \bar{y}_2) \pm z_{\alpha/2}\sqrt{\frac{\sigma_1^2}{n_1} + \frac{\sigma_2^2}{n_2}}$$

where σ_1^2 is estimated with $s_1^2 = 422^2$ and σ_2^2 is estimated with $s_2^2 = 271^2$

For confidence coefficient .90, $\alpha = 1 - .90 = .10$ and $\alpha/2 = .10/2 = .05$. From Table 4, Appendix II, $z_{.05} = 1.645$. The 90% confidence interval is

$$(1312 - 1352) \pm 1.645\sqrt{\frac{422^2}{100} + \frac{271^2}{47}} \Rightarrow -40 \pm 95.118 \Rightarrow (-135.118, 55.118)$$

We are 90% confident the difference between the mean shear strengths of epoxy-repaired truss joints for the two species of wood is between -135.118 and 55.118.

8.37   a.   Let μ_1 = mean percent yield after fertigation and μ_2 = mean percent yield before fertigation. The large sample confidence interval for $\mu_1 - \mu_1$ is

$$(\bar{y}_1 - \bar{y}_2) \pm z_{\alpha/2}\sqrt{\frac{\sigma_1^2}{n_1} + \frac{\sigma_2^2}{n_2}}$$

where σ_1^2 is estimated by $s_1^2 = 6^2$ and σ_2^2 is estimated by $s_2^2 = 8^2$

For confidence coefficient .90, $\alpha = 1 - .90 = .10$ and $\alpha/2 = .10/2 = .05$. From Table 4, Appendix II, $z_{.05} = 1.645$. The 90% confidence interval is

$$(75 - 40) \pm 1.645\sqrt{\frac{6^2}{100} + \frac{8^2}{100}} \Rightarrow 35 \pm 1.645 \Rightarrow (33.355, 36.645)$$

b.   We are 90% confident the difference between the mean percent yields before and after fertigation is between 33.355% and 36.645%.

8.39 a. First, we must compute the differences, d_i, between the diazinon residue during the day and at night.

Date	Day	Night	d_i
Jan. 11	5.4	24.3	−18.9
12	2.7	16.5	−13.8
13	34.2	47.2	−13.0
14	19.9	12.4	7.5
15	2.4	24.0	−21.6
16	7.0	21.6	−14.6
17	6.1	104.3	−98.2
18	7.7	96.9	−89.2
19	18.4	105.3	−86.9
20	27.1	78.7	−51.6
21	16.9	44.6	−27.7

$$\bar{d} = \frac{\sum d_i}{n} = \frac{-428}{11} = -38.909$$

$$s_d^2 = \frac{\sum d^2 - \frac{(\sum d)^2}{n}}{n-1} = \frac{30{,}033.96 - \frac{(-428)^2}{11}}{11-1} = 1{,}338.086909$$

$$s_d = \sqrt{1{,}338.086909} = 36.5799$$

Let μ_1 = diazinon residue during the day and μ_2 = mean diazinon residue at night. Then $\mu_d = \mu_1 - \mu_2$ is the difference between the mean diazinon residue for day and night. The small sample confidence interval for μ_d is

$$\bar{d} \pm t_{\alpha/2} s_d / \sqrt{n}$$

For confidence coefficient .90, $\alpha = 1 - .90 = .10$ and $\alpha/2 = .10/2 = .05$. From Table 7, Appendix II, with df = $n - 1 = 11 - 1 = 10$, $t_{.05} = 1.812$. The 90% confidence interval is

$$-38.909 \pm 1.812 \frac{(36.5799)}{\sqrt{11}} \Rightarrow -38.909 \pm 19.985 \Rightarrow (-58.894, -18.924)$$

We are 90% confident that the difference in mean diazinon residue between day and night is between −58.894 and −18.924.

b. We must assume the population of differences is normal.

c. Since the confidence interval in part a does not contain 0, there is a difference in the mean diazinon residue between day and night. Since the interval contains only negative numbers, the mean diazinon residue for night is greater than the mean diazinon residue for day.

8.41 **a.** Let μ_1 = mean reaction time score for restrained drivers and μ_2 = mean reaction time score for unrestrained drivers. Then $\mu_d = \mu_1 - \mu_2$ is the difference between mean reaction time scores for the restrained and unrestrained drivers. The small sample confidence interval for μ_d is

$$\bar{d} \pm t_{\alpha/2} s_d / \sqrt{n}$$

For confidence coefficient .90, $\alpha = 1 - .90 = .10$ and $\alpha/2 = .10/2 = .05$. From Table 7, Appendix II, $t_{.05} = 1.761$ with $n - 1 = 15 - 1 = 14$ degrees of freedom. The 90% confidence interval is

$$1.18 \pm 1.761 \frac{1.191}{\sqrt{15}} \Rightarrow 1.18 \pm .542 \Rightarrow (.638, 1.722)$$

We are 90% confident the difference in mean reaction time scores for the restrained and unrestrained drivers is between .638 and 1.722.

b. We must assume the population of differences is normally distributed and the sample paired observations are randomly selected from the target population of paired observations.

c. Since 0 is not in the confidence interval, we can infer that the mean reaction time scores for restrained drivers is greater than that for unrestrained drivers.

8.43 First, we compute the differences:

Veteran	d = Plasma − Fat
1	−2.4
2	−2.8
3	−2.3
4	−3.4
5	−3.9
6	−2.4
7	−4.0
8	−2.5
9	−5.0
10	.3
11	−.1
12	.4
13	0
14	.2
15	−.5
16	.7
17	9.0
18	−.5
19	.2
20	1.6

$$\bar{d} = \frac{\sum d}{n} = \frac{-17.4}{20} = -.87$$

$$s_d^2 = \frac{\sum d^2 - \frac{(\sum d)^2}{n}}{n - 1} = \frac{183.56 - \frac{(-17.4)^2}{20}}{20 - 1}$$

$$= \frac{168.422}{19}$$

$$= 8.864$$

The small sample confidence interval for μ_d is

$$\bar{d} \pm t_{\alpha/2} s_d/\sqrt{n}$$

For confidence coefficient .95, $\alpha = 1 - .95 = .05$ and $\alpha/2 = .05/2 = .025$. From Table 7, Appendix II, $t_{.025} = 2.093$ with $n - 1 = 20 - 1 = 19$ degrees of freedom. The 95% confidence interval is

$$-.87 \pm 2.093 \sqrt{\frac{8.864}{20}} \Rightarrow -.87 \pm 1.39 \Rightarrow (-2.26, .52)$$

We are 90% confident the difference in mean TCDD level in plasma and fat tissue is between $(-2.26$ and .52$)$. Since 0 is contained in the interval, there is no evidence of a difference in the mean TCDD levels between plasma and fat tissue.

8.45   Let p = proportion of recruiters who find it toughest to fill engineering positions.

$$\hat{p} = \frac{95}{285} = .333 \text{ and } \hat{q} = 1 - \hat{p} = 1 - .333 = .667$$

The confidence interval for p is

$$\hat{p} \pm z_{\alpha/2} \sqrt{\frac{\hat{p}\hat{q}}{2}}$$

For confidence coefficient .99, $\alpha = 1 - .99 = .01$ and $\alpha/2 = .01/2 = .005$. From Table 4, Appendix II, $z_{.005} = 2.575$. The 99% confidence interval is

$$.333 \pm 2.575 \sqrt{\frac{.333(.667)}{285}} \Rightarrow .333 \pm .072 \Rightarrow (.261, .405)$$

To change the proportions to percentages, multiply the end points by 100%. The interval is (26.1%, 40.5%).

We are 99% confident the true percentage of recruiters who find it toughest to fill engineering positions is between 26.1% and 40.5%.

In order for the above interval to be valid, $n\hat{p} \geq 4$ and $n\hat{q} \geq 4$

$$n\hat{p} = 285(.333) = 94.905 \geq 4 \text{ and } n\hat{q} = 285(.667) = 190.095 \geq 4$$

Thus, the interval is valid.

8.47   a.   Let p = proportion of all inquires that are due to falls.

$$\hat{p} = y/n = .23 \text{ and } \hat{q} = 1 - \hat{p} = 1 - .23 = .77$$

The confidence interval for p is

$$\hat{p} \pm z_{\alpha/2} \sqrt{\frac{\hat{p}\hat{q}}{n}}$$

For confidence coefficient .95, $\alpha = 1 - .95 = .05$ and $\alpha/2 = .05/2 = .025$. From Table 4, Appendix II, $z_{.025} = 1.96$. The 95% confidence interval is

$$.23 \pm 1.96 \sqrt{\frac{.23(.77)}{2514}} \Rightarrow .23 \pm .0165 \Rightarrow (.2135, .2465)$$

We are 95% confident the proportion of all injuries due to falls is between .2135 and .2465.

In order for the above interval to be valid, $n\hat{p}$ and $n\hat{q}$ must be greater than or equal to 4.

$$n\hat{p} = 2514(.23) = 578.22 \geq 4 \text{ and } n\hat{q} = 2514(.77) = 1935.78 \geq 4$$

Thus, the interval is valid.

b. Let p = proportion of all inquires that are due to burns or scalds.

$$\hat{p} = y/n = .20 \text{ and } \hat{q} = 1 - .20 = .80$$

The 95% confidence interval for p is

$$.2 \pm 1.96 \sqrt{\frac{.2(.8)}{2514}} \Rightarrow .2 \pm .0156 \Rightarrow (.1844, .2156)$$

We are 95% confident the proportion of all injuries due to burns or scalds is between .1844 and .2156.

In order for the above interval to be valid, $n\hat{p} \geq 4$ and $n\hat{q} \geq 4$.

$$n\hat{p} = 2514(.2) = 502.8 \geq 4 \text{ and } n\hat{q} = 2514(.8) = 2011.2 \geq 4$$

Thus, the interval is valid.

8.49 a. Let p = proportion of subjects who use the bright color level.

$$\hat{p} = \frac{58}{90} = .644 \text{ and } \hat{q} = 1 - .644 = .356$$

The confidence interval for p is

$$\hat{p} \pm z_{\alpha/2} \sqrt{\frac{\hat{p}\hat{q}}{n}}$$

For confidence coefficient .95, $\alpha = 1 - .95 = .05$ and $\alpha/2 = .05/2 = .025$. From Table 4, Appendix II, $z_{.025} = 1.96$. The 95% confidence interval is

$$.644 \pm 1.96 \sqrt{\frac{.644(.356)}{90}} \Rightarrow .644 \pm .099 \Rightarrow (.546, .743)$$

We are 95% confident the proportion of subjects who use the bright color level is between .545 and .743.

b. Since both values of the confidence interval are greater than .5, we can infer that a majority of subjects would select bright color levels over dark color levels as a cue.

In order for the above interval to be valid, $n\hat{p} \geq 4$ and $n\hat{q} \geq 4$.

$$n\hat{p} = 90(.644) = 57.96 \geq 4 \text{ and } n\hat{q} = 90(.356) = 32.04$$

Thus, the interval is valid.

8.51 a. Let p_1 = proportion of control cells that exhibited altered growth and p_2 = proportion of cells exposed to E2F1 that exhibited altered growth.

$$\hat{p}_1 = \frac{15}{158} = .095 \quad \hat{q}_1 = 1 - \hat{p}_1 = 1 - .095 = .905$$

$$\hat{p}_2 = \frac{41}{92} = .446 \quad \hat{q}_2 = 1 - \hat{p}_2 = 1 - .446 = .554$$

The confidence interval for the difference in two proportions is

$$(\hat{p}_1 - \hat{p}_2) \pm z_{\alpha/2} \sqrt{\frac{\hat{p}_1\hat{q}_1}{n_1} + \frac{\hat{p}_2\hat{q}_2}{n_2}}$$

For confidence coefficient .90, $\alpha = 1 - .90 = .10$ and $\alpha/2 = .10/2 = .05$. From Table 4, Appendix II, $z_{.05} = 1.645$. The 90% confidence interval is

$$(.095 - .446) \pm 1.645 \sqrt{\frac{.095(.905)}{158} + \frac{.446(.554)}{92}} \Rightarrow -.351 \pm .093$$

$$\Rightarrow (-.444, -.258)$$

In order for the above interval to be valid

$$n_1\hat{p}_1 \geq 4 \text{ and } n_1\hat{q}_1 \geq 4; \ n_2\hat{p}_2 \geq 4 \text{ and } n_2\hat{q}_2 \geq 4$$

$$n_1\hat{p}_1 = 158(.095) = 15.01 \geq 4 \text{ and } n_1\hat{q}_1 = 158(.905) = 142.99 \geq 4;$$
$$n_2\hat{p}_2 = 92(.446) = 41.03 \geq 4 \text{ and } n_2\hat{q}_2 = 92(.554) = 50.97 \geq 4$$

Since all conditions are met, the interval is valid.

b. We are 90% confident that the difference in the proportion of cells that exhibited altered growth between the control group and the group exposed to E2F1 is between $-.444$ and $-.258$. Since the confidence interval contains only negative values, there is evidence to indicate the proportion of control cells that exhibited altered growth is less than the proportion of cells exposed to E2F1 that exhibited altered growth.

8.53   Let p_1 = proportion of brill captured in the Mediterranean Sea that are infected and p_2 = proportion of brill captured in the Atlantic that are infected.

$$\hat{p}_1 = \frac{211}{588} = .359 \quad \text{and } \hat{q}_1 = 1 - .359 = .641$$

$$\hat{p}_2 = \frac{26}{123} = .211 \quad \text{and } \hat{q}_2 = 1 - .211 = .789$$

The confidence interval for the difference in two proportions is

$$(\hat{p}_1 - \hat{p}_2) \pm z_{\alpha/2} \sqrt{\frac{\hat{p}_1 \hat{q}_1}{n_1} + \frac{\hat{p}_2 \hat{q}_2}{n_2}}$$

For confidence coefficient .90, $\alpha = 1 - .90 = .10$ and $\alpha/2 = .10/2 = .05$. From Table 4, Appendix II, $z_{.05} = 1.645$. The 90% confidence interval is

$$(.359 - .211) \pm 1.645 \sqrt{\frac{.359(.641)}{588} + \frac{.211(.789)}{123}} \Rightarrow .148 \pm .069 \Rightarrow (.079, .217)$$

We are 90% confident the difference in the proportion of infected brill between the Mediterranean Sea and the Atlantic Ocean is between .079 and .217.

In order for the above interval to be valid,

$$n_1 \hat{p}_1 \geq 4 \text{ and } n_1 \hat{q}_1 \geq 4 \text{ or } 588(.359) = 211.092 \geq 4 \text{ and } 588(.641) = 376.908 \geq 4$$

and $n_2 \hat{p}_2 \geq 4$ and $n_2 \hat{q}_2 \geq 4$ or $123(.211) = 25.953 \geq 4$ and $123(.789) = 97.047 \geq 4$

Since all conditions are met, the interval is valid.

8.55   Let p_1 = proportion of large firms who indicate they have no need for outside consulting engineering services and p_2 = proportion of small firms who indicate they have no need for outside consulting engineering services.

$$\hat{p}_1 = \frac{12}{20} = .6 \text{ and } \hat{q}_1 = 1 - \hat{p}_1 = 1 - .6 = .4$$

$$\hat{p}_2 = \frac{6}{20} = .3 \text{ and } \hat{q}_2 = 1 - \hat{p}_2 = 1 - .3 = .7$$

The confidence interval for the difference in two proportions is

$$(\hat{p}_1 - \hat{p}_2) \pm z_{\alpha/2} \sqrt{\frac{\hat{p}_1 \hat{q}_1}{n_1} + \frac{\hat{p}_2 \hat{q}_2}{n_2}}$$

For confidence coefficient .90, $\alpha = 1 - .90 = .10$ and $\alpha/2 = .10/2 = .05$. From Table 4, Appendix II, $z_{.05} = 1.645$. The 90% confidence interval is

$$(.6 - .3) \pm 1.645 \sqrt{\frac{.6(.4)}{20} + \frac{.3(.7)}{20}} \Rightarrow (.3 \pm .247) \Rightarrow (.053, .547)$$

To obtain the percentages, multiply the endpoints by $100\% \Rightarrow (5.3\%, 54.7\%)$. We are 90% confident the difference in the percentage of large and small firms who indicate they have no need for outside consulting engineering services is between 5.3% and 54.7%.

In order for the above interval to be valid $n_1\hat{p}_1 \geq 4$ and $n_1\hat{q}_1 \geq 4$; $n_2\hat{p}_2 \geq 4$ and $n_2\hat{q}_2 \geq 4$

$n_1\hat{p}_1 = 20(.6) = 12 \geq 4$ and $n_1\hat{q}_1 = 20(.4) = 8 \geq 4$;
$n_2\hat{p}_2 = 20(.3) = 6 \geq 4$ and $n_2\hat{q}_2 = 20(.7) = 14 \geq 4$

Since all conditions are met, the interval is valid.

8.57   a.   The confidence interval for σ^2 is

$$\frac{(n-1)s^2}{\chi^2_{\alpha/2}} \leq \sigma^2 \leq \frac{(n-1)s^2}{\chi^2_{1-\alpha/2}}$$

For confidence coefficient .95, $\alpha = 1 - .95 = .05$ and $\alpha/2 = .05/2 = .025$. From Table 8, Appendix II, $\chi^2_{.025} = 30.1910$ and $\chi^2_{.975} = 7.56418$ with $n - 1 = 18 - 1 = 17$ degrees of freedom. The 95% confidence interval is

$$\frac{(18-1)6.3^2}{30.1910} \leq \sigma^2 \leq \frac{(18-1)6.3^2}{7.56418} \Rightarrow (22.349 \leq \sigma^2 \leq 89.201)$$

To get a confidence interval for σ, take the square root of the endpoints of the above interval:

$$\Rightarrow \left(\sqrt{22.349} \leq \sigma \leq \sqrt{89.201}\right) \Rightarrow (4.727 \leq \sigma \leq 9.445)$$

b.   No. Since 7 is contained in the above interval, there is no evidence that the true standard deviation is less than 7.

8.59   The confidence interval for σ^2 is

$$\frac{(n-1)s^2}{\chi^2_{\alpha/2}} \leq \sigma^2 \leq \frac{(n-1)s^2}{\chi^2_{1-\alpha/2}}$$

For confidence coefficient .90, $\alpha = 1 - .90 = .10$ and $\alpha/2 = .10/2 = .05$. From Table 8, Appendix II, $\chi^2_{.05} = 9.48773$ and $\chi^2_{.95} = 0.710721$ with $n - 1 = 5 - 1 = 4$ df. The 90% confidence interval is

$$\frac{(5-1)(.45)^2}{9.48773} \leq \sigma^2 \leq \frac{(5-1)(.45)^2}{0.710721} \Rightarrow (.0854 \leq \sigma^2 \leq 1.1397)$$

8.61   a. $s^2 = \dfrac{\sum y^2 - \dfrac{\sum y^2}{n}}{n-1} = \dfrac{29{,}345.78 - \dfrac{342.6^2}{4}}{4-1} = \dfrac{2.09}{3} = .6967$

For confidence coefficient .99, $\alpha = 1 - .99 = .01$ and $\alpha/2 = .01/2 = .005$. For df $= n - 1 = 4 - 1 = 3$, $\chi^2_{.005} = 12.8381$ and $\chi^2_{.995} = .0717212$ from Table 8. The 99% confidence interval is

$$\frac{(n-1)s^2}{\chi^2_{.005}} \le \sigma^2 \le \frac{(n-1)s^2}{\chi^2_{.995}} \Rightarrow \frac{(4-1).6967}{12.8381} \le \sigma^2 \le \frac{(4-1).6967}{.0717212}$$

$$\Rightarrow (.163 \le \sigma^2 \le 29.142)$$

b.   We must assume the observations come from a normal distribution.

8.63   Using Table 10, Appendix II,

a.   For $\nu_1 = 7$ and $\nu_2 = 25$, $F_{.05} = 2.40$

b.   For $\nu_1 = 10$ and $\nu_2 = 8$, $F_{.05} = 3.35$

c.   For $\nu_1 = 30$ and $\nu_2 = 60$, $F_{.05} = 1.65$

d.   For $\nu_1 = 15$ and $\nu_2 = 4$, $F_{.05} = 5.86$

8.65   The confidence interval for σ_1^2/σ_2^2 is

$$\frac{s_1^2}{s_2^2} \cdot \frac{1}{F_{\alpha/2(\nu_1,\nu_2)}} \le \frac{\sigma_1^2}{\sigma_2^2} \le \frac{s_1^2}{s_2^2} F_{\alpha/2(\nu_2,\nu_1)}$$

$\nu_1 = n_1 - 1 = 100 - 1 = 99$ and $\nu_2 = n_2 - 1 = 47 - 1 = 46$

$s_1^2 = 422^2 = 178{,}084$ and $s_2^2 = 271^2 = 73{,}441$

For confidence coefficient .90, $\alpha = 1 - .90 = .10$ and $\alpha/2 = .10/2 = .05$. From Table 10, Appendix II, $F_{.05(99, 46)} \approx 1.60$ and $F_{.05(46, 99)} \approx 1.50$. The 90% confidence interval is

$$\frac{178084}{73441} \cdot \frac{1}{1.60} \le \frac{\sigma_1^2}{\sigma_2^2} \le \frac{178084}{73441} \cdot 1.50 \Rightarrow (1.52 \le \frac{\sigma_1^2}{\sigma_2^2} \le 3.64)$$

We are 90% confident the ratio of the shear stress variances for the two species of wood is between 1.52 and 3.64. Since this interval does not contain 1 (the ratio of $\sigma_1^2/\sigma_2^2 = 1$ if $\sigma_1^2 = \sigma_2^2$), there is evidence to indicate the shear stress variances differ.

8.67 The confidence interval for σ_1^2/σ_2^2 is

$$\frac{s_1^2}{s_2^2} \cdot \frac{1}{F_{\alpha/2(\nu_1, \nu_2)}} \le \frac{\sigma_1^2}{\sigma_2^2} \le \frac{s_1^2}{s_2^2} F_{\alpha/2(\nu_2, \nu_1)}$$

From Exercise 8.36, $s_1 = 1.66$ and $s_2 = 1.56$. Thus, $s_1^2 = 1.66^2 = 2.7556$ with $\nu_1 = n_1 - 1 = 10 - 1 = 9$

$s_2^2 = 1.56^2 = 2.4336$ with $\nu_2 = n_2 - 1 = 10 - 1 = 9$

For confidence coefficient .95, $\alpha = 1 - .95 = .05$ and $\alpha/2 = .05/2 = .025$. From Table 10, Appendix II, $F_{.025(9, 9)} = 4.03$. The 95% confidence interval is

$$\frac{2.7556}{2.4336} \cdot \frac{1}{4.03} \le \frac{\sigma_1^2}{\sigma_2^2} \le \frac{2.7556}{2.4336} \cdot 4.03 \Rightarrow \left(.281 \le \frac{\sigma_1^2}{\sigma_2^2} \le 4.563\right)$$

We are 95% confident the ratio of the variances of cancer death rate increases for the two groups of cities is between .281 and 4.566. Since this interval does contain 1 (the ratio of $\sigma_1^2/\sigma_2^2 = 1$ if $\sigma_1^2 = \sigma_2^2$), there is no evidence to indicate variances are different.

8.69 To choose the sample size for estimating a population mean, we use the formula

$$n = \left[\frac{z_{\alpha/2}\sigma}{H}\right]^2$$

For confidence coefficient .90, $\alpha = 1 - .90 = .10$ and $\alpha/2 = .10/2 = .05$. From Table 4, Appendix II, $z_{.05} = 1.645$. Using $\sigma = 24.66$ and $H = 5$, the sample size is

$$n = \left[\frac{z_{\alpha/2}\sigma}{H}\right]^2 = \left[\frac{1.645(24.66)}{5}\right]^2 = 65.8 \approx 66$$

8.71 **a.** To choose the sample size for estimating a population proportion, we use the formula

$$n = \left[\frac{z_{\alpha/2}}{H}\right]^2 pq$$

For confidence coefficient .90, $\alpha = 1 - .90 = .10$ and $\alpha/2 = .10/2 = .05$. From Table 4, Appendix II, $z_{.05} = 1.645$. Using $p = .60$ and $q = 1 - p = 1 - .60 = .40$, and $H = .03$, the sample size is

$$n = \left[\frac{1.645}{.03}\right]^2 (.6)(.4) = 721.6 \approx 722$$

b. Using $\sigma = 2$ and confidence coefficient .90, the formula for choosing the sample size for estimating a population mean is

$$n = \left[\frac{z_{\alpha/2}\sigma}{H}\right]^2 = \left[\frac{1.645(2)}{.25}\right]^2 = 173.19 \approx 174$$

8.73 To choose the sample size for estimating the differences between two population means, we use the formula

$$n_1 = n_2 = \left[\frac{z_{\alpha/2}}{H}\right]^2 (\sigma_1^2 + \sigma_2^2)$$

For confidence coefficient .95, $\alpha = 1 - .95 = .05$ and $\alpha/2 = .05/2 = .025$. From Table 4, Appendix II, $z_{.025} = 1.96$. Using $\sigma = 3/4$ and $H = .05$, the sample size is

$$n_1 = n_2 = \left[\frac{1.96}{.05}\right]^2 \left[\left[\frac{3}{4}\right]^2 + \left[\frac{3}{4}\right]^2\right] = 1728.72 \approx 1729$$

8.75 The formula for determining sample size for estimating a population proportion is

$$n = \left[\frac{z_{\alpha/2}}{H}\right]^2 pq = \left[\frac{z_{\alpha/2}}{H}\right]^2 p(1 - p)$$

To show that n is maximized when $p = .5$, we take the derivative of n with respect to p, set it equal to 0, and solve.

$$\frac{dn}{dp} = \frac{d\left[\left[\frac{z_{\alpha/2}}{H}\right]^2 p(1 - p)\right]}{dp} = \left[\frac{z_{\alpha/2}}{H}\right]^2 [p(-1) + (1 - p)(1)] = \left[\frac{z_{\alpha/2}}{H}\right]^2 (1 - 2p)$$

Set this equal to 0 and solve:

$$\left[\frac{z_{\alpha/2}}{H}\right]^2 (1 - 2p) = 0 \Rightarrow 1 - 2p = 0 \Rightarrow p = \frac{1}{2} = .5$$

8.77 a. To compare precision, we need to compare the population variances. The confidence interval for σ_1^2/σ_2^2 is

$$\frac{s_1^2}{s_2^2} \cdot \frac{1}{F_{\alpha/2(\nu_1, \nu_2)}} \leq \frac{\sigma_1^2}{\sigma_2^2} \leq \frac{s_1^2}{s_2^2} \cdot F_{\alpha/2(\nu_2, \nu_1)}$$

$$s_1^2 = \frac{\sum y_1^2 - \frac{\left(\sum y_1\right)^2}{n_1}}{n_1 - 1} = \frac{10251 - \frac{225^2}{5}}{5 - 1} = \frac{126}{4} = 31.5$$

with $v_1 = n_1 - 1 = 5 - 1 = 4$ and

$$s_2^2 = \frac{\sum y_2^2 - \frac{\left(\sum y_2\right)^2}{n_2}}{n_2 - 1} = \frac{10351 - \frac{227^2}{5}}{5 - 1} = \frac{45.2}{4} = 11.3$$

with $v_2 = n_2 - 1 = 5 - 1 = 4$

For confidence coefficient .90, $\alpha = 1 - .90 = .10$ and $\alpha/2 = .10/2 = .05$. From Table 10, Appendix II, $F_{.05(4, 4)} = 6.39$. The 90% confidence interval is

$$\frac{31.5}{11.3} \cdot \frac{1}{6.39} \le \frac{\sigma_1^2}{\sigma_2^2} \le \frac{31.5}{11.3}(6.39) \Rightarrow \left(.436 \le \frac{\sigma_1^2}{\sigma_2^2} \le 17.813\right)$$

We are 90% confident the ratio of the variances for the two instruments is between .436 and 17.813.

b. Since 1 is in the interval $(\sigma_1^2/\sigma_2^2 = 1 \Rightarrow \sigma_1^2 = \sigma_2^2)$, there is no evidence to indicate the two variances are different. Thus, we could infer the variances are equal.

c. We must assume the two samples are independent and the two populations are normal.

8.79 To choose the sample size for estimating a population proportion, we use the formula

$$n = \left[\frac{z_{\alpha/2}}{H}\right]^2 pq$$

For confidence coefficient .95, $\alpha = 1 - .95 = .05$ and $\alpha/2 = .05/2 = .025$. From Table 4, Appendix II, $z_{.025} = 1.96$. Using $p = .16$ and $q = 1 - .16 = .84$, and $H = .03$, the sample size is

$$n = \left[\frac{z_{\alpha/2}}{H}\right]^2 pq = \left[\frac{1.96}{.03}\right]^2 (.16)(.84) = 573.68 \approx 574$$

8.81 The confidence interval for σ^2 is

$$\frac{(n - 1)s^2}{\chi_{\alpha/2}^2} \le \sigma^2 \le \frac{(n - 1)s^2}{\chi_{1-\alpha/2}^2}$$

For confidence coefficient .95, $\alpha = 1 - .95 = .05$ and $\alpha/2 = .05/2 = .025$. From Table 8, Appendix II, $\chi^2_{.025} = 32.8523$ and $\chi^2_{.975} = 8.90655$ with $n - 1 = 20 - 1 = 19$ degrees of freedom. The 95% confidence interval is

$$\frac{(20 - 1)(.07)^2}{32.8523} \le \sigma^2 \le \frac{(20 - 1)(.07)^2}{8.90655} \Rightarrow (.0028 \le \sigma^2 \le .0105)$$

We are 95% confident the variance of the fill is between .0028 and .0105.

8.83   To choose the sample size for estimating the difference between two means, we use

$$n_1 = n_2 = \left[\frac{z_{\alpha/2}}{H}\right]^2 (\sigma_1^2 + \sigma_2^2)$$

For confidence coefficient .95, $\alpha = 1 - .95 = .05$ and $\alpha/2 = .05/2 = .025$. From Table 4, Appendix II, $z_{.025} = 1.96$. We will estimate σ by $\sigma \approx \dfrac{\text{Range}}{4} = \dfrac{4}{4} = 1$. Thus, we estimate $\sigma_1^2 = \sigma_2^2 = 1^2 = 1$. For $H = .2$, the sample size is

$$n_1 = n_2 = \left[\frac{1.96}{.2}\right]^2 (1 + 1) = 192.08 \approx 193$$

8.85   a.   Let μ_1 = mean load capacity for Alloy A and μ_2 = mean load capacity for Alloy B. The confidence interval for $(\mu_1 - \mu_2)$, small sample, is

$$\bar{y}_1 - \bar{y}_2 \pm t_{\alpha/2} \sqrt{s_p^2 \left[\frac{1}{n_1} + \frac{1}{n_2}\right]}$$

where $s_p^2 = \dfrac{(n_1 - 1)s_1^2 + (n_2 - 1)s_2^2}{n_1 + n_2 - 2} = \dfrac{(11 - 1)24.4 + (17 - 1)19.9}{11 + 17 - 2} = 21.6308$

For confidence coefficient .99, $\alpha = 1 - .99 = .01$ and $\alpha/2 = .01/2 = .005$. From Table 7, Appendix II, $t_{.005} = 2.779$ with df $= n_1 + n_2 - 2 = 11 + 17 - 2 = 26$. The 99% confidence interval is

$$(43.7 - 48.5) \pm 2.779 \sqrt{21.6308 \left[\frac{1}{11} + \frac{1}{17}\right]} \Rightarrow -4.8 \pm 5.001$$
$$\Rightarrow (-9.801, .201)$$

b.   The necessary assumptions are:

Both populations are normal, $\sigma_1^2 = \sigma_2^2 = \sigma^2$, and the samples are independent.

c. We are 90% confident the difference between the true average load capacities for the two alloys is between -9.801 and $.201$.

Since 0 (if $\mu_1 = \mu_2$, then $\mu_1 - \mu_2 = 0$) is in the interval, there is insufficient evidence to conclude the average load capacities are different.

d. To choose the sample size for estimating $\mu_1 - \mu_2$, we use:

$$n_1 = n_2 = \left[\frac{z_{\alpha/2}}{H}\right]^2 (\sigma_1^2 + \sigma_2^2)$$

For confidence coefficient .99, $\alpha = 1 - .99 = .01$ and $\alpha/2 = .01/2 = .005$. From Table 4, Appendix II, $z_{.005} = 2.58$. We will use s_1^2 and s_2^2 to estimate σ_1^2 and σ_2^2, respectively. For $H = .2$, the sample size is

$$n_1 = n_2 = \left[\frac{2.58}{2}\right]^2 (24.4. + 19.9) = 73.7 \approx 74$$

8.87   a.   First, we must find $E(y)$.

$$E(y) = \int_{-\infty}^{\infty} yf(y)dy = \int_{\theta}^{\theta+1} y \, dy = \left.\frac{y^2}{2}\right]_{\theta}^{\theta+1} = \frac{(\theta + 1)^2}{2} - \frac{\theta^2}{2}$$

$$= \frac{\theta^2 + 2\theta + 1 - \theta^2}{2} = \frac{2\theta + 1}{2} = \theta + \frac{1}{2}$$

Thus, $E(\bar{y}) = E\left[\dfrac{y_1 + y_2 + \cdots + y_n}{n}\right] = \dfrac{1}{n}[E(y_1) + E(y_2) + \cdots + E(y_n)]$

$$= \frac{1}{n}\left[\left(\theta + \frac{1}{2}\right) + \left(\theta + \frac{1}{2}\right) + \cdots + \left(\theta + \frac{1}{2}\right)\right]$$

$$= \frac{1}{n}n\left(\theta + \frac{1}{2}\right) = \theta + \frac{1}{2}$$

The bias is $\dfrac{1}{2}$.

b. First, we must find $V(y)$. We know $V(y) = E(y^2) - [E(y)]^2$.

$$E(y^2) = \int_{-\infty}^{\infty} y^2 f(y)dy = \int_{\theta}^{\theta+1} y^2 \, dy = \left.\frac{y^3}{3}\right]_{\theta}^{\theta+1} = \frac{(\theta + 1)^3}{3} - \frac{\theta^3}{3}$$

$$= \frac{3\theta^2 + 3\theta + 1}{3} = \theta^2 + \theta + \frac{1}{3}$$

$$V(y) = E(y^2) - [E(y)]^2 = \theta^2 + \theta + \frac{1}{3} - \left(\theta + \frac{1}{2}\right)^2$$

$$= \theta^2 + \theta + \frac{1}{3} - \left(\theta^2 + \theta + \frac{1}{4}\right) = \frac{1}{3} - \frac{1}{4} = \frac{4}{12} - \frac{3}{12} = \frac{1}{12}$$

$$V(\bar{y}) = V\left[\frac{y_1 + y_2 + \cdots + y_n}{n}\right] = \frac{1}{n^2}[V(y_1) + V(y_2) + \cdots + V(y_n)]$$

$$= \frac{1}{n^2}\left[\frac{1}{12} + \frac{1}{12} + \cdots + \frac{1}{12}\right] = \frac{1}{n^2}\left(\frac{n}{12}\right) = \frac{1}{12n}$$

c.   If $E(\bar{y}) = \theta + \frac{1}{2}$, then $\bar{y} - \frac{1}{2}$ would be an unbiased estimator of θ.

$$E\left[\bar{y} - \frac{1}{2}\right] = \left(\theta + \frac{1}{2}\right) - \frac{1}{2} = \theta$$

8.89   a.   Let $w = 2y/\beta$. If y has a gamma distribution with $\alpha = 1$ and arbitrary β, the density function for y is

$$f(y) = \begin{cases} \frac{1}{\beta}e^{-y/\beta} & y > 0 \\ 0 & \text{elsewhere} \end{cases}$$

If the range for y is $y > 0$, then the range for w is $w > 2(0)/\beta$ or $w > 0$.

$$P(w \le w_0) = G(w_0) = F(y_0) = P(y \le y_0) \text{ where } w_0 = 2y_0/\beta$$

which implies $y_0 = \beta w_0/2$

Thus, $F(y_0) = F(\beta w_0/2) = \int_0^{\beta w_0/2} \frac{1}{\beta}e^{-y/\beta}dy = -e^{-y/\beta}\Big|_0^{\beta w_0/2}$

$$= -e^{-w_0/2} - (-1) = 1 - e^{-w_0/2}$$

The cumulative distribution function for w is $F(w) = 1 - e^{-w/2}$

The density function for w is

$$\frac{dF(w)}{dw} = \frac{d\left[1 - e^{-w/2}\right]}{dw} = -\left(-\frac{1}{2}e^{-w/2}\right) = \frac{1}{2}e^{-w/2}$$

Thus, $f(w) = \begin{cases} \frac{1}{2}e^{-w/2} & w > 0 \\ 0 & \text{elsewhere} \end{cases}$

Thus, density function indicates w has a gamma distribution with $\alpha = 1$ and $\beta = 2$.

b. From Section 5.6, a chi-square random variable has a gamma distribution with $\alpha = \nu/2$ and $\beta = 2$, where ν is the degrees of freedom. Thus, any gamma distribution with $\beta = 2$ can be transformed to a chi-square distribution with $\nu = 2\alpha$.

From part a, $2y/\beta$ then has a chi-square distribution with $\nu = 2(\alpha) = 2(1) = 2$.

c. Using the pivotal statistic $w = 2y/\beta$, the confidence interval for β is

$$P\left(\chi_{.975}^2 \le w \le \chi_{.025}^2\right) = P\left(\chi_{.975}^2 \le 2y/\beta \le \chi_{.025}^2\right)$$

$$= P\left[\frac{1}{\chi_{.975}^2} \ge \beta/2y \ge \frac{1}{\chi_{.025}^2}\right]$$

$$= P\left[\frac{2y}{\chi_{.975}^2} \ge \beta \ge \frac{2y}{\chi_{.025}^2}\right] = .95$$

or

$$= P\left[\frac{2y}{\chi_{.025}^2} \le \beta \le \frac{2y}{\chi_{.975}^2}\right] = .95$$

where the critical χ^2 values have 2 df.

8.91 a. The midpoint of the interval $\bar{y} - t_{\alpha/2}\left(\dfrac{s}{\sqrt{n}}\right) \le \mu \le \bar{y} + t_{\alpha/2}\left(\dfrac{s}{\sqrt{n}}\right)$ is $\bar{y}$.

If $E(y_i) = \mu$ for all i,

$$E(\bar{y}) = E\left[\frac{y_1 + y_2 + \cdots + y_n}{n}\right] = \frac{1}{n}[E(y_1) + E(y_2) + \cdots + E(y_n)]$$

$$= \frac{1}{n}(n\mu) = \mu$$

Since the expected value of the midpoint of the confidence interval is μ, the confidence interval is unbiased.

b. The midpoint of the interval $\dfrac{(n-1)s^2}{\chi_{\alpha/2}^2} \le \sigma^2 \le \dfrac{(n-1)s^2}{\chi_{1-\alpha/2}^2}$

is $\dfrac{\dfrac{(n-1)s^2}{\chi_{\alpha/2}^2} + \dfrac{(n-1)s^2}{\chi_{1-\alpha/2}^2}}{2} = \dfrac{(n-1)s^2}{2}\left[\dfrac{1}{\chi_{\alpha/2}^2} + \dfrac{1}{\chi_{1-\alpha/2}^2}\right]$

We know $E(s^2) = \sigma^2$.

Therefore, $E\left\{\dfrac{(n-1)s^2}{2}\left[\dfrac{1}{\chi_{\alpha/2}^2} + \dfrac{1}{\chi_{1-\alpha/2}^2}\right]\right\} = \dfrac{(n-1)\sigma^2}{2}\left[\dfrac{1}{\chi_{\alpha/2}^2} + \dfrac{1}{\chi_{1-\alpha/2}^2}\right] \ne \sigma^2$

Since the expected value of the midpoint of the interval is not σ^2, the interval is biased.

Estimation

CHAPTER NINE

....

Tests of Hypotheses

9.1   **a.** α = probability of committing a Type I error or probability of rejecting H_0 when H_0 is true.

        b. β = probability of committing a Type II error or probability of accepting H_0 when H_0 is false.

9.3   **a.**   A false negative would be accepting H_0 when it is false, which is a Type II error.

        b.   A false positive would be rejecting H_0 when it is true, which is a Type I error.

        c.   According to Dunnett, a false positive or Type I error would be more serious. Much money and time would be spent with further testing of an ineffective drug.

9.5   **a.** α = probability of rejecting H_0 when H_0 is true

$$= P(y \geq 6 \text{ if } p = .1) = 1 - P(y \leq 5) = 1 - \sum_{y=0}^{5} p(y) = 1 - .9666 = .0334$$

        ($P(y \leq 5)$ is found using Table 1, Appendix II, with $n = 25$ and $p = .1$)

        b. β = probability of accepting H_0 when it is false

$$= P(y \leq 5 \text{ if } p = .2) = \sum_{y=0}^{5} p(y) = .6167$$

        ($P(y \leq 5)$ is found using Table 1, Appendix II, with $n = 25$ and $p = .2$). The power of the test = $1 - \beta = 1 - .6167 = .3833$.

        c. β = probability of accepting H_0 when it is false

$$= P(y \leq 5 \text{ if } p = .4) = \sum_{y=0}^{5} p(y) = .0294$$

        ($P(y \leq 5)$ is found using Table 1, Appendix II, with $n = 25$ and $p = .4$). The power of the test = $1 - \beta = 1 - .0294 = .9706$.

9.7 $L(\mu) = \prod_{i=1}^{n} f(y_i) = \prod_{i=1}^{n} f(y_i) \frac{1}{\sqrt{2\pi}} e^{-(y_i - \mu)^2/2} = \left[\frac{1}{\sqrt{2\pi}} \right]^n e^{-\sum (y_i - \mu)^2/2}$

....

9.9 From Exercise 9.8, $\lambda = e^{-n\bar{y}^2/2}$

The rejection region is given as $\lambda \leq \lambda_\alpha$

$$\Rightarrow e^{-n\bar{y}^2/2} \leq \lambda_\alpha \Rightarrow -n\bar{y}^2/2 \leq \ln(\lambda_\alpha) \Rightarrow -\bar{y}^2 \leq \frac{2}{n}\ln(\lambda_\alpha) \Rightarrow \bar{y}^2 \geq -\frac{2}{n}\ln(\lambda_\alpha)$$

$$\Rightarrow \bar{y}^2 \geq \bar{y}_a \quad \text{where } \bar{y}_a = -\frac{2}{n}\ln(\lambda_a)$$

9.11 Since it is desired to determine if the mean breaking strength is less than 22 pounds, we test:

H_0: $\mu = 22$
H_a: $\mu < 22$

9.13 Let μ_1 = mean Datapro rating for the software vendor and let μ_2 = mean Datapro rating for the rival vendor.

Since the software vendor wants to determine if its product has a higher mean, we test:

H_0: $\mu_1 - \mu_2 = 0$
H_a: $\mu_1 - \mu_2 > 0$

9.15 Let μ_1 = mean number of items produced by method 1 and let μ_2 = mean number of items produced by method 2.

Since we want to determine if the mean number of items produced differ for the two methods, we test:

H_0: $\mu_1 - \mu_2 = 0$
H_a: $\mu_1 - \mu_2 \neq 0$

9.17 Let μ = mean throughput per 40 hour week.

Since it is desired to see if the true mean throughput is less than 1920 parts, we test:

H_0: $\mu = 1920$
H_a: $\mu < 1920$

The rejection region for a small sample, one-tailed test requires $\alpha = .05$ in the lower tail of the t distribution with df $= n - 1 = 5 - 1 = 4$. From Table 7, Appendix II, $t_{.05} = 2.132$ with 4 df. The rejection region is $t < -2.132$.

The test statistic is $t = \dfrac{\bar{y} - \mu_0}{s/\sqrt{n}} = \dfrac{1908.8 - 1920}{18/\sqrt{5}} = -1.39$

Since the observed value of the test statistic does not fall in the rejection region ($t = -1.39 \not< -2.132$), H_0 is not rejected. There is insufficient evidence to indicate the true mean throughput per 40-hour week for the system is less than 1920 parts at $\alpha = .05$.

9.19   Since it is desired to determine if the mean blood lead content in young black children is greater than 14, we use the hypotheses:

        H_0: $\mu = 14$
        H_a: $\mu > 14$

The rejection region for a large-sample, one-tailed test with $\alpha = .01$ requires $\alpha = .01$ in the upper tail of the z distribution.

From Table 4, Appendix II, $z_{.01} = 2.33$. The rejection region is $z > 2.33$.

The test statistic is $z = \dfrac{\bar{y} - \mu_0}{\sigma_{\bar{y}}} \approx \dfrac{\bar{y} - \mu_0}{s/\sqrt{n}} = \dfrac{21 - 14}{10/\sqrt{200}} = 9.90$

Since the observed value of the test statistic falls in the rejection region ($z = 9.9 > 2.33$), H_0 is rejected. There is sufficient evidence to indicate the mean blood lead concentration in young black children is greater than 14 at $\alpha = .01$.

9.21   To determine if the mean drill chip length differs from 75 mm, we test:

        H_0: $\mu = 75$
        H_a: $\mu \neq 75$

The rejection region for a large-sample, two-tailed test requires $\alpha/2 = .01/2 = .005$ in each tail of the z distribution. From Table 4, $z_{.005} = 2.575$. The rejection region is $z < -2.575$ or $z > 2.575$.

The test statistic is $z = \dfrac{\bar{y} - \mu_0}{\sigma_{\bar{y}}} \approx \dfrac{\bar{y} - \mu_0}{s/\sqrt{n}} = \dfrac{81.2 - 75}{50.2/\sqrt{50}} = .87$

Since the observed value of the test statistic does not fall in the rejection region ($z = .87 \not> 2.575$), H_0 is not rejected. There is insufficient evidence to indicate the mean drill chip length differs from 75 mm at $\alpha = .01$.

9.23   Since it is desired to determine if the mean breaking strength is more than 2500 pounds per linear foot, we test the hypotheses:

        H_0: $\mu = 2500$
        H_a: $\mu > 2500$

The rejection region for a small-sample, one-tailed test requires $\alpha = .10$ in the upper tail of the t distribution with df $= n - 1 = 7 - 1 = 6$. Table 7, Appendix II, gives $t_{.10} = 1.440$ with 6 df. The rejection region is $t > 1.440$.

$$\bar{y} = \frac{\sum y}{n} = \frac{18,000}{7} = 2571.429$$

$$s^2 = \frac{\sum y^2 - \frac{(\sum y)^2}{n}}{n - 1} = \frac{46,365,200 - \frac{18,000^2}{7}}{6} = \frac{79,485.715}{6} = 13,247.619$$

$$s = \sqrt{13,247.619} = 115.098$$

The test statistic is $t = \dfrac{\bar{y} - \mu_0}{s/\sqrt{n}} = \dfrac{2571.429 - 2500}{115.098/\sqrt{7}} = \dfrac{71.429}{43.503} = 1.64$

Since the observed value of the test statistic falls in the rejection region ($t = 1.64 > 1.440$), H_0 is rejected. There is sufficient evidence to indicate the mean breaking strength is greater than 2500 pounds per lineal foot at $\alpha = .10$. Yes, the pipe meets specifications.

9.25 a. From Example 9.9, the rejection region is $t > 1.729$. In terms of $\bar{y}$:

$$\bar{y}_0 = \mu_0 + t_\alpha \left[\frac{s}{\sqrt{n}} \right] = 1 + 1.729 \left[\frac{1.7}{\sqrt{20}} \right] = 1.657$$

The rejection region in terms of $\bar{y}$ is $\bar{y} > 1.657$.

β = probability of accepting H_0 when H_0 is false
$\quad = P(\bar{y} \leq 1.657 \mid \mu_a = 1.015)$

$$= P\left[t \leq \frac{1.657 - 1.015}{1.7/\sqrt{20}} \right] = P(t \leq 1.689)$$

Using Table 7, with df $= n - 1 = 20 - 1 = 19$, $.90 < P(t \leq 1.689) < .95$. Thus, $.90 < \beta < .95$.

b. β = probability of accepting H_0 when H_0 is false
$\quad = P(\bar{y} \leq 1.657 \mid \mu_a = 1.045)$

$$= P\left[t \leq \frac{1.657 - 1.045}{1.7/\sqrt{20}} \right] = P(t \leq 1.610)$$

Using Table 7, with df $= n - 1 = 20 - 1 = 19$, $.90 < P(t \leq 1.689) < .95$. Thus, $.90 < \beta < .95$.

Power $= 1 - \beta \Rightarrow .05 <$ Power $< .10$

9.27 The p-values associated with each test statistic are found in Table 4, Appendix II.

a. p-value $= P(z > 1.96) = .5 - P(0 \le z \le 1.96) = .5 - .4750 = .0250$

b. p-value $= P(z > 1.645) = .5 - P(0 \le z \le 1.645) = .5 - .4500 = .05$

c. p-value $= P(z > 2.67) = .5 - P(0 \le z \le 2.67) = .5 - .4962 = .0038$

d. p-value $= P(z > 1.25) = .5 - P(0 \le z \le 1.25) = .5 - .3944 = .1056$

9.29 The test statistic for Example 9.8 is $z = 4.03$. For a two-tailed test, the p-value $= P(z \ge 4.03) + P(z \le -4.03)$. From Table 4, Appendix II, $P(z \ge 4.03) + P(z \le -4.03) \approx 0$.

9.31 The p-value $= P(t \ge 1.642) = .1517/2 = .07585$. The probability of observing our test statistic or anything more unusual is .076. Since this p-value is less than $\alpha = .10$, we reject H_0. This agrees with the inference in Exercise 9.23.

9.35 Let μ_1 = mean oxon/thion ratio of foggy days and let μ_2 = mean oxon/thion ratio of cloudy/clear days.

Some preliminary calculations:

$$\bar{y}_1 = \frac{\sum y_1}{n_1} = \frac{2.19}{8} = .27375 \qquad s_1^2 = \frac{\sum y_1^2 - \frac{\left(\sum y_1\right)^2}{n_1}}{n_1 - 1} = \frac{.698 - \frac{2.19^2}{8}}{8 - 1} = .01407$$

$$\bar{y}_2 = \frac{\sum y_2}{n_2} = \frac{1.809}{4} = .45225 \qquad s_2^2 = \frac{\sum y_2^2 - \frac{\left(\sum y_2\right)^2}{n_2}}{n_1 - 1} = \frac{.922455 - \frac{1.809^2}{4}}{4 - 1} = .03478$$

$$s_p^2 = \frac{(n_1 - 1)s_1^2 + (n_2 - 1)s_2^2}{n_1 + n_2 - 2} = \frac{(8 - 1).01407 + (4 - 1)(.03478)}{8 + 4 - 2} = .020283$$

To determine if the mean oxon/thion ratios of foggy and clear/cloudy days are different, we test:

H_0: $\mu_1 - \mu_2 = 0$
H_a: $\mu_1 - \mu_2 \ne 0$

The rejection region for a small-sample, two-tailed test requires $\alpha/2 = .05/2 = .025$ in each tail of the t distribution with df $= n_1 + n_2 - 2 = 8 + 4 - 2 = 10$. From Table 7, Appendix , $t_{.025} = 2.228$. The rejection region is $t > 2.228$ or $t < -2.228$.

The test statistic is $t = \dfrac{\bar{y}_1 - \bar{y}_2 - (\mu_2 - \mu_2)}{\sqrt{s_p^2\left[\dfrac{1}{n_1} + \dfrac{1}{n_2}\right]}} = \dfrac{(.27375 - .45225) - 0}{\sqrt{.020283\left[\dfrac{1}{8} + \dfrac{1}{4}\right]}} = -2.05$

Since the observed value of the test statistic does not fall in the rejection region ($t = -2.05 \not< -2.228$), H_0 is not rejected. There is insufficient evidence to indicate the mean oxon/thion ratios of foggy and clear/cloudy days are different at $\alpha = .05$.

9.37 a. Let μ_1 = mean perception of managers at less automated firms and μ_2 = mean perception of managers at highly automated firms.

Since we want to determine if there is a difference in the mean perception between managers of highly automated and less automated firms, we test:

$$H_0: \mu_1 - \mu_2 = 0$$
$$H_a: \mu_1 - \mu_2 \neq 0$$

The rejection region for the small-sample, two-tailed test requires $\alpha/2 = .01/2 = .005$ in each tail of the t distribution with degrees of freedom = $n_1 + n_2 - 2 = 17 + 8 - 2 = 23$. From Table 7, Appendix II, $t_{.005} = 2.807$. The rejection region is $t < -2.807$ or $t > 2.807$.

Assuming the variances are equal, we find the pooled variance.

$$s_p^2 = \frac{(n_1 - 1)s_1^2 + (n_2 - 1)s_2^2}{n_1 + n_2 - 2} = \frac{(17 - 1).762^2 + (8 - 1).721^2}{17 + 8 - 2} = .562$$

The test statistic is $t = \dfrac{(\bar{y}_1 - \bar{y}_2) - D_0}{\sqrt{s_p^2\left[\dfrac{1}{n_1} + \dfrac{1}{n_2}\right]}} = \dfrac{(3.274 - 3.280) - 0}{\sqrt{.562\left[\dfrac{1}{17} + \dfrac{1}{8}\right]}} = -.0187$

Since the observed value of the test statistic does not fall in the rejection region ($t = -.0187 \not< -2.807$), H_0 is not rejected. There is insufficient evidence to indicate a difference in the mean perceptions between managers of highly automated and less automated firms at $\alpha = .01$.

b. If the variances are not equal, the test statistic is

$$t = \frac{(\bar{y}_1 - \bar{y}_2) - D_0}{\sqrt{\dfrac{s_1^2}{n_1} + \dfrac{s_2^2}{n_2}}} = \frac{(3.274 - 3.280) - 0}{\sqrt{\dfrac{.762^2}{17} + \dfrac{.721^2}{8}}} = -.019$$

The degrees of freedom are:

$$\nu = \frac{\left(s_1^2/n_1 + s_2^2/n_2\right)^2}{\dfrac{\left(s_1^2/n_1\right)^2}{n_1 - 1} + \dfrac{\left(s_2^2/n_2\right)^2}{n_2 - 1}} = \frac{(.762^2/17 + .721^2/8)^2}{\dfrac{(.762^2/17)^2}{17 - 1} + \dfrac{(.721^2/8)^2}{8 - 1}} = \frac{.009828}{.000676114}$$

$$= 14.5 \approx 14$$

The rejection region requires $\alpha/2 = .01/2 = .005$ in each tail of the t distribution with 14 df. From Table 7, Appendix II, $t_{.005} = 2.977$. The rejection region is $t < -2.977$ and $t > 2.977$.

Since the observed value of the test statistic does not fall in the rejection region ($t = -.019 \not< -2.977$), H_0 is not rejected. The conclusion is the same as in part a.

9.39   Let μ_1 = average number of seedlings germinating from the seed caches of desert rodents and μ_2 = average number of seedlings germinating from the control areas.

Since it is desired to determine if the average number of seedlings germinating is higher from the seed caches than from the control areas, we test:

$$H_0: \mu_1 - \mu_2 = 0$$
$$H_a: \mu_1 - \mu_2 > 0$$

The test statistic is $z = \dfrac{(\bar{y}_1 - \bar{y}_2) - D_0}{\sqrt{\dfrac{\sigma_1^2}{n_1} + \dfrac{\sigma_2^2}{n_2}}} \approx \dfrac{(\bar{y}_1 - \bar{y}_2) - D_0}{\sqrt{\dfrac{s_1^2}{n_1} + \dfrac{s_2^2}{n_2}}} = \dfrac{(5.3 - 2.7) - 0}{\sqrt{\dfrac{1.3^2}{40} + \dfrac{.7^2}{40}}} = \dfrac{2.6}{.23345}$

$$= 11.14$$

The rejection region for a large-sample, one-tailed test requires $\alpha = .05$ in the upper tail of the z distribution.

From Table 4, Appendix II, $z_{.05} = 1.645$. The rejection region is $z > 1.645$.

Since the observed value of the test statistic falls in the rejection region ($z = 11.14 > 1.645$), H_0 is rejected. There is sufficient evidence to indicate the average number of seedlings germinating is higher from the seed caches of desert rodents than from control areas ($\mu_1 - \mu_2 > 0$) at $\alpha = .05$.

9.41   Let μ_1 = mean percentage of correct syllables under condition S + F + A and μ_2 = mean percentage of correct syllables under condition S.

Since it is desired to determine if the mean percentage is higher under condition S + F + A, we test:

$$H_0: \mu_1 - \mu_2 = 0$$
$$H_a: \mu_1 - \mu_2 > 0$$

The test statistic is $t = \dfrac{\bar{d} - D_0}{s_d/\sqrt{n}} = \dfrac{20.4 - 0}{17.44/\sqrt{10}} = \dfrac{20.4}{5.515} = 3.70$

The rejection region for this small-sample, one-tailed test requires $\alpha = .05$ to be in the upper tail of the t distribution with df $= n - 1 = 10 - 1 = 9$. Table 7, Appendix II, gives $t_{.05} = 1.833$ with 9 df.

The rejection region is $t > 1.833$.

Since the observed value of the test statistic falls in the rejection region ($t = 3.7 > 1.833$), H_0 is rejected. There is sufficient evidence to indicate the mean percentage under condition S + F + A is higher ($\mu_1 - \mu_2 > 0$) at $\alpha = .05$.

9.43 Let μ_1 = mean number of swims required for male rat pups and let μ_2 = mean number of swims required for female rat pups.

To determine if the mean number of swims required by male and female rat pups differ, we test:

H_0: $\mu_1 - \mu_2 = 0$
H_a: $\mu_1 - \mu_2 \neq 0$

The rejection region for a small-sample, two-tailed test, requires $\alpha/2 = .10/2 = .05$ in each tail of the t distribution with df $= n - 1 = 19 - 1 = 18$. From Table 7, Appendix II, $t_{.05} = 1.734$. The rejection region is $t > 1.734$ or $t < -1.734$.

The test statistic is $t = \dfrac{\bar{y} - D_0}{s_d/\sqrt{n}} = .46$ (from printout)

Since the observed value of the test statistic does not fall in the rejection region ($t = .46 \not> 1.734$), H_0 is not rejected. There is insufficient evidence to indicate that the mean number of swims required by male and female rat pups differ at $\alpha = .10$.

We could also make our decision based on the given p-value. The p-value is .65. Since the p-value is greater than $\alpha = .10$, H_0 is not rejected.

9.45 To determine if the true success rate of the feeder exceeds .90, we test:

H_0: $p = .90$
H_a: $p > .90$

The rejection region for a large-sample, one-tailed test requires $\alpha = .10$ in the upper tail of the z distribution. From Table 4, Appendix II, $z_{.10} = 1.28$. The rejection region is $z > 1.28$.

The test statistic is $z = \dfrac{\hat{p} - p_0}{\sqrt{\dfrac{p_0 q_0}{n}}} = \dfrac{.94 - .90}{\sqrt{\dfrac{.90(.10)}{100}}} = 1.33$

Since the observed value of the test statistic falls in the rejection region ($z = 1.33 > 1.28$), H_0 is rejected. There is sufficient evidence to indicate the true success rate of the feeder exceeds .90 at $\alpha = .10$.

9.47 Since it is desired to test the manufacturer's claim that the proportion of data lost with the new controller in place is .01, we will test:

H_0: $p = .01$
H_a: $p > .01$

$$\hat{p} = \frac{6}{200} = .03$$

The rejection region for the large-sample, one-tailed test requires $\alpha = .05$ in the upper tail of the z distribution. From Table 4, Appendix II, $z_{.05} = 1.645$. The rejection region is $z > 1.645$.

The test statistic is $z = \dfrac{\hat{p} - p_0}{\sqrt{\dfrac{p_0 q_0}{n}}} = \dfrac{.03 - .01}{\sqrt{\dfrac{.01(.99)}{200}}} = 2.84$

Since the observed value of the test statistic does fall in the rejection region ($z = 2.84 > 1.645$), H_0 can be rejected. There is sufficient evidence to indicate the proportion of errors with the new controller in place exceeds .01 at $\alpha = .05$.

9.49 Since we want to see if the mock weapon detection rate at LAX is less than the national rate of .80,

H_0: $p = .80$
H_a: $p < .80$

$$\hat{p} = \frac{72}{100} = .72$$

The rejection region for a large sample, one-tailed test requires $\alpha = .05$ in the lower tail of the z distribution. From Table 4, Appendix II, $z_{.05} = 1.645$. The rejection region is $z < -1.645$.

The test statistic is $z = \dfrac{\hat{p} - p_0}{\sqrt{\dfrac{p_0 q_0}{n}}} = \dfrac{.72 - .80}{\sqrt{\dfrac{.80(.20)}{100}}} = -2.00$

Since the value of the test statistic falls in the rejection region ($z = -2.00 < -1.645$), H_0 is rejected. There is sufficient evidence at $\alpha = .05$ to indicate the mock weapon detection rate at LAX is less than the national rate.

In order for the above test to be valid, $n\hat{p} \geq 4$ and $n\hat{q} \geq 4$.

$n\hat{p} = 100(.72) = 72 \geq 4$ and $n\hat{q} = 100(.28) = 28 \geq 4$. Thus, the test is valid.

9.51   Since it is desired to determine if the proportion of A-E firms that rely heavily on public-sector clients has declined, we test:

$$H_0: \ p = 1/3 = .333$$
$$H_a: \ p < 1.3 = .333$$

$$\hat{p} = \frac{10}{60} = \frac{1}{6} = .167$$

The test statistic is $z = \dfrac{\hat{p} - p_0}{\sqrt{p_0 q_0/n}} = \dfrac{.167 - .333}{\sqrt{.333(1 - .333)/60}} = \dfrac{-.167}{.0609} = -2.74$

The rejection region for a large-sample, one-tailed test requires $\alpha = .05$ in the lower tail of the z distribution.

From Table 4, Appendix II, $z_{.05} = 1.645$. The rejection region is $z < -1.645$.

Since the observed value of the test statistic falls in the rejection region ($z = -2.74 < -1.645$), H_0 is rejected. There is sufficient evidence to indicate the proportion of A-E firms that rely heavily on public-sector clients has declined at $\alpha = .05$.

In order for the above test to be valid, $n\hat{p} \geq 4$ and $n\hat{q} \geq 4$.

$n\hat{p} = 60(1/6) = 10 \geq 4$ and $n\hat{q} = 60(5/6) = 50 \geq 4$. Thus, the test is valid.

9.53   Let p_1 = proportion of Verapamil users who had recurring heart attacks and p_2 = proportion of the control group who had recurring heart attacks.

To determine if the calcium blocker (Verapamil) is effective in reducing heart attacks, we test:

$$H_0: \ p_1 - p_2 = 0$$
$$H_a: \ p_1 - p_2 < 0$$

$$\hat{p}_1 = \frac{146}{897} = .163 \qquad \hat{p}_2 = \frac{180}{878} = .205 \qquad \hat{p} = \frac{146 + 180}{897 + 878} = .184$$

The rejection region for this large-sample, one-tailed test requires $\alpha = .01$ in the lower tail of the z distribution. From Table 4, Appendix II, $z_{.01} = 2.33$. The rejection region is $z < -2.33$.

The test statistic is $z = \dfrac{(\hat{p}_1 - \hat{p}_2) - D_0}{\sqrt{\hat{p}\hat{q}\left[\dfrac{1}{n_1} + \dfrac{1}{n_2}\right]}} = \dfrac{(.163 - .205) - 0}{\sqrt{.184(.816)\left[\dfrac{1}{878} + \dfrac{1}{897}\right]}} = -2.30$

Since the observed value of the test statistic does not fall in the rejection region ($z = -2.30 \not< -2.33$), H_0 cannot be rejected. There is insufficient evident to indicate the calcium blockers are effective in reducing the risk of heart attack at $\alpha = .01$.

9.55 **a.** Let p_1 = proportion of MFWS users who rely on the computer as their information source and p_2 = proportion of non-MFWS users who rely on the computer as their information source.

Some preliminary calculations:

$$\hat{p}_1 = \frac{4}{12} = .333 \qquad \hat{p}_2 = \frac{2}{25} = .08 \qquad \hat{p} = \frac{4+2}{12+25} = .162$$

To determine if a difference exists between the proportion of MFWS users and non-MFWS users who rely on the computer as their major information source, we test:

$H_0: p_1 - p_2 = 0$
$H_a: p_1 - p_2 \neq 0$

The rejection region for a large-sample, two-tailed test requires $\alpha/2 = .10/2 = .05$ in each tail of the z distribution. From Table 4, Appendix II, $z_{.05} = 1.645$. The rejection region is $z > 1.645$ or $z < -1.645$.

The test statistic is $z = \dfrac{(\hat{p}_1 - \hat{p}_2) - D_0}{\sqrt{\hat{p}\hat{q}\left[\dfrac{1}{n_1} + \dfrac{1}{n_2}\right]}} = \dfrac{(.333 - .08) - 0}{\sqrt{.162(.838)\left[\dfrac{1}{12} + \dfrac{1}{25}\right]}} = 1.96$

Since the observed value of the test statistic falls in the rejection region ($z = 1.96 > 1.645$), H_0 is rejected. There is sufficient evident to indicate that a difference exists between the proportion of MFWS users and non-MFWS users who rely on the computer as their major information source at $\alpha = .10$.

b. The sample sizes are large enough if $n_1\hat{p}_1 \geq 4$ and $n_1\hat{q}_1 \geq 4$; $n_2\hat{p}_2 \geq 4$ and $n_2\hat{q}_2 \geq 4$.

$n_1\hat{p}_1 = 12(.333) = 4 \geq 4$ and $n_1\hat{q}_1 = 12(.667) = 8 \geq 4$
$n_2\hat{p}_2 = 25(.08) = 2 \not\geq 4$ and $n_2\hat{q}_2 = 25(.92) = 23 \geq 4$

Since $n_2\hat{p}_2 = 25(.08) = 2 \not\geq 4$, the sample sizes are not large enough.

9.57 Let p_1 = proportion of firms with computerized external market data in 1982 and p_2 = proportion of firms with computerized external market data in 1987.

Since it is desired to test for a significant increase in the percentage of firms with computerized external market data over the 5-year period, we test:

$H_0: p_2 - p_1 = 0$
$H_a: p_2 - p_1 > 0$

The sample proportions are $\hat{p}_1 = .25$ and $\hat{p}_2 = .33$.

$$\hat{p} = \frac{371(.25) + 459(.33)}{371 + 459} = .294 \qquad \hat{q} = -\hat{p} = 1 - .294 = .706$$

The test statistic is $z = \dfrac{(\hat{p}_2 - \hat{p}_1) - D_0}{\sqrt{\hat{p}\hat{q}\left[\dfrac{1}{n_1} + \dfrac{1}{n_2}\right]}} = \dfrac{.33 - .25 - 0}{\sqrt{.294(.706)\left[\dfrac{1}{371} + \dfrac{1}{459}\right]}} = \dfrac{.08}{.0318} = 2.52$

The rejection region for a large-sample, one-tailed test requires $\alpha = .05$ in the upper tail of the z distribution.

From Table 4, Appendix II, $z_{.05} = 1.645$. The rejection region is $z > 1.645$.

Since the observed value of the test statistic falls in the rejection region ($z = 2.52 > 1.645$), H_0 is rejected. There is sufficient evidence to indicate there was a significant increase in percentage of firms with computerized external market data from 1982 to 1987 at $\alpha = .05$.

In order for the above test to be valid,

$$n_1\hat{p}_1 \geq 4 \text{ and } n_1\hat{q}_1 \geq 4;\ n_2\hat{p}_2 \geq 4 \text{ and } n_2\hat{q}_2 \geq 4.$$

$n_1\hat{p}_1 = 371(.25) = 92.75 \geq 4$ and $n_1\hat{q}_1 = 371(.75) = 278.25 \geq 4$
$n_2\hat{p}_2 = 459(.33) = 151.47 \geq 4$ and $n_2\hat{q}_2 = 459(.67) = 307.53 \geq 4$

Since all conditions are met, the test is valid.

9.59   a.   Since we wish to determine if the SNR variance exceeds .54, we test:

$$H_0:\ \sigma^2 = .54$$
$$H_a:\ \sigma^2 > .54$$

b.   Using the normal population assumption, we estimate that

Range $\approx 4\sigma$

$$\Rightarrow 2.97 \approx 4\sigma \Rightarrow s = \hat{\sigma} = \dfrac{2.97}{4} = .7425$$

c.   The test statistic is $\chi^2 = \dfrac{(n-1)s^2}{\sigma_0^2} = \dfrac{(41-1)(.7425)^2}{.54} = 40.8375$

The rejection region requires $\alpha = .10$ in the upper tail of the χ^2 distribution with $n - 1$ $= 41 - 1 = 40$ df. From Table 8, Appendix II, $\chi^2_{.10} = 51.8050$. The rejection region is $\chi^2 > 51.8050$. Since the observed value of the test statistic does not fall in the rejection region ($\chi^2 = 40.8375 \not> 51.8050$), H_0 cannot be rejected. There is insufficient evidence to indicate the SNR variance exceeds .54 at $\alpha = .10$.

9.61   a.   Since we desire to test if the variance of the amount of rubber cement dispensed into the cans exceeds .3, we test:

$$H_0: \sigma^2 = .30$$
$$H_a: \sigma^2 > .30$$

The rejection region requires $\alpha = .05$ in the upper tail of the χ^2 distribution with $n - 1 = 10 - 1 = 9$ df. From Table 8, Appendix II, $\chi^2_{.05} = 16.9190$. The rejection region is $\chi^2 > 16.9190$.

The test statistic is $\chi^2 = \dfrac{(n-1)s^2}{\sigma_0^2} = \dfrac{(10-1)(.48)^2}{.30} = 6.912$

Since the observed value of the test statistic does not fall in the rejection region ($\chi^2 = 6.912 \not> 16.9190$), H_0 cannot be rejected. There is insufficient evidence to indicate the dispensing machines are in need of adjustment at $\alpha = .05$.

b.   The assumption necessary is that the distribution of the amount of rubber cement dispensed is approximately normal.

9.63   Let σ_1^2 = variance under foggy conditions and σ_2^2 = variance under clear/cloudy conditions.

From Exercise 9.35, $s_1^2 = .01407$ and $s_2^2 = .03478$ with $n_1 = 8$ and $n_2 = 4$.

To determine if the variances are different, we test:

$$H_0: \frac{\sigma_1^2}{\sigma_2^2} = 1$$

$$H_a: \frac{\sigma_1^2}{\sigma_2^2} \neq 1$$

The rejection region for this two-tailed test requires $\alpha/2 = .05/2 = .02$ in the upper tail of the F distribution with $\nu_1 = n_2 - 1 = 4 - 1 = 3$ and $\nu_2 = n_1 - 1 = 8 - 1 = 7$ degrees of freedom. From Table 11, Appendix II, $F_{.025} = 5.42$. The rejection region is $F > 5.42$.

The test statistic is $F = \dfrac{\text{Larger sample variance}}{\text{Smaller sample variance}} = \dfrac{s_2^2}{s_1^2} = \dfrac{.03478}{.01407} = 2.472$

Since the observed value of the test statistic does not fall in the rejection region ($F = 2.472 \not> 5.42$), H_0 is not rejected. There is insufficient evidence to indicate the variances are different at $\alpha = .05$.

9.65 $s_1^2 = \dfrac{\sum y_1^2 - \dfrac{(\sum y_1)^2}{n_1}}{n_1 - 1} = \dfrac{891.12 - \dfrac{82^2}{8}}{8 - 1} = \dfrac{50.62}{7} = 7.23143$

$s_2^2 = \dfrac{\sum y_2^2 - \dfrac{(\sum y_2)^2}{n_2}}{n_2 - 1} = \dfrac{973.29 - \dfrac{86.1^2}{8}}{8 - 1} = \dfrac{46.63875}{7} = 6.66268$

a. To determine if there is a difference in the variation in the cracking torsion moments of the two types of T-beams, we test:

$$H_0: \frac{\sigma_1^2}{\sigma_2^2} = 1$$

$$H_a: \frac{\sigma_1^2}{\sigma_2^2} \neq 1$$

The test statistic is $F = \dfrac{\text{Larger sample variance}}{\text{Smaller sample variance}} = \dfrac{s_1^2}{s_2^2} = \dfrac{7.23143}{6.66268} = 1.085$

The rejection region for this two-tailed test requires $\alpha/2 = .10/2 = .05$ in the upper tail of the F distribution with numerator df $= n_1 - 1 = 8 - 1 = 7$ and denominator df $= n - 1 = 8 - 1 = 7$. From Table 10, Appendix II, $F_{.05} = 3.79$. The rejection region is $F > 3.79$.

Since the observed value of the test statistic does not fall in the rejection region ($F = 1.085 \not> 3.79$), H_0 is not rejected. There is insufficient evidence to indicate a difference in the variation in the cracking torsion moments of the two types of T-beams at $\alpha = .10$.

b. The assumptions necessary are that the two populations being sampled from are normal and the samples are independently and randomly selected.

9.67 Let σ_1^2 = variance in the percentage of correct syllables for inexperienced speedreaders and σ_2^2 = variance in the percentage of correct syllables for experienced speedreaders. To determine if a difference exists between the two variances, we test:

$$H_0: \frac{\sigma_1^2}{\sigma_2^2} = 1$$

$$H_a: \frac{\sigma_1^2}{\sigma_2^2} \neq 1$$

The rejection region for this two-tailed test requires $\alpha/2 = .05$ in the upper tail of the F distribution with $\nu_1 = n_2 - 1 = 11$ and $\nu_2 = n_1 - 1 = 23$ degrees of freedom. From Table 10, Appendix II, $F_{.05} \approx 2.27$. Reject H_0 if $F > 2.27$.

The test statistic is $F = \dfrac{\text{Larger sample variance}}{\text{Smaller sample variance}} = \dfrac{s_2^2}{s_1^2} = \dfrac{(12.4)^2}{(8.7)^2} = 2.03$

Since the observed value of the test statistic does not fall in the rejection region ($F= 2.03 \ngtr 2.27$), H_0 is not rejected. There is insufficient evidence that a difference exists between the two variances at $\alpha = .10$.

9.69   From the hint,

$$P\left[\frac{\text{Larger sample variance}}{\text{Smaller sample variance}} > F_{\alpha/2}\right]$$

$$= P\left[\frac{s_1^2}{s_2^2} > F_{\alpha/2} \text{ or } \frac{s_2^2}{s_1^2} > F_{\alpha/2}\right]$$

$$= P\left[\frac{s_1^2}{s_2^2} > F_{\alpha/2}\right] + P\left[\frac{s_2^2}{s_1^2} > F_{\alpha/2}\right] \qquad \left(\text{because } \frac{s_1^2}{s_2^2} \text{ and } \frac{s_2^2}{s_1^2} \text{ cannot both be greater than } F_{\alpha/2} \text{ at the same time}\right)$$

$$= \frac{\alpha}{2} + \frac{\alpha}{2} \qquad \text{(from Exercise 9.68)}$$

$$= \alpha$$

9.71   a. α = probability of rejecting H_0 when H_0 is true
        $= P(y \leq 1 \text{ or } y \geq 8 \text{ if } p = .5) = P(y \leq 1) + [1 - P(y \leq 7)]$
        $= .0107 + (1 - .9453) = .0654$

        where $p(y \leq 1)$ and $p(y \leq 7)$ are found using Table 1, Appendix II, with $n = 10$ and $p = .5$

     b. β = probability of accepting H_0 when H_0 is false
        $= P(2 \leq y \leq 7 \text{ if } p = .4) = p(y \leq 7) - p(y \leq 1)$
        $= .9877 - .0464 = .9413$

        where $p(y \leq 7)$ and $p(y \leq 1)$ are found using Table 1, Appendix II, with $n = 10$ and $p = .4$

        Power $1 - \beta = 1 - .9413 = .0587$

c. β = probability of accepting H_0 when H_0 is false

$$= P(2 \leq y \leq 7 \text{ if } p = .8) = p(y \leq 7) - p(y \leq 1) = .3222 - 0 = .3222$$

where $p(y \leq 7)$ and $p(y \leq 1)$ are found using Table 1, Appendix II, with $n = 10$ and $p = .8$

Power $= 1 - \beta = 1 - .3222 = .6778$

9.73 a. Let μ_1 = mean of SPT flow times and μ_2 = mean of proposed rule flow times. To determine if the average in flow time is less under the proposed scheduling rule than under the SPT approach, we test:

H_0: $\mu_1 - \mu_2 = 0$
H_a: $\mu_1 - \mu_2 > 0$

The test statistic is $z = \dfrac{(\bar{y}_1 - \bar{y}_2) - D_0}{\sqrt{\dfrac{\sigma_1^2}{n_1} + \dfrac{\sigma_2^2}{n_2}}} = \dfrac{(158.28 - 117.07) - 0}{\sqrt{\dfrac{8532.80}{32} + \dfrac{5208.53}{32}}} = 1.99$

The rejection region for this large-sample, one-tailed test requires $\alpha = .05$ in the upper tail of the z distribution. From Table 4, Appendix II, $z_{.05} = 1.645$. The rejection region is $z > 1.645$.

Since the observed value of the test statistic falls in the rejection region ($z = 1.99 > 1.645$), H_0 is rejected. There is sufficient evidence to indicate the average flow time is less under the proposed scheduling rule than under the SPT approach at $\alpha = .05$.

b. Let μ_1 = mean of SPT tardiness scores and μ_2 = mean of proposed rule tardiness scores. To determine if the proposed scheduling rule will lead to a reduction in the average in the tardiness of tire tests, we test:

H_0: $\mu_1 - \mu_2 = 0$
H_a: $\mu_1 - \mu_2 > 0$

The test statistic is $z = \dfrac{(\bar{y}_1 - \bar{y}_2) - D_0}{\sqrt{\dfrac{\sigma_1^2}{n_1} + \dfrac{\sigma_2^2}{n_2}}} \approx \dfrac{(5.26 - 4.52) - 0}{\sqrt{\dfrac{452.09}{32} + \dfrac{319.41}{32}}} = .15$

The rejection region is $z > 1.645$ (see part a).

Since the observed value of the test statistic does not fall in the rejection region ($z = .15 \not> 1.645$), H_0 is not rejected. There is insufficient evidence to indicate the proposed scheduling rule will lead to a reduction in the average tardiness of tire tests at $\alpha = .05$.

9.75   a.   Since we wish to determine if the mean level of radiation is less than 5 picocuries per liter of water, we test:

$$H_0: \mu = 5$$
$$H_a: \mu < 5$$

The rejection region for this small-sample, one-tailed test requires $\alpha = .01$ in the lower tail of the t distribution with $n - 1 = 24 - 1 = 23$ df. From Table 7, Appendix II, $t_{.01} = 2.500$. The rejection region is $t < -2.500$.

The test statistic is $t = \dfrac{\bar{y} - \mu_0}{s/\sqrt{n}} = \dfrac{4.61 - 5}{.87/\sqrt{24}} = -2.196$

Since the observed value of the test statistic does not fall in the rejection region ($t = -2.196 \not< -2.500$), H_0 cannot be rejected. There is insufficient evidence to indicate that the mean level of radiation is safe at $\alpha = .01$.

b.   We want our chance of making a Type I error to be small. That is, we want the probability of saying the mean level of radiation is safe when it really is unsafe to be small.

c. β = probability of accepting H_0 when H_0 is false.

Accepting H_0 will happen if $t > -2.500$. In terms of $\bar{y}$:

$$t = \frac{\bar{y} - \mu_0}{s/\sqrt{n}} = \frac{\bar{y} - 5}{.87/\sqrt{24}} > -2.500$$

$$\Rightarrow \bar{y} - 5 > -.44397 \Rightarrow \bar{y} > 4.55603$$

$$\beta = P(\bar{y} > 4.55603 \text{ if } \mu_a = 4.5)$$

$$= P\left(t > \frac{4.55603 - 4.5}{.87/\sqrt{24}}\right) = P(t > .3155)$$

$$\Rightarrow .5 > P(t > .3155) > .10 \Rightarrow .10 < \beta < .50$$

d. p-value $= P(t \leq -2.196)$
$\Rightarrow P(t < -2.069) > P(t \leq -2.196) > P(t < -2.500)$
$\Rightarrow .025 > p\text{-value} > .010$
$\Rightarrow .01 < p\text{-value} < .025$

9.77   Let μ_1 = mean central memory access time for 16 PE's and μ_2 = mean central memory access time for 48 PE's.

To detect a difference in the mean central memory access time for the 16 and 48 PE's, we test:

$$H_0: \mu_1 - \mu_2 = 0$$
$$H_a: \mu_1 - \mu_2 \neq 0$$

The rejection region for this large-sample, two tailed test requires $\alpha/2 = .05/2 = .025$ in both tails of the z distribution. From Table 4, Appendix II, $z_{.025} = 1.96$. The rejection region is $z > 1.96$ or $z < -1.96$.

The test statistic is $z = \dfrac{(\bar{y}_1 - \bar{y}_2) - D_0}{\sqrt{\dfrac{\sigma_1^2}{n_1} + \dfrac{\sigma_2^2}{n_2}}} = \dfrac{(8.94 - 8.83) - 0}{\sqrt{\dfrac{(3.10)^2}{1000} + \dfrac{(3.50)^2}{1000}}} = .744$

Since the observed value of the test statistic does not fall in the rejection region ($z = .744 \not> 1.96$), H_0 cannot be rejected. There is insufficient evidence to indicate a difference in the mean central memory access times of instructions processed with 16 and 48 PE's using $\alpha = .05$.

9.79   a.   Let $p_1 =$ proportion of employees in experimental group who experience some type of physical discomfort at the end of the day and $p_2 =$ proportion of employees in control group who experience some type of physical discomfort at the end of the day. We want to test:

$$H_0:\ p_1 - p_2 = 0$$
$$H_a:\ p_1 - p_2 < 0$$

The rejection region for the large-sample, one-tailed test requires $\alpha = .03$ in the lower tail of the z distribution. Using Table 4, $z_{.03} = 1.88$, so our rejection region is $z < -1.88$.

$$\hat{p}_1 = \frac{3}{50} = .06 \qquad \hat{p}_2 = \frac{12}{50} = .24 \qquad \hat{p} = \frac{3 + 12}{50 + 50} = .15$$

The test statistic is $z = \dfrac{(\hat{p}_1 - \hat{p}_2) - D_0}{\sqrt{\hat{p}\hat{q}\left[\dfrac{1}{n_1} + \dfrac{1}{n_2}\right]}} = \dfrac{.06 - .24 - 0}{\sqrt{(.15)(.85)\left[\dfrac{1}{50} + \dfrac{1}{50}\right]}} = -2.521$

Since the observed value of the test statistic falls in the rejection region ($z = -2.521 < -1.88$), H_0 is rejected. There is sufficient evidence to indicate the proportion of employees in the experimental group who experience some type of physical discomfort at the end of the day is significantly less than the corresponding proportion in the control group at $\alpha = .03$.

b.   The p-value for the test is $p = P(z \le -2.52) = .5 - .4941 = .0059$

9.81 Let σ_1^2 = variance of Coastal Douglas firs and σ_2^2 = variance of Southern Pine.

Since we want to detect a difference in the variances, we test:

$$H_0: \frac{\sigma_1^2}{\sigma_2^2} = 1$$

$$H_a: \frac{\sigma_1^2}{\sigma_2^2} \neq 1$$

The rejection region for the two-tailed test requires $\alpha/2 = .02/2 = .01$ in the upper tail of the F distribution with $\nu_1 = n_1 - 1 = 118 - 1 = 117$ and $\nu_2 = n_2 - 1 = 147 - 1 = 146$ df.

Using Table 12, Appendix II, $F_{.01} = 1.53$ (using $\nu_1 = \nu_2 = 120$). The rejection region is $F > 1.53$.

The test statistic is $F = \dfrac{\text{Larger sample variance}}{\text{Smaller sample variance}} = \dfrac{s_1^2}{s_2^2} = \dfrac{644.62^2}{611.72^2} = 1.11$

Since the observed value of the test statistic does not fall in the rejection region ($F = 1.11 \not> 1.53$), H_0 cannot be rejected. We have insufficient evidence to indicate the variation in wooden pole strengths of the two tree types differ at $\alpha = .02$.

9.83 Since we wish to determine if the machine is producing bolts with a mean length differing from 1 inch, we test:

$$H_0: \mu = 1$$
$$H_a: \mu \neq 1$$

The rejection region for a large-sample, two-tailed test requires $\alpha/2 = .01/2 = .005$ in both tails of the z distribution. From Table 4, Appendix II, $z_{.005} = 2.576$. The rejection region is $z > 2.576$ or $z < -2.576$.

The test statistic is $z = \dfrac{\bar{y} - \mu_0}{s/\sqrt{n}} = \dfrac{1.02 - 1}{.04/\sqrt{50}} = 3.54$

Since the observed value of the test statistic falls in the rejection region ($z = 3.54 > 2.576$), H_0 is rejected. There is sufficient evidence to indicate the process is out of control at $\alpha = .01$.

CHAPTER TEN

....

Categorical Data Analysis

10.1   a.   Let p_1 = proportion of companies that do not take action.

$$\hat{p}_1 = \frac{n_1}{n} = \frac{10}{121} = .083 \text{ and } \hat{q}_1 = 1 - \hat{p}_1 = 1 - .083 = .917$$

The confidence interval for p_1 is $\hat{p}_1 \pm z_{\alpha/2} \sqrt{\dfrac{\hat{p}_1 \hat{q}_1}{n}}$

For confidence coefficient .90, $\alpha = 1 - .90 = .10$ and $\alpha/2 = .10/2 = .05$. From Table 4, Appendix II, $z_{.05} = 1.645$. The 90% confidence interval for p_1 is

$$.083 \pm 1.645 \sqrt{\frac{.083(.971)}{121}} \Rightarrow .083 \pm .041 \Rightarrow (.042, .124)$$

We are 90% confident the proportion of all companies that do not take action is between .042 and .124.

   b.   Let p_2 = proportion of companies that do internal audits.

$$\hat{p}_2 = \frac{n_2}{n} = \frac{49}{121} = .405 \text{ and } \hat{q}_2 = 1 - \hat{p}_2 = 1 - .405 = .595$$

The confidence interval for p_2 is

$$.405 \pm 1.645 \sqrt{\frac{.405(.595)}{121}} \Rightarrow .405 \pm .073 \Rightarrow (.332, .478)$$

We are 90% confident the proportion of all companies that do internal audits is between .332 and .478.

   c.   Let p_3 = proportion of companies that use the honor system.

$$\hat{p}_3 = \frac{n_3}{n} = \frac{28}{121} = .231 \text{ and } \hat{q}_3 = 1 - \hat{p}_3 = 1 - .231 = .769$$

The confidence interval for $p_2 - p_3$ is

$$(\hat{p}_2 - \hat{p}_3) \pm z_{\alpha/2} \sqrt{\frac{\hat{p}_2(1 - \hat{p}_2) + \hat{p}_3(1 - \hat{p}_3) + 2\hat{p}_2\hat{p}_3}{n}}$$

....

The 90% confidence interval for $p_2 - p_3$ is

$$(.405 - .231) \pm 1.645 \sqrt{\frac{.405(.595) + .231(.769) + 2(.405)(.231)}{121}} \Rightarrow .174 \pm .116$$

$$\Rightarrow (.058, .290)$$

We are 90% confident that the difference in the proportion of companies that do internal audits and the proportion of companies that use the honor system is between .058 and .290.

d. Let p_5 = proportion of companies that use other methods.

$$\hat{p}_5 = \frac{n_5}{n} = \frac{22}{121} = .182 \text{ and } \hat{q}_5 = 1 - \hat{p}_5 = 1 - .182 = .818$$

The 90% confidence interval for $p_1 - p_5$ is

$$(.083 - .182) \pm 1.645 \sqrt{\frac{.083(.917) + .182(.818) + 2(.083)(.182)}{121}} \Rightarrow -.099 \pm .076$$

$$\Rightarrow (-.175, -.023)$$

We are 90% confident that the difference in the proportion of companies that do not take any action and the proportion of companies that use the other methods is between $-.175$ and $-.023$.

10.3 a. The form of the confidence interval for p_2 is

$$\hat{p}_2 \pm z_{\alpha/2} \sqrt{\frac{\hat{p}_2(1 - \hat{p}_2)}{n}}$$

$$\hat{p}_2 = \frac{n_2}{n} = \frac{11}{40} = .275$$

For confidence coefficient .99, $\alpha = 1 - .99 = .01$ and $\alpha/2 = .01/2 = .005$. From Table 4, Appendix II, $z_{.005} = 2.58$. The confidence interval is

$$.275 \pm 2.58 \sqrt{\frac{.275(1 - .275)}{40}} \Rightarrow .275 \pm .182 \Rightarrow (.093, .457)$$

b. The form of the confidence interval for $(p_2 - p_1)$ is

$$(\hat{p}_2 - \hat{p}_1) \pm z_{\alpha/2} \sqrt{\frac{\hat{p}_2(1 - \hat{p}_2) + \hat{p}_1(1 - \hat{p}_1) + 2\hat{p}_2\hat{p}_1}{n}}$$

$$\hat{p}_2 = .275$$

$$\hat{p}_1 = \frac{n_1}{n} = \frac{6}{40} = .150$$

Using the information from part a, the confidence interval is

$$(.275 - .150) \pm 2.58 \sqrt{\frac{.275(1 - .275) + .150(1 - .150) + 2(.275)(.150)}{40}}$$

$$\Rightarrow .125 \pm .261 \Rightarrow (-.136, .386)$$

10.5   a.   The form of the confidence interval for p_C is

$$\hat{p}_C \pm z_{\alpha/2} \sqrt{\frac{\hat{p}_C(1 - \hat{p}_C)}{n}}$$

$$\hat{p}_C = \frac{n_C}{n} = \frac{22}{100} = .22$$

For confidence coefficient .90, $\alpha = 1 - .90 = .10$ and $\alpha/2 = .10/2 = .05$. From Table 4, Appendix II, $z_{.05} = 1.645$. The confidence interval is

$$.22 \pm 1.645 \sqrt{\frac{.22(1 - .22)}{100}} \Rightarrow .22 \pm .068 \Rightarrow (.152, .288)$$

b.   The form of the confidence interval for $(p_E - p_B)$ is

$$(\hat{p}_E - \hat{p}_B) \pm z_{\alpha/2} \sqrt{\frac{\hat{p}_E(1 - \hat{p}_E) + \hat{p}_B(1 - \hat{p}_B) + 2\hat{p}_E\hat{p}_B}{n}}$$

$$\hat{p}_E = \frac{n_E}{n} = \frac{19}{100} = .19$$

$$\hat{p}_B = \frac{n_B}{n} = \frac{27}{100} = .27$$

Using the information from part a, the confidence interval is

$$(.19 - .27) \pm 1.645 \sqrt{\frac{.19(1 - .19) + .27(1 - .27) + 2(.19)(.27)}{100}}$$

$$\Rightarrow -.08 \pm .111 \Rightarrow (-.191, .031)$$

c. $\hat{p}_A = \frac{n_A}{n} = \frac{17}{100} = .17$

$$\hat{p}_D = \frac{n_D}{n} = \frac{15}{100} = .15$$

Using the information from part **b**, the confidence interval is

$$(.17 - .15) \pm 1.645 \sqrt{\frac{.17(1 - .17) + .15(1 - .15) + 2(.17)(.15)}{100}}$$

$$\Rightarrow .02 \pm .093 \Rightarrow (-.073, .095)$$

10.7 The probability function for the multinomial is

$$p(n_1, n_2, \ldots, n_k) = \frac{n!}{n_1! n_2! \cdots n_k!} \, p_1^{n_1} p_2^{n_2} \cdots p_k^{n_k}$$

where $n = n_1 + n_2 + \cdots + n_k$ and $p_1 + p_2 + \cdots + p_k = 1$

Since this is a probability function, we know if we sum over all possible values of $n_1, n_2, \ldots, n_k$, the result is 1.

$$\sum_{n_1=0}^{n} \sum_{n_2=0}^{n} \cdots \sum_{n_k=0}^{n} \frac{n!}{n_1! n_2! \cdots n_k!} \, p_1^{n_1} p_2^{n_2} \cdots p_k^{n_k} = 1 \text{ where } n = n_1 + n_2 + \cdots + n_k$$

Without loss of generality, let $i = 1$ and $j = 2$.

$$E(n_1 n_2) = \sum_{n_1} \sum_{n_2} \cdots \sum_{n_k} n_1 n_2 \frac{n!}{n_1! n_2! \cdots n_k!} \, p_1^{n_1} p_2^{n_2} \cdots p_k^{n_k}$$

$$= n(n - 1) p_1 p_2 \sum_{n_1} \sum_{n_2} \cdots \sum_{n_k} \frac{n - 2!}{(n_1 - 1)!(n_2 - 1)! \, n_3! \cdots n_k!} \, p_1^{n_1 - 1} p_2^{n_2 - 1} p_3^{n_3} \cdots p_k^{n_k}$$

$$= n(n - 1) p_1 p_2$$

$$\text{Cov}(n_i, n_j) = E(n_i n_j) - E(n_i) E(n_j) = n(n - 1) p_i p_j - n p_i (n p_j) = n p_i p_j (n - 1 - n) = -n p_i p_j$$

10.9 Let $p_1 =$ proportion of butterfly hotspots, $p_2 =$ proportion of butterfly coldspots, and $p_3 =$ proportion of neutral area.

To test the theory, we test:

H_0: $p_1 = .05$, $p_2 = .05$, and $p_3 = .90$
H_a: At least one of the proportions differ from its theoretical value

The expected numbers in each category are:

$E(n_1) = n p_1 = 2588(.05) = 129.4$
$E(n_2) = n p_2 = 2588(.05) = 129.4$
$E(n_3) = n p_3 = 2588(.90) = 2329.2$

The observed and expected counts are:

	Butterfly Hotspots	Butterfly Coldspots	Neutral Areas
Observed	123	147	2,318
Expected	129.4	129.4	2,329.2

The test statistic is

$$\chi^2 = \sum \frac{[n_i - E(n_i)]^2}{E(n_i)} = \frac{(123 - 129.4)^2}{129.4} + \frac{(147 - 129.4)^2}{129.4} + \frac{(2318 - 2329.2)^2}{2329.2}$$
$$= 2.764$$

The rejection region requires $\alpha = .01$ in the upper tail of the χ^2 distribution with df $= k - 1 = 3 - 1 = 2$. From Table 8, Appendix II, $\chi^2_{.01} = 9.21034$. The rejection region is $\chi^2 > 9.21034$.

Since the observed value of the test statistic does not fall in the rejection region ($\chi^2 = 2.764 \ngtr 9.21034$), H_0 is not rejected. There is insufficient evidence to indicate the proportions differ from their theoretical values at $\alpha = .01$.

10.11 To determine if the incidence of gastrointestinal disease during the epidemic is related to water consumption, we test:

H_0: $p_1 = p_2 = p_3 = p_4 = .25$
H_a: At least two of the proportions are not equal

The expected number in each category is:

$$E(n_1) = E(n_2) = E(n_3) = E(n_4) = np_{i,0} = 40(.25) = 10$$

The observed and expected counts are:

	Water Consumption			
	0	1-2	3-4	5 or more
Observed	6	11	13	10
Expected	10	10	10	10

The test statistic is

$$\chi^2 = \sum \frac{[n_i - E(n_i)]^2}{E(n_i)} = \frac{(6 - 10)^2}{10} + \frac{(11 - 10)^2}{10} + \frac{(13 - 10)^2}{10} + \frac{(10 - 10)^2}{10} = 2.6$$

The rejection region requires $\alpha = .01$ in the upper tail of the χ^2 distribution with df $= k - 1$ $= 4 - 1 = 3$. From Table 8, Appendix II, $\chi_{.01}^2 = 11.3449$. The rejection region is $\chi^2 > 11.3449$.

Since the observed value of the test statistic does not fall in the rejection region ($\chi^2 = 2.6 \not> 11.3449$), H_0 is not rejected. There is insufficient evidence to indicate the incidence of gastrointestinal disease during the epidemic is related to water consumption at $\alpha = .01$.

10.13   To determine if a preference for one or more of the five water management strategies exists, we test:

H_0: $p_1 = p_2 = p_3 = p_4 = p_5 = .20$
H_a: At least two of the proportions are not equal

The expected number in each category is:

$E(n_i) = np_{i,0} = 100(.20) = 20$

The observed and expected counts are:

	Strategy				
	A	B	C	D	E
Observed	17	27	22	15	19
Expected	20	20	20	20	20

$$\chi^2 = \sum \frac{[n_i - E(n_i)]^2}{E(n_i)}$$

$$= \frac{(17 - 20)^2}{20} + \frac{(27 - 20)^2}{20} + \frac{(22 - 20)^2}{20} + \frac{(15 - 20)^2}{20} + \frac{(19 - 20)^2}{20} = 4.4$$

The rejection region requires $\alpha = .05$ in the upper tail of the χ^2 distribution with df $= k - 1$ $= 5 - 1 = 4$. From Table 8, Appendix II, $\chi_{.05}^2 = 9.48773$. The rejection region is $\chi^2 > 9.48773$.

Since the observed value of the test statistic does not fall in the rejection region ($\chi^2 = 4.4 \not> 9.48773$), H_0 is not rejected. There is insufficient evidence to indicate a preference for one or more of the five other water management strategies exists at $\alpha = .05$.

10.15   For $k = 2$:

$$\chi^2 = \sum_{i=1}^{2} \frac{(n_i - np_i)^2}{np_i} = \frac{(n_1 - np_1)^2}{np_1} + \frac{(n_2 - np_2)^2}{np_2}$$

For a binomial experiment, $n_1 = y$, $n_2 = n - y$, $p_1 = p$, and $p_2 = (1 - p)$

$$\chi^2 = \frac{(y - np)^2}{np} + \frac{[(n - y) - n(1 - p)]^2}{n(1 - p)}$$

$$= \frac{y^2 - 2ynp + n^2p^2}{np} + \frac{(n - y)^2 - 2n(n - y)(1 - p) + n^2(1 - p)^2}{n(1 - p)}$$

$$= \frac{y_2 - 2ynp + n^2p^2}{np} + \frac{n_2 - 2ny + y^2 - 2n^2 + 2n^2p + 2ny - 2npy + n^2 - 2n^2p + n^2p^2}{n(1 - p)}$$

$$= \frac{y^2(1 - p) - 2ynp(1 - p) + n^2p^2(1 - p) + n^2p - 2nyp + y^2p - 2n^2p + 2n^2p^2 + 2nyp - 2nyp^2 + n^2p - 2n^2p^2 + n^2p^3}{np(1 - p)}$$

$$= \frac{y^2 - y^2p - 2ynp + 2ynp^2 + n^2p^2 - n^2p^3 + y^2p + 2n^2p^2 - 2nyp^2 - 2n^2p^2 + n^2p^3}{np(1 - p)}$$

$$= \frac{y^2 - 2ynp + n^2p^2}{np(1 - p)} = \frac{(y - np)^2}{np(1 - p)} = \frac{(y - np)^2}{npq} = z^2$$

10.17 a. Yes, the sampling appears to satisfy the assumptions of a multinomial experiment. The experiment contains 120 trials and $2(4) = 8$ categories. Since the 120 rats were randomly selected, the trials are considered independent and the probabilities are considered constant.

b. $\hat{E}(n_{ij}) = \dfrac{n_i \cdot n_j}{n}$

$\hat{E}(n_{11}) = \dfrac{80(30)}{120} = 20 \qquad \hat{E}(n_{21}) = \dfrac{40(30)}{120} = 10$

$\hat{E}(n_{12}) = \dfrac{80(30)}{120} = 20 \qquad \hat{E}(n_{22}) = \dfrac{40(30)}{120} = 10$

$\hat{E}(n_{13}) = \dfrac{80(30)}{120} = 20 \qquad \hat{E}(n_{23}) = \dfrac{40(30)}{120} = 10$

$\hat{E}(n_{14}) = \dfrac{80(30)}{120} = 20 \qquad \hat{E}(n_{24}) = \dfrac{40(30)}{120} = 10$

c. $\chi^2 = \sum\sum \dfrac{[n_{ij} - \hat{E}(n_{ij})]^2}{\hat{E}(n_{ij})}$

$$= \frac{(27 - 20)^2}{20} + \frac{(20 - 20)^2}{20} + \frac{(19 - 20)^2}{20} + \frac{(14 - 20)^2}{20} + \frac{(3 - 10)^2}{10}$$

$$+ \frac{(10 - 10)^2}{10} + \frac{(11 - 10)^2}{10} + \frac{(16 - 10)^2}{10} = 12.9$$

d. To determine if diet and presence/absence of cancer are independent, we test:

H_0: Diet and presence/absence of cancer are independent
H_a: Diet and presence/absence of cancer are dependent

The test statistic is $\chi^2 = 12.9$.

The rejection region requires $\alpha = .05$ in the upper tail of the χ^2 with df = $(r-1)(c-1) = (2-1)(4-1) = 3$. From Table 8, Appendix II, $\chi^2_{.05} = 5.99147$. The rejection region is $\chi^2 > 5.99147$.

Since the observed value of the test statistic falls in the rejection region ($\chi^2 = 12.9 > 5.99147$), H_0 is rejected. There is sufficient evidence to indicate that diet and presence/absence of cancer are not independent at $\alpha = .05$.

e. Let p_1 = proportion of rats on high fat/no fiber diet with cancer and let p_2 = proportion of rats on high fat/fiber diet with cancer.

$$\hat{p}_1 = \frac{27}{30} = .9 \qquad \hat{p}_2 = \frac{20}{30} = .667$$

The confidence interval for the difference between two proportions is

$$(\hat{p}_1 - \hat{p}_2) \pm z_{\alpha/2} \sqrt{\frac{\hat{p}_1\hat{q}_1}{n_1} + \frac{\hat{p}_2\hat{q}_2}{n_2}}$$

For confidence coefficient .95, $\alpha = 1 - .95 = .05$ and $\alpha/2 = .05/2 = .025$. From Table 4, Appendix II, $z_{.025} = 1.96$. The 95% confidence interval is

$$(.90 - .667) \pm 1.645 \sqrt{\frac{.9(.1)}{30} + \frac{.667(.333)}{30}} \Rightarrow .233 \pm .2 \Rightarrow (.033, .433)$$

To obtain the confidence interval for the percentage, multiply the endpoints by 100%. The interval is (3.3%, 43.3%).

We are 95% confident that the difference in the percentage of rats with cancer between those on high fat/no fiber diets and those on high fat/fiber diets is between 3.3% and 43.3%.

Since the rats were divided into groups according to diets, we assume the groups are independent.

10.19   To determine if dose at time of exposure and sex are independent, we test:

H_0: Dose and sex are independent
H_a: Dose and sex are dependent

The expected category counts are:

$$\hat{E}(n_{ij}) = \frac{n_i.n_{.j}}{n}$$

$$\hat{E}(n_{11}) = \frac{19(17)}{58} = 5.569$$

$$\hat{E}(n_{12}) = \frac{19(41)}{58} = 13.431$$

$$\hat{E}(n_{21}) = \frac{26(17)}{58} = 7.621$$

$$\hat{E}(n_{22}) = \frac{26(41)}{58} = 18.379$$

$$\hat{E}(n_{31}) = \frac{13(17)}{58} = 3.810$$

$$\hat{E}(n_{32}) = \frac{13(41)}{58} = 9.190$$

The observed and expected category counts are:

Dose (rad)	Males	Females	Totals
Less than 1	6 (5.569)	13 (13.431)	19
1–10	8 (7.621)	18 (18.379)	26
11 or more	3 (3.810)	10 (9.190)	13
Totals	17	41	58

The test statistic is

$$\chi^2 = \sum\sum \frac{\left[n_{ij} - \hat{E}(n_{ij})\right]^2}{\hat{E}(n_{ij})}$$

$$= \frac{(6 - 5.569)^2}{5.569} + \frac{(13 - 13.431)^2}{13.431} + \frac{(8 - 7.621)^2}{7.621} + \frac{(18 - 18.379)^2}{18.379}$$

$$+ \frac{(3 - 3.810)^2}{3.810} + \frac{(10 - 9.190)^2}{9.190} = .3174$$

The rejection region requires $\alpha = .01$ in the upper tail of the χ^2 distribution with df = $(r - 1)(c - 1) = (3 - 1)(2 - 1) = 2$. From Table 8, Appendix II, $\chi^2_{.01} = 9.21034$. The rejection region is $\chi^2 > 9.21034$.

Since the observed value of the test statistic does not fall in the rejection region ($\chi^2 = .3174 \ngtr 9.21034$), H_0 is not rejected. There is insufficient evidence to indicate that dose and sex are dependent at $\alpha = .01$.

Note: One of the expected category counts is less than 5. The calculated x^2 statistic may not have a x^2 distribution.

10.21 To determine if defect rate and years of experience are independent, we test:

H_0: Defect rate and years of experience are independent
H_a: Defect rate and years of experience are dependent

The test statistic is $x^2 = 1.351$ (from printout).

The rejection region requires $\alpha = .05$ in the upper tail of the x^2 distribution with $df = (r - 1)$ $(c - 1) = (3 - 1)(3 - 1) = 4$. From Table 8, Appendix II, $x^2_{.05} = 9.48773$. The rejection region is $x^2 > 9.48773$.

Since the observed value of the test statistic does not fall in the rejection region ($x^2 = 1.351 \not> 9.48773$), H_0 cannot be rejected. There is insufficient evidence to indicate that years of experience and defect rate are dependent at $\alpha = .05$.

10.23 Some preliminary calculations are:

$$\hat{E}(n_{ij}) = \frac{n_i . n_{.j}}{n}$$

$$\hat{E}(n_{11}) = \frac{109(374)}{567} = 71.9 \qquad \hat{E}(n_{21}) = \frac{458(374)}{567} = 302.1$$

$$\hat{E}(n_{12}) = \frac{109(193)}{567} = 37.1 \qquad \hat{E}(n_{22}) = \frac{458(193)}{567} = 155.9$$

To determine if dose the proportion of patients taking Seldane-D who experience insomnia differs from the corresponding proportion for patients receiving the placebo, we test:

H_0: The proportion of patients experiencing insomnia is the same for those receiving Seldane-D and those receiving the placebo
H_a: The proportion of patients experiencing insomnia is not the same for those receiving Seldane-D and those receiving the placebo

The test statistic is

$$x^2 = \sum\sum \frac{\left[n_{ij} - \hat{E}(n_{ij})\right]^2}{\hat{E}(n_{ij})}$$

$$= \frac{(97 - 71.9)^2}{71.9} + \frac{(12 - 37.1)^2}{37.1} + \frac{(277 - 302.1)^2}{302.1} + \frac{(181 - 155.9)^2}{155.9} = 31.87$$

The rejection region requires $\alpha = .10$ in the upper tail of the x^2 distribution with $df = (r - 1)(c - 1) = (2 - 1)(2 - 1) = 1$. From Table 8, Appendix II, $x^2_{.10} = 2.70554$. The rejection region is $x^2 > 2.70554$.

Since the observed value of the test statistic falls in the rejection region ($x^2 = 31.87 > 2.70554$), H_0 is rejected. There is sufficient evidence to indicate that the proportion of patients taking Seldane-D who experience insomnia differs from the corresponding proportion for patients receiving the placebo at $\alpha = .10$.

10.25 a. To determine if the percentages of the different types of programming statements differ for the two languages, we test:

H_0: The proportions of the different types of programming statements are the same for the two languages

H_a: The proportions of the different types of programming statements are different for the two languages

The expected category counts are:

$$\hat{E}(n_{ij}) = \frac{n_{i\cdot}n_{\cdot j}}{n}$$

$$\hat{E}(n_{11}) = \frac{2170(10,412)}{19,882} = 1136.407$$

$$\hat{E}(n_{12}) = \frac{2170(9470)}{19,882} = 1033.593$$

$$\vdots$$

$$\hat{E}(n_{52}) = \frac{726(9470)}{19,882} = 345.801$$

The observed and expected category counts are:

	ALGOL		PASCAL		Totals
IF	125	(1136.407)	2045	(1033.593)	2170
FOR	968	(690.223)	350	(627.777)	1318
IO	135	(1037.953)	1847	(944.047)	1982
IF ASSIGNMENT	8293	(7167.218)	4763	(6518.782)	13686
Other	261	(380.199)	465	(345.801)	726
Totals	10,412		9470		19,882

The test statistic is

$$x^2 = \sum\sum \frac{\left[n_{ij} - \hat{E}(n_{ij})\right]^2}{\hat{E}(n_{ij})}$$

$$= \frac{(125 - 1136.407)^2}{1136.407} + \frac{(2045 - 1033.593)^2}{1033.593} + \cdots + \frac{(465 - 345.801)^2}{345.801}$$

$$= 4755.1933$$

The rejection region requires $\alpha = .05$ in the upper tail of the χ^2 distribution with df = $(r-1)(c-1) = (5-1)(2-1) = 4$. From Table 8, Appendix II, $\chi^2_{.05} = 9.48773$. The rejection region is $\chi^2 > 9.48773$.

Since the observed value of the test statistic falls in the rejection region ($\chi^2 = 4755.1993 > 9.48773$), H_0 is rejected. There is sufficient evidence to indicate the percentages of the different types of programming statements differ for the two languages at $\alpha = .05$.

b.   The form of the confidence interval for $(p_A - p_P)$ is

$$(p_A - p_P) \pm z_{\alpha/2} \sqrt{\frac{\hat{p}_A(1 - \hat{p}_A)}{n_A} + \frac{\hat{p}_P(1 - \hat{p}_P)}{n_P}}$$

$$\hat{p}_A = \frac{X_A}{n_A} = \frac{8923}{10,412} = .857 \qquad \hat{p}_P = \frac{X_P}{n_P} = \frac{4763}{9470} = .503$$

For confidence coefficient .95, $\alpha = 1 - .95 = .05$ and $\alpha/2 = .05/2 = .025$. From Table 4, Appendix II, $z_{.025} = 1.96$. The confidence interval is

$$(.857 - .503) \pm 1.96 \sqrt{\frac{.857(1 - .857)}{10412} + \frac{.503(1 - .503)}{9470}} \Rightarrow .354 \pm .0121$$
$$\Rightarrow (.3419, .3661)$$

10.27   To determine if the hourly employees have a preference for one of the work schedules, we test:

$H_0: p_1 = p_2 = p_3 = p_4 = .25$
$H_a:$ At least two of the proportions differ

The expected number of each category is:

$E(n_i) = np_{i,0} = 671(.25) = 167.75 \quad (i = 1, 2, 3, 4)$

The observed and expected category counts are:

	8-hr fixed	8-hr rotating	12-hr fixed	12-hr rotating
Observed	389	54	208	20
Expected	(167.75)	(167.75)	(167.75)	(167.75)

The test statistic is

$$\chi^2 = \sum \frac{[n_i - E(n_i)]^2}{E(n_i)} = \frac{(389 - 167.75)^2}{167.75} + \frac{(54 - 167.75)^2}{167.75} + \frac{(208 - 167.75)^2}{167.75}$$
$$+ \frac{(20 - 167.75)^2}{167.75} = 508.7377$$

The rejection region requires $\alpha = .01$ in the upper tail of the χ^2 distribution with df $= k - 1 = 4 - 1 = 3$. From Table 8, Appendix II, $\chi^2_{.05} = 11.3449$. The rejection region is $\chi^2 > 11.3449$.

Since the observed value of the test statistic falls in the rejection region ($\chi^2 = 508.7377 > 11.3449$), H_0 is rejected. There is sufficient evidence to indicate the employees have a preference for one of the work schedules at $\alpha = .05$.

10.29 a. To determine if street characteristics and crime rates are dependent for the white middle income pair, we test:

H_0: Crime rate and street characteristics are independent
H_a: Crime rate and street characteristics are dependent

The expected category counts are:

$$\hat{E}(n_{ij}) = \frac{n_i . n_{.j}}{n}$$

$$\hat{E}(n_{11}) = \frac{29(48)}{81} = 17.185$$

$$\hat{E}(n_{12}) = \frac{29(33)}{81} = 11.815$$

$$\hat{E}(n_{21}) = \frac{16(48)}{81} = 9.481$$

$$\hat{E}(n_{22}) = \frac{16(33)}{81} = 6.519$$

$$\hat{E}(n_{31}) = \frac{36(48)}{81} = 21.333$$

$$\hat{E}(n_{32}) = \frac{36(33)}{81} = 14.667$$

The observed and expected category counts are:

	High	Low	Totals
Major	20 (17.185)	9 (11.815)	29
Small	7 (9.481)	9 (6.519)	16
Other	21 (21.333)	15 (14.667)	36
Totals	48	33	81

The test statistic is

$$\chi^2 = \sum\sum \frac{\left[n_{ij} - \hat{E}(n_{ij})\right]^2}{\hat{E}(n_{ij})}$$

$$= \frac{(20 - 17.185)^2}{17.185} + \frac{(9 - 11.815)^2}{11.815} + \frac{(7 - 9.481)^2}{9.481} + \frac{(9 - 6.519)^2}{6.519}$$

$$+ \frac{(21 - 21.333)^2}{21.333} + \frac{(15 - 14.667)^2}{14.667} = 2.7380$$

The rejection region requires $\alpha = .05$ in the upper tail of the χ^2 distribution with df $= (r - 1)(c - 1) = (3 - 1)(2 - 1) = 2$. From Table 8, Appendix II, $\chi^2_{.05} = 5.99147$. The rejection region is $\chi^2 > 5.99147$.

Since the observed value of the test statistic does not fall in the rejection region ($\chi^2 = 2.7380 \not> 5.99147$), H_0 is not rejected. There is insufficient evidence to indicate crime rate and street characteristics for the white middle income pair are dependent at $\alpha = .05$.

To determine if street characteristics and crime rates are dependent for the black lower middle income pair, we test:

H_0: Crime rate and street characteristics are independent
H_a: Crime rate and street characteristics are dependent

The expected category counts are:

$$\hat{E}(n_{ij}) = \frac{n_i . n_{.j}}{n}$$

$$\hat{E}(n_{11}) = \frac{26(86)}{128} = 17.469$$

$$\hat{E}(n_{12}) = \frac{26(42)}{128} = 8.531$$

$$\hat{E}(n_{21}) = \frac{52(86)}{128} = 34.938$$

$$\hat{E}(n_{22}) = \frac{52(42)}{128} = 17.062$$

$$\hat{E}(n_{31}) = \frac{50(86)}{128} = 33.594$$

$$\hat{E}(n_{32}) = \frac{50(42)}{128} = 16.406$$

The observed and expected category counts are:

	High	Low	Totals
Major	25 (17.469)	1 (8.531)	26
Small	25 (34.938)	27 (17.062)	52
Other	36 (33.594)	14 (16.406)	50
Totals	86	42	128

The test statistic is

$$\chi^2 = \sum\sum \frac{\left[n_{ij} - \hat{E}(n_{ij})\right]^2}{\hat{E}(n_{ij})}$$

$$= \frac{(25 - 17.469)^2}{17.469} + \frac{(1 - 8.531)^2}{8.531} + \frac{(25 - 34.938)^2}{34.938} + \frac{(27 - 17.062)^2}{17.062}$$

$$+ \frac{(36 - 33.594)^2}{33.594} + \frac{(14 - 16.406)^2}{16.406} = 19.0354$$

The rejection region requires $\alpha = .05$ in the upper tail of the χ^2 distribution with df $= (r - 1)(c - 1) = (3 - 1)(2 - 1) = 2$. From Table 8, Appendix II, $\chi^2_{.05} = 5.99147$. The rejection region is $\chi^2 > 5.99147$.

Since the observed value of the test statistic falls in the rejection region ($\chi^2 = 19.0354 > 5.99147$), H_0 is rejected. There is sufficient evidence to indicate crime rate and street characteristics for the black lower middle income pair are dependent at $\alpha = .05$.

To determine if street characteristics and crime rates are dependent for the black lower income pair, we test:

H_0: Crime rate and street characteristics are independent
H_a: Crime rate and street characteristics are dependent

The expected category counts are:

$$\hat{E}(n_{ij}) = \frac{n_i \cdot n_{\cdot j}}{n}$$

$$\hat{E}(n_{11}) = \frac{52(33)}{128} = 13.406$$

$$\hat{E}(n_{12}) = \frac{52(95)}{128} = 38.594$$

$$\hat{E}(n_{21}) = \frac{50(33)}{128} = 12.891$$

$$\hat{E}(n_{22}) = \frac{50(95)}{128} = 37.109$$

$$\hat{E}(n_{31}) = \frac{26(33)}{128} = 6.703$$

$$\hat{E}(n_{32}) = \frac{26(95)}{128} = 19.297$$

The observed and expected category counts are:

	High	Low	Totals
Major	22 (13.406)	30 (38.594)	52
Small	8 (12.891)	42 (37.109)	50
Other	3 (6.703)	23 (19.297)	26
Totals	33	95	128

The test statistic is

$$\chi^2 = \sum\sum \frac{\left[n_{ij} - \hat{E}(n_{ij})\right]^2}{\hat{E}(n_{ij})}$$

$$= \frac{(22 - 13.406)^2}{13.406} + \frac{(30 - 38.594)^2}{38.594} + \frac{(8 - 12.891)^2}{12.891} + \frac{(42 - 37.109)^2}{37.109}$$

$$+ \frac{(3 - 6.703)^2}{6.703} + \frac{(23 - 19.297)^2}{19.297} = 12.6795$$

The rejection region requires $\alpha = .05$ in the upper tail of the χ^2 distribution with df = $(r - 1)(c - 1) = (3 - 1)(2 - 1) = 2$. From Table 8, Appendix II, $\chi^2_{.05} = 5.99147$. The rejection region is $\chi^2 > 5.99147$.

Since the observed value of the test statistic falls in the rejection region ($\chi^2 = 12.6795 > 5.99147$), H_0 is rejected. There is sufficient evidence to indicate crime rate and street characteristics for the black lower income pair are dependent at $\alpha = .05$.

b. The form of the confidence interval for $\left(p_{1_H} - p_{1_L}\right)$ is

$$\left(\hat{p}_{1_H} - \hat{p}_{1_L}\right) \pm z_{\alpha/2} \sqrt{\frac{\hat{p}_{1_H}\left(1 - \hat{p}_{1_H}\right)}{n_H} + \frac{\hat{p}_{1_L}\left(1 - \hat{p}_{1_L}\right)}{n_L}}$$

For the white middle income pair,

$$\hat{p}_{1_H} = \frac{n_{1_H}}{n_H} = \frac{20}{48} = .417 \qquad \hat{p}_{1_L} = \frac{n_{1_L}}{n_L} = \frac{9}{33} = .273$$

For confidence coefficient .95, $\alpha = 1 - .95 = .05$ and $\alpha/2 = .05/2 = .025$. From Table 4, Appendix II, $z_{.025} = 1.96$. The confidence interval is

$$(.417 - .273) \pm 1.96 \sqrt{\frac{.417(1 - .417)}{48} + \frac{.273(1 - .273)}{33}} \Rightarrow .144 \pm .2063$$

$$\Rightarrow (-.0623, .3503)$$

For the black lower middle income pair,

$$\hat{p}_{1_H} = \frac{n_{1_H}}{n_H} = \frac{25}{86} = .291 \qquad \hat{p}_{1_L} = \frac{n_{1_L}}{n_L} = \frac{1}{42} = .024$$

The confidence interval is

$$(.291 - .024) \pm 1.96 \sqrt{\frac{.291(1 - .291)}{86} + \frac{.024(1 - .024)}{42}} \Rightarrow .267 \pm .1066$$
$$\Rightarrow (.1604, .3736)$$

For the black lower income pair,

$$\hat{p}_{1_H} = \frac{n_{1_H}}{n_H} = \frac{23}{33} = .667 \qquad \hat{p}_{1_L} = \frac{n_{1_L}}{n_L} = \frac{30}{95} = .316$$

The confidence interval is

$$(.667 - .316) \pm 1.96 \sqrt{\frac{.667(1 - .667)}{33} + \frac{.316(1 - .316)}{95}} \Rightarrow .351 \pm .1860$$
$$\Rightarrow (.1650, .5370)$$

10.31   To determine if the proportion of times the model conditions are satisfied is identical for the two groups, we test:

H_0: Proportion of times the model conditions are satisfied is identical for the two groups
H_a: Proportion of times the model conditions are satisfied is not identical for the two groups

The expected category counts are:

$$\hat{E}(n_{ij}) = \frac{n_i \cdot n_{\cdot j}}{n}$$

$$\hat{E}(n_{11}) = \frac{20(41)}{82} = 10$$

$$\hat{E}(n_{12}) = \frac{20(41)}{82} = 10$$

$$\hat{E}(n_{21}) = \frac{62(41)}{82} = 31$$

$$\hat{E}(n_{22}) = \frac{62(41)}{82} = 31$$

The observed and expected category counts are:

	Receivers	Producers	Totals
Satisfied	20 (10)	0 (10)	20
Not Satisfied	21 (31)	41 (31)	62
Totals	41	41	82

The test statistic is

$$\chi^2 = \sum\sum \frac{\left[n_{ij} - \hat{E}(n_{ij})\right]^2}{\hat{E}(n_{ij})}$$

$$= \frac{(20 - 10)^2}{10} + \frac{(0 - 10)^2}{10} + \frac{(21 - 31)^2}{31} + \frac{(41 - 31)^2}{31} = 26.4516$$

The rejection region requires $\alpha = .01$ in the upper tail of the χ^2 distribution with df = $(r - 1)(c - 1) = (2 - 1)(2 - 1) = 1$. From Table 8, Appendix II, $\chi^2_{.01} = 6.63490$. The rejection region is $\chi^2 > 6.63490$.

Since the observed value of the test statistic falls in the rejection region ($\chi^2 = 26.4516 > 6.63490$), H_0 is rejected. There is sufficient evidence to indicate the proportions of times the model conditions are satisfied are not identical at $\alpha = .01$.

10.33   a.   To determine if the type if commercial and recall of brand name are dependent, we test:

H_0: Type of commercial and brand name recall are independent
H_a: Type of commercial and brand name recall are dependent

The expected category counts are:

$$\hat{E}(n_{ij}) = \frac{n_i.n._j}{n}$$

$$\hat{E}(n_{11}) = \frac{57(57)}{200} = 16.245$$

$$\hat{E}(n_{12}) = \frac{57(74)}{200} = 21.090$$

$$\hat{E}(n_{13}) = \frac{57(69)}{200} = 19.665$$

$$\hat{E}(n_{21}) = \frac{143(57)}{200} = 40.755$$

$$\hat{E}(n_{22}) = \frac{143(74)}{200} = 52.910$$

$$\hat{E}(n_{23}) = \frac{143(69)}{200} = 49.335$$

The observed and expected category counts are:

	Normal	Version 1	Version 2	Totals
Yes	15 (16.245)	32 (21.090)	10 (19.665)	57
No	42 (40.755)	42 (52.910)	59 (49.335)	143
Totals	57	74	69	200

The test statistic is

$$\chi^2 = \sum\sum \frac{\left[n_{ij} - \hat{E}(n_{ij})\right]^2}{\hat{E}(n_{ij})}$$

$$= \frac{(15 - 16.245)^2}{16.245} + \frac{(32 - 21.090)^2}{21.090} + \cdots + \frac{(59 - 49.335)^2}{49.335} = 14.6705$$

The rejection region requires $\alpha = .05$ in the upper tail of the χ^2 distribution with df $= (r - 1)(c - 1) = (2 - 1)(3 - 1) = 2$. From Table 8, Appendix II, $\chi^2_{.05} = 5.99147$. The rejection region is $\chi^2 > 5.99147$.

Since the observed value of the test statistic falls in the rejection region ($\chi^2 = 14.6705 > 5.99147$), H_0 is rejected. There is sufficient evidence to indicate the type of commercial and brand name recall are dependent at $\alpha = .05$.

b.   The form of the confidence interval for $(p_{11} - p_{12})$ is

$$(\hat{p}_{11} - \hat{p}_{12}) \pm z_{\alpha/2}\sqrt{\frac{\hat{p}_{11}(1 - \hat{p}_{11})}{n_1} + \frac{\hat{p}_{12}(1 - \hat{p}_{12})}{n_2}}$$

$$\hat{p}_{11} = \frac{n_{11}}{n_1} = \frac{15}{57} = .263 \qquad \hat{p}_{12} = \frac{n_{12}}{n_2} = \frac{32}{74} = .432$$

For confidence coefficient .95, $\alpha = 1 - .95 = .05$ and $\alpha/2 = .05/2 = .025$. From Table 4, Appendix II, $z_{.025} = 1.96$. The confidence interval is

$$(.263 - .432) \pm 1.96\sqrt{\frac{.263(1 - .263)}{57} + \frac{.432(1 - .432)}{74}} \Rightarrow -.169 \pm .1606$$

$$\Rightarrow (-.3296, -.0084)$$

10.35   To determine if the proportions of problems are different among the four DSS components, we test:

H_0: $p_1 = p_2 = p_3 = p_4 = .25$
H_a: At least two of the category proportions are different

The expected category counts are:

$$E(n_i) = np_i = 151(.25) = 37.75 \ (i = 1, 2, 3, 4)$$

The observed and expected category counts are:

	1	2	3	4
Observed	31	28	45	47
Expected	(37.75)	(37.75)	(37.75)	(37.75)

The test statistic is

$$\chi^2 = \sum \frac{[n_i - E(n_i)]^2}{E(n_i)} = \frac{(31 - 37.75)^2}{37.75} + \frac{(28 - 37.75)^2}{37.75} + \frac{(45 - 37.75)^2}{37.75}$$
$$+ \frac{(47 - 37.75)^2}{37.75} = 7.3841$$

The rejection region requires $\alpha = .05$ in the upper tail of the χ^2 distribution with df $= k - 1$ $= 4 - 1 = 3$. From Table 8, Appendix II, $\chi^2_{.05} = 7.81473$. The rejection region is $\chi^2 > 7.81473$.

Since the observed value of the test statistic does not fall in the rejection region ($\chi^2 = 7.3841$ $\not> 7.81473$), H_0 is not rejected. There is insufficient evidence to indicate the proportions of problems are different for at least two of the four DSS components at $\alpha = .05$.

10.37 To determine if a difference among the proportions of transactions assigned to the five memory location exists, we test:

H_0: $p_1 = p_2 = p_3 = p_4 = p_5 = .25$
H_a: At least two of the proportions are different

The expected category counts are:

$$E(n_i) = np_{i,0} = 425(.20) = 85 \ (i = 1, 2, 3, 4, 5)$$

The observed and expected category counts are:

	1	2	3	4	5
Observed	90	78	100	72	85
Expected	(85)	(85)	(85)	(85)	(85)

The test statistic is

$$\chi^2 = \sum \frac{[n_i - E(n_i)]^2}{E(n_i)} = \frac{(90 - 85)^2}{85} + \frac{(78 - 85)^2}{85} + \frac{(100 - 85)^2}{85} + \frac{(72 - 85)^2}{85}$$
$$+ \frac{(85 - 85)^2}{85} = 5.5059$$

The rejection region requires $\alpha = .025$ in the upper tail of the χ^2 distribution with df $= k - 1$ $= 5 - 1 = 4$. From Table 8, Appendix II, $\chi^2_{.025} = 11.1433$. The rejection region is $\chi^2 > 11.1433$.

Since the observed value of the test statistic does not fall in the rejection region ($\chi^2 = 5.5059$ $\not> 11.1433$), H_0 is not rejected. There is insufficient evidence to indicate a difference in the location proportions at $\alpha = .025$.

CHAPTER ELEVEN

· ·

Simple Linear Regression

11.1   a.

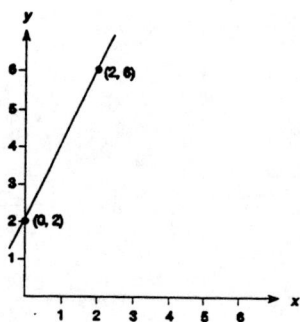

b.

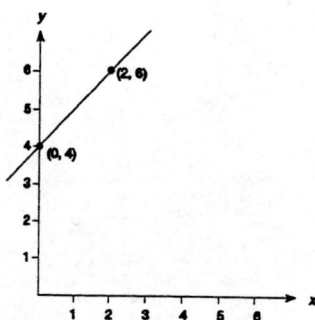

c.

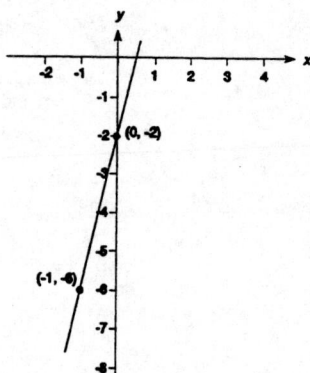

d.

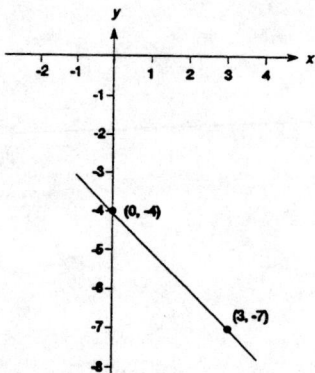

11.3   Find the equation of the line $y = \beta_0 + \beta_1 x$, where

$$\beta_0 = y\text{-intercept and } \beta_1 = \text{slope} = \frac{y_1 - y_2}{x_1 - x_2} \quad \text{(deterministic line)}$$

a.   Using the points $(0, 2)$ and $(2, 6)$:

$\beta_0 = 2$ (point where line crosses the y axis or when $x = 0$)

$\beta_1 = \dfrac{2 - 6}{0 - 2} = 2$

Thus, $y = \beta_0 + \beta_1 x = 2 + 2x$

b. Using the points $(0, 4)$ and $(2, 6)$:

$$\beta_0 = 4 \text{ (point where line crosses the } y \text{ axis or when } x = 0)$$

$$\beta_1 = \frac{4 - 6}{0 - 2} = 1$$

Thus, $y = \beta_0 + \beta_1 x = 4 + x$

c. Using the points $(0, -2)$ and $(-1, -6)$:

$$\beta_0 = -2 \text{ (point where line crosses the } y \text{ axis or when } x = 0)$$

$$\beta_1 = \frac{-2 - (-6)}{0 - (-1)} = 4$$

Thus, $y = \beta_0 + \beta_1 x = -2 + 4x$

d. Using the points $(0, -4)$ and $(3, -7)$:

$$\beta_0 = -4 \text{ (point where line crosses the } y \text{ axis or when } x = 0)$$

$$\beta_1 = \frac{-4 - (-7)}{0 - 3} = -1$$

Thus, $y = \beta_0 + \beta_1 x = -4 - x$

11.5 $y = \beta_0 + \beta_1 x$ where $\beta_0 = y\text{-intercept}$ and $\beta_1 = \text{slope of the line (deterministic line)}$

a. $\beta_0 = 3, \beta_1 = 2$

b. $\beta_0 = 1, \beta_1 = 1$

c. $\beta_0 = -2, \beta_1 = 3$

d. $\beta_0 = 0, \beta_1 = 5$

e. $\beta_0 = 4, \beta_1 = -2$

11.7   a.   The plot of the data is:

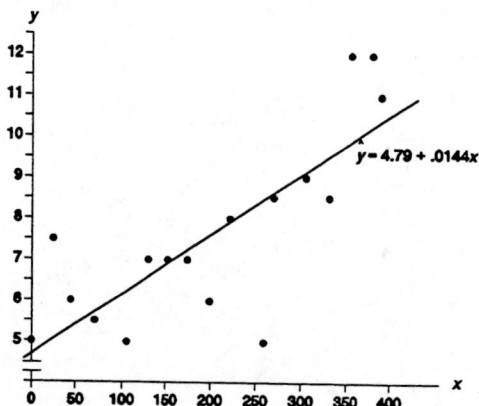

b.   Some preliminary calculations are:

$$\sum x = 3,395 \qquad \sum x^2 = 931,025$$

$$\sum y = 130.27 \qquad \sum xy = 29,656.15 \qquad \sum y^2 = 1081.3975$$

$$SS_{xy} = \sum xy - \frac{\sum x \sum y}{n} = 29,656.15 - \frac{3,395(130.27)}{17} = 3,640.46471$$

$$SS_{xx} = \sum x^2 - \frac{(\sum x)^2}{n} = 931,025 - \frac{3,395^2}{17} = 253,023.5294$$

$$\bar{x} = \frac{\sum x}{n} = \frac{3,395}{17} = 199.7058824 \qquad \bar{y} = \frac{\sum y}{n} = \frac{130.27}{17} = 7.662941176$$

$$\hat{\beta}_1 = \frac{SS_{xy}}{SS_{xx}} = \frac{3,640.46471}{253,023.5294} = 0.01438785 \approx 0.0144$$

$$\hat{\beta}_0 = \bar{y} - \hat{\beta}_1 \bar{x} = 7.662941176 - 0.01438785(199.7058824) = 4.7896028 \approx 4.79$$

The least squares prediction equation is $\hat{y} = 4.79 + 0.0144x$.

c.   See the graph in part a.

d. $\hat{\beta}_0 = 4.79$ Since $x = 0$ is in the observed range, we estimate the mean time to drill 5 feet starting at depth 0 feet is 4.79 minutes.

       $\hat{\beta}_1 = 0.0144$ We estimate that the mean time to drill 5 feet will increase by .0144 minutes for each additional foot of depth that drilling begins.

11.9   a.

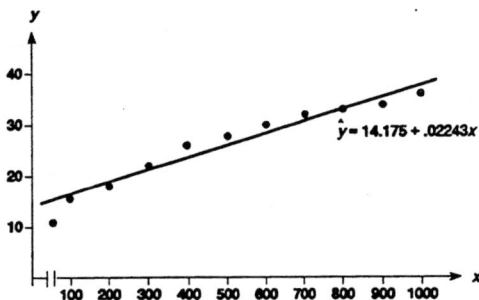

It does appear that x and y are linearly related. As x tends to increase, so does y.

b.   Some preliminary calculations are:

$$\sum x = 5550 \qquad \sum x^2 = 3,852,500$$

$$\sum y = 280.4 \qquad \sum xy = 165,075$$

$$SS_{xy} = \sum xy - \frac{\sum x \sum y}{n} = 165,075 - \frac{5550(280.4)}{11} = 23600.45455$$

$$SS_{xx} = \sum x^2 - \frac{(\sum x)^2}{n} = 3,852,500 - \frac{5550^2}{11} = 1052272.727$$

$$\bar{x} = \frac{\sum x}{n} = \frac{5550}{11} = 504.54545 \qquad \bar{y} = \frac{\sum y}{n} = \frac{280.4}{11} = 25.490909$$

$$\hat{\beta}_1 = \frac{SS_{xy}}{SS_{xx}} = \frac{23600.45455}{1052272.727} = .02243$$

$$\hat{\beta}_0 = \bar{y} - \hat{\beta}_1 \bar{x} = 25.490909 - .022428(504.54545) = 14.175$$

The least squares line is $\hat{y} = 14.175 + .02243x$.

c.   Two points on the least squares line are (100, 16.418) and (800, 32.119). See the line plotted on the graph in part a.

d. $\hat{\beta}_0 = 14.175$ This has no meaning since $x = 0$ is not in the observed range. It is the y-intercept.

$\hat{\beta}_1 = .02243$ We estimate that the mean site attenuation will increase by .02243 decibels for each 1 megahertz increase in transmission frequency.

11.11 a.

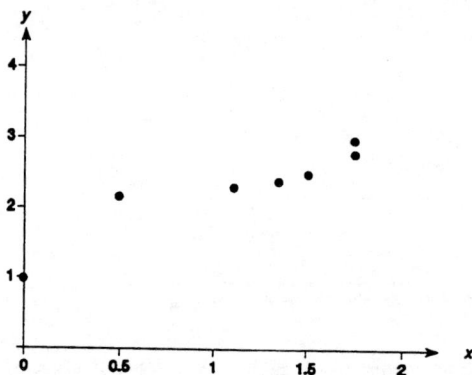

The relationship between shear strength and precompression stress appear to be linear. As precompression stress increases, shear strength tends to increase.

b. Some preliminary calculations are:

$$\sum x = 8.06 \qquad \sum x^2 = 11.7388$$
$$\sum y = 16.3 \qquad \sum xy = 21.195$$

$$SS_{xy} = \sum xy - \frac{\sum x \sum y}{n} = 21.195 - \frac{8.06(16.3)}{7} = 2.42671429$$

$$SS_{xx} = \sum x^2 - \frac{(\sum x)^2}{n} = 11.7388 - \frac{8.06^2}{7} = 2.458285714$$

$$\bar{x} = \frac{\sum x}{n} = \frac{8.06}{7} = 1.1514286 \qquad \bar{y} = \frac{\sum y}{n} = \frac{16.3}{7} = 2.3285714$$

$$\hat{\beta}_1 = \frac{SS_{xy}}{SS_{xx}} = \frac{2.42671429}{2.458285714} = .987157138 \approx .987$$

$$\hat{\beta}_0 = \bar{y} - \hat{\beta}_1 \bar{x} = 2.3285714 - .987157138(1.1514286) = 1.19193 \approx 1.192$$

c. $\hat{\beta}_0 = 1.192$ We estimate the mean shear strength is 1.192 when the precompression stress is 0.

$\hat{\beta}_1 = .987$ We estimate that the mean shear strength will increase by .987 for each 1 unit increase in precompression stress.

11.13 The two linear equations are:

$$n\hat{\beta}_0 + \hat{\beta}_1 \sum x_i = \sum y_i$$
$$\hat{\beta}_0 \sum x_i + \hat{\beta}_1 \sum x_i^2 = \sum x_i y_i$$

From the first equation, divide every term by n.

$$\hat{\beta}_0 + \hat{\beta}_1 \bar{x} = \bar{y}$$
$$\Rightarrow \hat{\beta}_0 = \bar{y} - \hat{\beta}_1 \bar{x}$$

Now substitute the value of $\hat{\beta}_0$ into the second equation.

$$(\bar{y} - \hat{\beta}_1 \bar{x})\sum x_i + \hat{\beta}_1 \sum x_i^2 = \sum x_i y_i$$

$$\Rightarrow \left[\frac{\sum y_i}{n} - \hat{\beta}_1 \frac{\sum x_i}{n}\right]\sum x_i + \hat{\beta}_1 \sum x_i^2 = \sum x_i y_i$$

$$\Rightarrow \frac{\sum x_i \sum y_i}{n} - \hat{\beta}_1 \frac{(\sum x_i)^2}{n} + \hat{\beta}_1 \sum x_i^2 = \sum x_i y_i$$

$$\Rightarrow \hat{\beta}_1 \left[\sum x_i^2 - \frac{(\sum x_i)^2}{n}\right] = \sum x_i y_i - \frac{\sum x_i y_i}{n}$$

$$\Rightarrow \hat{\beta}_1 SS_{xx} = SS_{xy}$$

$$\Rightarrow \hat{\beta}_1 = \frac{SS_{xy}}{SS_{xx}}$$

11.15 Show that $E(\hat{\beta}_0) = \beta_0$

First, the model is $y_i = \beta_0 + \beta_1 x_i + \epsilon_i$. $E(y_i) = E(\beta_0 + \beta_1 x_i + \epsilon_i) = \beta_0 + \beta_1 x_i$

$$E(\hat{\beta}_0) = E(\bar{y} - \hat{\beta}_1 \bar{x}) = E(\bar{y}) - \beta_1 \bar{x} \qquad \text{(from Exercise 11.14)}$$

$$= \frac{1}{n}E\left(\sum y_i\right) - \beta_1 \frac{\sum x_i}{n} = \frac{1}{n}\sum(\beta_0 + \beta_1 x_i) - \beta_1 \frac{\sum x_i}{n} = \frac{1}{n}\left(n\beta_0 + \beta_1 \sum x_i\right) - \beta_1 \frac{\sum x_i}{n}$$

$$= \beta_0$$

11.17 Some preliminary calculations are:

$$\sum x = 15 \qquad \sum x^2 = 55 \qquad \sum xy = 37$$
$$\sum y = 10 \qquad \sum y^2 = 26$$

$$SS_{xy} = \sum xy - \frac{\sum x \sum y}{n} = 37 - \frac{15(10)}{5} = 7$$

$$SS_{xx} = \sum x^2 - \frac{(\sum x)^2}{n} = 55 - \frac{15^2}{5} = 10$$

$$SS_{yy} = \sum y^2 - \frac{(\sum y)^2}{n} = 26 - \frac{10^2}{5} = 6$$

$$\hat{\beta}_1 = \frac{SS_{xy}}{SS_{xx}} = \frac{7}{10} = .7$$

$$SSE = SS_{yy} - \hat{\beta}_1 SS_{xy} = 6 - .7(7) = 1.10$$

. . . .
Simple Linear Regression

11.19 a.

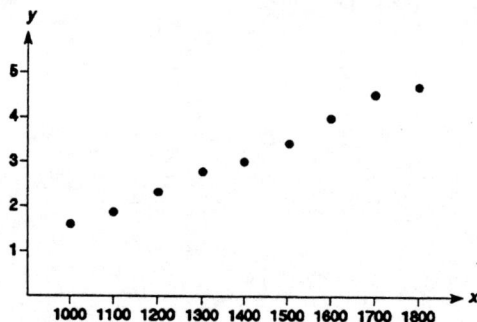

b.   Some preliminary calculations are:

$$\sum x = 12,600 \qquad \sum x^2 = 18,240,000 \qquad \sum xy = 40,968$$

$$\sum y = 27.68 \qquad \sum y^2 = 93.3448$$

$$SS_{xy} = \sum xy - \frac{\sum x \sum y}{n} = 40,968 - \frac{12,600(27.68)}{9} = 2216$$

$$SS_{xx} = \sum x^2 - \frac{(\sum x)^2}{n} = 18,240,000 - \frac{12,600^2}{9} = 600,000$$

$$SS_{yy} = \sum y^2 - \frac{(\sum y)^2}{n} = 93.3448 - \frac{27.68^2}{9} = 8.21342222$$

$$\bar{x} = \frac{\sum x}{n} = \frac{12,000}{9} = 1400 \qquad \bar{y} = \frac{\sum y}{n} = \frac{27.68}{9} = 3.0755556$$

$$\hat{\beta}_1 = \frac{SS_{xy}}{SS_{xx}} = \frac{2216}{600,000} = .003693$$

$$\hat{\beta}_0 = \bar{y} - \hat{\beta}_1 \bar{x} = 3.0755556 - (.003693)1400 = -2.095$$

The least squares line is $\hat{y} = -2.095 + .003693x$.

$\hat{\beta}_0 = -2.095$ Since $x = 0$ is not in the observed region, $\hat{\beta}_0$ has no meaning. It is the
y-intercept.

$\hat{\beta}_1 = .003693$ We estimate the mean oxygen diffusivity increases by .003693 for each
degree increase in temperature.

c. $SSE = SS_{yy} - \hat{\beta}_1 SS_{xy} = 8.21342222 - .003693333(2216) = .028996$
$s^2 = SSE/(n - 2) = .028996/(9 - 2) = .00414$

d. $s = \sqrt{.00414} = .0643$

We expect most of the observed oxygen diffusivity values to lie within $2s$ or $2(.0643)$ or $.1286$ of their respective least squares predicted values, $\hat{y}_i$.

11.21   By Theorem 11.1,

$$\chi^2 = \frac{(n-2)s^2}{\sigma^2}$$

$$V(\chi^2) = V\left[\frac{(n-2)s^2}{\sigma^2}\right] = 2\nu = 2(n-2)$$

$$\Rightarrow \frac{(n-2)^2}{\sigma^4}V(s^2) = 2(n-2) \Rightarrow V(s^2) = \frac{2\sigma^4}{(n-2)}$$

11.23   a.   Some preliminary calculations are:

Let x = TCDD levels in plasma and y = TCDD levels in fat tissue.

$$\sum x = 119.8 \qquad \sum x^2 = 1,972.82$$

$$\sum y = 137.2 \qquad \sum y^2 = 2,302.86 \quad \sum xy = 2,046.06$$

$$SS_{xy} = \sum xy - \frac{\sum x \sum y}{n} = 2,046.06 - \frac{119.8(137.2)}{20} = 1,224.232$$

$$SS_{xx} = \sum x^2 - \frac{(\sum x)^2}{n} = 1,972.82 - \frac{119.8^2}{20} = 1,255.218$$

$$SS_{yy} = \sum y^2 - \frac{(\sum y)^2}{n} = 2,302.86 - \frac{137.2^2}{20} = 1,361.668$$

$$\bar{x} = \frac{\sum x}{n} = \frac{119.8}{20} = 5.99 \quad \bar{y} = \frac{\sum y}{n} = \frac{137.2}{20} = 6.86$$

Using the TCDD level in plasma as the independent variable, the parameter estimates are:

$$\hat{\beta}_1 = \frac{SS_{xy}}{SS_{xx}} = \frac{1,224.232}{1,255.218} = 0.975314248 \approx 0.9753$$

$$\hat{\beta}_0 = \bar{y} - \hat{\beta}_1\bar{x} = 6.86 - 0.975314248(5.99) = 1.017867653 \approx 1.0179$$

The least squares prediction equation is $\hat{y} = 1.0179 + 0.9753x$.

Using the TCDD level in fat tissue as the independent variable, the parameter estimates are:

$$\hat{\beta}_1 = \frac{SS_{xy}}{SS_{yy}} = \frac{1,224.232}{1,361.668} = 0.899067907 \approx 0.8991$$

$$\hat{\beta}_0 = \bar{x} - \hat{\beta}_1 \bar{y} = 5.99 - 0.899067907(6.86) = -0.177605848 \approx -0.1776$$

The least squares prediction equation is $\hat{x} = -0.1776 + 0.8991y$.

b. If we want to see if fat tissue level (y) is a useful linear predictor of blood plasma level (x), we will use the model:

$$x = \beta_0 + \beta_1 y + \epsilon$$

We must first calculate SSE, s^2, and s.

$$SSE = SS_{xx} - \hat{\beta}_1 SS_{xy} = 1,255.218 - 0.899067907(1,224.232) = 154.550298$$

$$s^2 = \frac{SSE}{n-2} = \frac{154.550298}{20-2} = 8.586127667 \qquad s = \sqrt{8.586127667} = 2.93021$$

To determine of fat tissue level (y) is a useful linear predictor of blood plasma level (x), we test:

H_0: $\beta_1 = 0$
H_a: $\beta_1 \neq 0$

The test statistic is $t = \dfrac{\hat{\beta}_1}{s/\sqrt{SS_{yy}}} = \dfrac{.8991}{2.93021/\sqrt{1,361.668}} = 11.32$

The rejection region for this small-sample, two-tailed test requires $\alpha/2 = .05/2 = .025$ in each tail of the t distribution with df $= n - 2 = 20 - 2 = 18$. From Table 7, Appendix II, $t_{.025} = 2.101$. The rejection region is $t < -2.101$ or $t > 2.101$.

Since the observed value of the test statistic falls in the rejection region ($t = 11.32 > 2.101$), H_0 is rejected. There is sufficient evidence to indicate that fat tissue level (y) is a useful predictor of blood plasma level (x) at $\alpha = .05$.

c. If we want to see if blood plasma level (x) is a useful linear predictor of fat tissue level (y), we will use the model:

$$y = \beta_0 + \beta_1 x + \epsilon$$

We must first calculate SSE, s^2, and s.

$$SSE = SS_{yy} - \hat{\beta}_1 SS_{xy} = 1,361.668 - 0.975314248(1,224.232) = 167.657088$$

$$s^2 = \frac{SSE}{n-2} = \frac{167.657088}{20-2} = 9.314282667 \qquad s = \sqrt{9.314282667} = 3.05193$$

To determine of blood plasma level (x) is a useful linear predictor of fat tissue level (y), we test:

$$H_0: \beta_1 = 0$$
$$H_a: \beta_1 \neq 0$$

The test statistic is $t = \dfrac{\hat{\beta}_1}{s/\sqrt{SS_{xx}}} = \dfrac{.9753}{3.05193/\sqrt{1,255.218}} = 11.32$

The rejection region is $t < -2.101$ or $t > 2.101$ (from part **b** above).

Since the observed value of the test statistic falls in the rejection region $(t = 11.32 > 2.101)$, H_0 is rejected. There is sufficient evidence to indicate that blood plasma level (x) is a useful predictor of fat tissue level (y) at $\alpha = .05$.

11.25 From Exercise 11.8. $\hat{\beta}_1 = 2.4264$ and $SS_{xx} = 3.9532$. From Exercise 11.18c, $s = .454$

The confidence interval for β_1 is $\hat{\beta}_1 \pm t_{\alpha/2} s_{\hat{\beta}_1}$ where $s_{\hat{\beta}_1} = \dfrac{s}{\sqrt{SS_{xx}}}$

For confidence coefficient .95, $\alpha = 1 - .95 = .05$ and $\alpha/2 = .05/2 = .025$. From Table 7 with df $= n - 2 = 24 - 2 = 22$, $t_{.025} = 2.074$. The 95% confidence interval is

$$2.4264 \pm 2.074 \frac{.454}{\sqrt{3.9532}} \Rightarrow 2.4264 \pm .4736 \Rightarrow (1.9528, 2.9000)$$

We are 95% confident the mean heat transfer enhancement will increase from between 1.9528 and 2.9000 for each 1 unit increase in unflooded area ratio.

11.27 From Exercise 11.11, $\hat{\beta}_1 = .987$ and $SS_{xx} = 2.4583$

From Exercise 11.18f, $s = .224$

To test if the slope of the line is positive, we test:

$$H_0: \beta_1 = 0$$
$$H_a: \beta_1 > 0$$

The test statistic is $t = \dfrac{\hat{\beta}_1}{s/\sqrt{SS_{xx}}} = \dfrac{.987}{.224/\sqrt{2.4583}} = 6.91$

For a small-sample, one-tailed test, the rejection region requires $\alpha = .10$ in the upper tail of the t distribution with df $= n - 2 = 7 - 2 = 5$. From Table 7, Appendix II, $t_{.10} = 1.476$. The rejection region is $t > 1.476$.

Since the observed value of the test statistic falls in the rejection region ($t = 6.91 > 1.476$), H_0 is rejected. There is sufficient evidence to indicate β_1 is greater than 0 or that the slope of the line is positive at $\alpha = .10$.

11.29 From Exercise 11.20, $\hat{\beta}_1 = 6.91175$, $SS_{xx} = 4.111$, and $s = .6537$

The form of the confidence interval for β_1 is:

$$\hat{\beta}_1 \pm t_{\alpha/2}\frac{s}{\sqrt{SS_{xx}}}$$

For confidence coefficient .99, $\alpha = 1 - .99 = .01$ and $\alpha/2 = .01/2 = .005$. From Table 7, Appendix II, with df $= n - 2 = 10 - 2 = 8$, $t_{.005} = 3.355$. The 99% confidence interval is

$$6.912 \pm 3.355\frac{.6537}{\sqrt{4.111}} \Rightarrow 6.912 \pm 1.082 \Rightarrow (5.830, 7.994)$$

Since 0 is not contained in the interval and the interval contains only positive values, there is sufficient evidence to indicate the amount of oxidative compounds measured using the TG technique is positively linearly related to the total percentage of oxidation products determined by the standard method.

11.31 a. Some preliminary calculations are:

$$\sum x = 273.9 \qquad\qquad \sum y = 31.9$$
$$\sum x^2 = 13{,}223.11 \qquad\qquad \sum y^2 = 275.69 \qquad\qquad \sum xy = 1694.23$$

$$SS_{xx} = \sum x^2 - \frac{(\sum x)^2}{n} = 13{,}223.11 - \frac{(273.9)^2}{6} = 719.575$$

$$SS_{yy} = \sum y^2 - \frac{(\sum y)^2}{n} = 275.69 - \frac{(31.9)^2}{6} = 106.08833$$

$$SS_{xy} = \sum xy - \frac{\sum x \sum y}{n} = 1694.23 - \frac{(273.9)(31.9)}{6} = 237.995$$

$$\hat{\beta}_1 = \frac{SS_{xy}}{SS_{xx}} = \frac{237.995}{719.575} = .33074384 \approx .3307$$

$$SSE = SS_{yy} - \hat{\beta}_1 SS_{xy} = 106.08833 - (.33074384)(237.995) = 27.37295$$

$$s^2 = \frac{SSE}{n-2} = \frac{27.37295}{6-2} = 6.8432 \qquad s = \sqrt{s^2} = \sqrt{6.8432} = 2.616$$

We want to test if cracking growth rate increases linearly with maximum load, we test:

$$H_0: \beta_1 = 0$$
$$H_a: \beta_1 > 0$$

The test statistic is $t = \dfrac{\hat{\beta}_1 - 0}{s/\sqrt{SS_{xx}}} = \dfrac{.3307 - 0}{2.616/\sqrt{719.575}} = 3.39$

For a one-tailed, small-sample test, the rejection region requires $\alpha = .10$ in the upper tail of the t distribution with df $= n - 2 = 6 - 2 = 4$. From Table 7, Appendix II, $t_{.10} = 1.533$. The rejection region is $t > 1.533$.

Since the observed value of the test statistic falls in the rejection region ($t = 3.39 > 1.533$), H_0 is rejected. There is sufficient evidence to indicate the cracking growth rate increases linearly with maximum load at $\alpha = .10$.

b. The form of the confidence interval is

$$\hat{\beta}_1 \pm t_{\alpha/2} s/\sqrt{SS_{xx}}$$

For confidence coefficient .90, $\alpha = 1 - .90 = .10$. $\alpha/2 = .10/2 = .05$. From Table 7, Appendix II, with df $= n - 2 = 4$, $t_{.05} = 2.132$. The confidence interval is

$$.3307 \pm 2.132(2.616)/\sqrt{719.575} \Rightarrow .3307 \pm .2079 \Rightarrow (.1128, .5386)$$

We are 90% confident that the mean increase in cracking growth rate for every 1-unit increase in maximum load will fall between .1128 and .5386.

11.33 We know that $t = \dfrac{z}{\sqrt{\chi^2/df}}$ has a t distribution with df degrees of freedom.

From Exercise 11.32, $z = \dfrac{\hat{\beta}_1 - \beta_1}{\sigma/\sqrt{SS_{xx}}}$ has a z distribution.

From Theorem 11.1, $\chi^2 = \dfrac{(n - 2)s^2}{\sigma^2}$ has a chi-square distribution with $n - 2$ degrees of freedom.

Then, $t = \dfrac{z}{\sqrt{\chi^2/df}} = \dfrac{\dfrac{\hat{\beta}_1 - \beta_1}{\sigma/\sqrt{SS_{xx}}}}{\sqrt{\dfrac{(n - 2)s^2}{\sigma^2}/n - 2}} = \dfrac{\dfrac{\hat{\beta}_1 - \beta_1}{\sigma/\sqrt{SS_{xx}}}}{s/\sigma} = \dfrac{\hat{\beta}_1 - \beta_1}{s/\sqrt{SS_{xx}}}$

has a t distribution with $n - 2$ df.

11.35 a. From Exercises 11.6 and 11.18:

$$SS_{xx} = 280 \qquad SS_{xy} = .900 \qquad SS_{yy} = .00357$$

$$r = \frac{SS_{xy}}{\sqrt{SS_{xx}SS_{yy}}} = \frac{.900}{\sqrt{280(.00357)}} = .900$$

$$r^2 = .900^2 = .81$$

There is strong positive linear relationship between x and y. 81% of the sum of squared deviations of the congestion times about their predicted values can be attributed to the linear relationship between the congestion times and the number of vehicles.

b. From Exercises 11.7 and 11.18:

$$SS_{xx} = 253,023.5294 \qquad SS_{xy} = 3,640.46471 \qquad SS_{yy} = 83.14615$$

$$r = \frac{SS_{xy}}{\sqrt{SS_{xx}SS_{yy}}} = \frac{3,640.46471}{\sqrt{253,023.5294(83.14615)}} = .794$$

$$r^2 = .794^2 = .630$$

There is a fairly strong positive linear relationship between x and y. 63% of the sum of squared deviations of the times to drill 5 feet about their predicted values can be attributed to the linear relationship between the times to drill 5 feet and the depth at which drilling begins.

c. From Exercises 11.8 and 11.18:

$$SS_{xx} = 3.9532 \qquad SS_{xy} = 9.592 \qquad SS_{yy} = 27.805$$

$$r = \frac{SS_{xy}}{\sqrt{SS_{xx}SS_{yy}}} = \frac{9.592}{\sqrt{3.9532 \cdot 27.805}} = .915$$

$$r^2 = .915^2 = .837$$

There is a strong positive linear relationship between x and y. 83.7% of the sum of squared deviations of the heat transfer enhancement values about their predicted values can be explained by the linear relationship between heat transfer enhancement and unflooded area ratio.

d. From Exercises 11.9 and 11.18:

$$SS_{xx} = 1,052,272.7 \qquad SS_{xy} = 23,600.45 \qquad SS_{yy} = 570.289$$

$$r = \frac{SS_{xy}}{\sqrt{SS_{xx}SS_{yy}}} = \frac{23,600.45}{\sqrt{1,052,272.7 \cdot 570.289}} = .9634$$

$$r^2 = .9634^2 = .9281$$

There is a strong positive linear relationship between x and y. 92.81% of the sum of squared deviations of the site attenuation values about their predicted values can be attributed by the linear relationship between site attenuation and transmission frequency.

e. From Exercises 11.10 and 11.18:

$$SS_{xx} = 8{,}780 \qquad SS_{xy} = 1{,}918 \qquad SS_{yy} = 534.8$$

$$r = \frac{SS_{xy}}{\sqrt{SS_{xx}SS_{yy}}} = \frac{1{,}918}{\sqrt{8{,}780(534.8)}} = .885$$

$$r^2 = .885^2 = .783$$

There is a fairly strong positive linear relationship between x and y. 78.3% of the sum of squared deviations of the passivation potential scores about their predicted values can be attributed to the linear relationship between the passivation potential scores and the annealing time.

f. From Exercises 11.11 and 11.18:

$$SS_{xx} = 2.4583 \qquad SS_{xy} = 2.4267 \qquad SS_{yy} = 2.6465$$

$$r = \frac{SS_{xy}}{\sqrt{SS_{xx}SS_{yy}}} = \frac{2.4267}{\sqrt{2.4583 \cdot 2.6465}} = .9514$$

$$r^2 = .9514^2 = .9052$$

There is a strong positive linear relationship between x and y. 90.52% of the sum of squared deviations of the precompression stress values about their predicted values can be explained by the linear relationship between precompression stress and shear strength.

11.37   a. From Exercise 11.19:

$$SS_{xy} = 2{,}216 \qquad SS_{xy} = 600{,}000 \qquad SS_{yy} = 8.2134222$$

$$r = \frac{SS_{xy}}{\sqrt{SS_{xx}SS_{yy}}} = \frac{2{,}216}{\sqrt{600{,}000(8.21342222)}} = \frac{2{,}216}{2219.92192} = .998233031$$

$$\approx .998$$

Since the value of r is very close to 1, it implies that a very strong positive linear relationship exists between oxygen diffusivity and temperature.

$$r^2 = .998^2 = .996$$

The proportion of the sum of squares of deviations of the oxygen diffusivity values about their predicted values that can be attributed to the linear relation between oxygen diffusivity and temperature is .996.

b.   To determine whether temperature and oxygen diffusivity are positively correlated, we test:

$$H_0: \rho = 0$$
$$H_a: \rho > 0$$

The test statistic is $t = \dfrac{r\sqrt{n-2}}{\sqrt{1-r^2}} = \dfrac{.998233\sqrt{9-2}}{\sqrt{1-.998233^2}} = 44.45$

For a small sample, one-tailed test, the rejection region requires $\alpha = .05$ in the upper tail of the t distribution. From Table 7, Appendix II, with df $= n - 2 = 9 - 2 = 7$, $t_{.05} = 1.895$. The rejection region is $t > 1.895$.

Since the observed value of the test statistic falls in the rejection region ($t = 44.45 > 1.895$), H_0 is rejected. There is sufficient evidence to indicate that temperature and oxygen diffusivity are positively correlated at $\alpha = .05$. The results do agree with the results of Exercise 11.28.

11.39   a. $r = -.50$ Since this value of r is half way between 0 and -1, it indicates a relatively weak negative linear relationship between a child's weight percentile and the number of cigarettes smoked per day in the child's home for girls.

      b. $p = .03$ The probability of observing the test statistic or anything more unusual is $p = .03$ if H_0 is true. Since this p-value is so small, we would reject H_0 and conclude that $\rho \neq 0$ or that there is a significant linear correlation between a child's weight percentile and the number of cigarettes smoked per day in the child's home for girls.

      c. $r = -.12$ Since this value of r is fairly close to 0, it indicates a very weak negative linear relationship between a child's weight percentile and the number of cigarettes smoked per day in the child's home for boys.

      d. $p = .57$ The probability of observing the test statistic or anything more unusual is $p = .57$ if H_0 is true. Since the p-value is not very small, we would not reject H_0. There is insufficient evidence to indicate $\rho \neq 0$ or that there is insufficient evidence to indicate a linear relationship between a child's weight percentile and the number of cigarettes smoked per day in the child's home for boys.

11.41   a. $\hat{\beta}_1 = .020$ We estimate that the mean log (base 10) of the average number of infections per leaf will increase by .020 for each additional 1% of leaves infected.

      b. $r^2 = .816$ 81.6% of the variability in the log (base 10) of the average number of infections per leaf values can be explained by the linear relationship between the log (base 10) values and the percent of leaves infected.

      c. $s = .288$ We expect most of the log (base 10) values will fall within $2s$ or $2(.288)$ or $(.576)$ units of their respective predicted values.

d. $r = \sqrt{.816} = .903$ Since this value of r is very close to 1, it indicates there is a fairly strong positive linear relationship between the log (base 10) of the average number of infections per leaf and the percent of leaves infected.

e.   To determine if the model is useful, we test:

$$H_0: \beta_1 = 0$$
$$H_a: \beta_1 \neq 0$$

The test statistic is $t = \dfrac{r\sqrt{n-2}}{\sqrt{1-r^2}} = \dfrac{\sqrt{.816}\sqrt{100-2}}{\sqrt{1-.816}} = 20.85$

The rejection region for this two-tailed test requires $\alpha/2 = .05/2 = .025$ in each tail of the t distribution with df $= n - 2 = 100 - 2 = 98$. From Table 7, Appendix II, $t_{.025} \approx 2.00$. The rejection region is $t > 2.00$ or $t < -2.00$.

Since the observed value of the test statistic falls in the rejection region ($t = 20.85 > 2.00$), H_0 is rejected. There is sufficient evidence to indicate the model is useful at $\alpha = .05$.

f. $\hat{y} = -.939 + .020x = -.939 + .020(80\%) = .661$

Antilog $(.661) = 4.58 =$ Severity of the disease when 80% of the leaves are infected.

11.43   Show $\dfrac{\hat{\beta}_1}{s/\sqrt{SS_{xx}}} = \dfrac{r\sqrt{n-2}}{\sqrt{1-r^2}}$

From Exercise 11.42, we know:

$$\hat{\beta}_1 = r\sqrt{\frac{SS_{yy}}{SS_{xx}}} \quad \text{and} \quad SSE = SS_{yy}(1 - r^2)$$

$$\Rightarrow \frac{\hat{\beta}_1}{s/\sqrt{SS_{xx}}} = \frac{r\sqrt{\dfrac{SS_{yy}}{SS_{xx}}}}{s/\sqrt{SS_{xx}}}$$

$$= \frac{r\sqrt{SS_{yy}}}{s} = \frac{r\sqrt{SS_{yy}}}{\sqrt{\dfrac{SS_{yy}(1 - r^2)}{(n-2)}}} \quad \left(\text{since } s = \sqrt{\frac{SSE}{n-2}}\right)$$

$$= \frac{r}{\sqrt{\dfrac{(1-r^2)}{n-2}}} = \frac{r\sqrt{n-2}}{\sqrt{1-r^2}}$$

11.45 Since the standard error contains the term $\dfrac{(x_p - \bar{x})^2}{SS_{xx}}$, the further x_p is from $\bar{x}$, the larger the standard error. This causes the confidence intervals to be wider for values of x_p further from $\bar{x}$. The implication is our best confidence intervals (narrowest) will be found when $x_p = \bar{x}$.

11.47 Some preliminary calculations are:

$$\sum x = 45.12 \qquad \sum x^2 = 88.7788$$

$$\sum y = 114.6 \qquad \sum xy = 225.04 \qquad \sum y^2 = 575.02$$

$$SS_{xy} = \sum xy - \frac{\sum x \sum y}{n} = 225.04 - \frac{45.12(114.6)}{24} = 9.592$$

$$SS_{xx} = \sum x^2 - \frac{(\sum x)^2}{n} = 88.7788 - \frac{45.12^2}{24} = 3.9532$$

$$SS_{yy} = \sum y^2 - \frac{(\sum y)^2}{n} = 575.02 - \frac{114.6^2}{24} = 27.805$$

$$\bar{x} = \frac{\sum x}{n} = \frac{45.12}{24} = 1.88 \qquad \bar{y} = \frac{\sum y}{n} = \frac{114.6}{24} = 4.775$$

$$\hat{\beta}_1 = \frac{SS_{xy}}{SS_{xx}} = \frac{9.592}{3.9532} = 2.426388748 \approx 2.4264$$

$$\hat{\beta}_0 = \bar{y} - \hat{\beta}_1 \bar{x} = 4.775 - 2.426388748(1.88) = 0.213389154 \approx .2134$$
$$SSE = SS_{yy} - \hat{\beta}_1 SS_{xy} = 27.805 - 2.426388748(9.592) = 4.53107913$$
$$s^2 = \frac{SSE}{n-2} = \frac{4.53107913}{24-2} = .205958142 \qquad s = \sqrt{.205958142} = .45383$$

The 99% prediction interval for the heat transfer coefficient of an integralfin tube with an unflooded area ratio of 1.95 is

$$\hat{y} \pm t_{\alpha/2} s \sqrt{1 + \frac{1}{n} + \frac{(x_p - \bar{x})^2}{SS_{xx}}}$$

$$\hat{y} = \hat{\beta}_0 + \hat{\beta}_1 x = .2134 + 2.4264(1.95) = 4.9449$$

For confidence coefficient .99, $\alpha = 1 - .99 = .01$ and $\alpha/2 = .01/2 = .005$. From Table 7, Appendix II, with df $= n - 2 = 24 - 2 = 22$, $t_{.005} = 2.819$. The 99% prediction interval is

$$4.9449 \pm 2.819(.45383) \sqrt{1 + \frac{1}{24} + \frac{(1.95 - 1.88)^2}{3.9532}} \Rightarrow 4.9449 \pm 1.3065$$
$$\Rightarrow (3.6384, 6.2514)$$

We are 99% confident that when the unflooded area ratio is 1.95, the actual value of the heat transfer coefficient of an integralfin tube is between 3.6384 and 6.2514.

11.49   Some preliminary calculations are:

$$\sum x = 285 \qquad \sum x^2 = 25{,}025$$

$$\sum y = -1964 \qquad \sum xy = -110{,}030 \qquad \sum y^2 = 771{,}994$$

$$SS_{xy} = \sum xy - \frac{\sum x \sum y}{n} = -110{,}030 - \frac{285(-1964)}{5} = 1918$$

$$SS_{xx} = \sum x^2 - \frac{(\sum x)^2}{n} = 25{,}025 - \frac{285^2}{5} = 8780$$

$$SS_{yy} = \sum y^2 - \frac{(\sum y)^2}{n} = 771{,}994 - \frac{(-1964)^2}{5} = 534.8$$

$$\bar{x} = \frac{\sum x}{n} = \frac{285}{5} = 57 \qquad \bar{y} = \frac{\sum y}{n} = \frac{-1964}{5} = -392.8$$

$$\hat{\beta}_1 = \frac{SS_{xy}}{SS_{xx}} = \frac{1918}{8780} = .218450125 \approx .21845$$

$$\hat{\beta}_0 = \bar{y} - \hat{\beta}_1 \bar{x} = -392.8 - .218451025(57) = -405.2517084 \approx -405.2517$$

$$SSE = SS_{yy} - \hat{\beta}_1 SS_{xy} = 534.8 - .218451025(1918) = 115.8109339$$

$$s^2 = \frac{SSE}{n-2} = \frac{115.8109339}{5-2} = 38.60364463 \qquad s = \sqrt{38.60364463} = 6.21318$$

The 95% confidence interval for the mean passivation potential of crystallized alloy when annealing time is 30 minutes is

$$\hat{y} \pm t_{\alpha/2} s \sqrt{\frac{1}{n} + \frac{(x_p - \bar{x})^2}{SS_{xx}}}$$

$$\hat{y} = \hat{\beta}_0 + \hat{\beta}_1 x = -405.2517 + .21845(30) = -398.6982$$

For confidence coefficient .95, $\alpha = 1 - .95 = .05$ and $\alpha/2 = .05/2 = .025$. From Table 7, Appendix II, with df $= n - 2 = 5 - 2 = 3$, $t_{.025} = 3.182$. The 95% confidence interval is

$$-398.6982 \pm 3.182(6.21318)\sqrt{\frac{1}{5} + \frac{(30-57)^2}{8780}} \Rightarrow -398.6982 \pm 10.5179$$

$$\Rightarrow (-409.2161, -388.1803)$$

We are 95% confident that when the annealing time is 30 minutes, the mean passivation potential of crystallized alloy is between -409.2161 and -388.1803.

11.51   **a.**   Some preliminary calculations are:

$$\sum x = 583 \qquad \sum x^2 = 10435.64 \quad \sum xy = 9260.4$$

$$\sum y = 517 \qquad \sum y^2 = 8271$$

$$SS_{xy} = \sum xy - \frac{\sum x \sum y}{n} = 9260.4 - \frac{583(517)}{33} = 126.733333$$

$$SS_{xx} = \sum x^2 - \frac{(\sum x)^2}{n} = 10435.64 - \frac{583^2}{33} = 135.973334$$

$$SS_{yy} = \sum y^2 - \frac{(\sum y)^2}{n} = 8271 - \frac{517^2}{33} = 171.333333$$

$$\bar{x} = \frac{\sum x}{n} = \frac{583}{33} = 17.666667 \qquad \bar{y} = \frac{\sum y}{n} = \frac{517}{33} = 15.666667$$

$$\hat{\beta}_1 = \frac{SS_{xy}}{SS_{xx}} = \frac{126.733333}{135.973334} = .93204549 \approx .93205$$

$$\hat{\beta}_0 = \bar{y} - \hat{\beta}_1 \bar{x} = 15.666667 - .93204549(17.666667) = -.7994703$$

The least squares prediction equation is $\hat{y} = -.7995 + .93205x$

b.   To determine if the model is useful for predicting y, we test:

$$H_0: \ \beta_1 = 0$$
$$H_a: \ \beta_1 \neq 0$$

$$SSE = SS_{yy} - \hat{\beta}_1 SS_{xy} = 171.333333 - .93204549(126.733333) = 53.21210155$$

$$s^2 = \frac{SSE}{n-2} = \frac{53.21210155}{33-2} = 1.7165194$$

The test statistic is $t = \dfrac{\hat{\beta}_1}{s/\sqrt{SS_{xx}}} = \dfrac{.93204549}{\sqrt{1.7165194/135.973334}} = 8.30$

For a two-tailed test, the rejection region requires $\alpha/2 = .10/2 = .05$ in each tail of the t distribution with df $= n - 2 = 33 - 2 = 31$. From Table 7, Appendix II, $t_{.05} \approx$ 1.699. The rejection region is $t < -1.699$ or $t > 1.699$.

Since the observed value of the test statistic falls in the rejection region ($t = 8.30 >$ 1.699), H_0 is rejected. There is sufficient evidence to indicate the model is useful for predicting y at $\alpha = .10$.

c. The 95% prediction interval for the instructor assigned grade y at $x = 17.5$ is:

$$\hat{y} \pm t_{\alpha/2}s\sqrt{1 + \frac{1}{n} + \frac{(x_p - \bar{x})^2}{SS_{xx}}}$$

$$\hat{y} = \hat{\beta}_0 + \hat{\beta}_1 x = -.7994703 + .93204549(17.5) = 15.5113$$

For confidence coefficient .95, $\alpha = 1 - .95 = .05$ and $\alpha/2 = .05/2 = .025$. From Table 7, Appendix II, with df $= n - 2 = 33 - 2 = 31$, $t_{.025} = 2.045$.

The prediction interval is

$$15.5113 \pm 2.045\sqrt{1.7165194}\sqrt{1 + \frac{1}{33} + \frac{(17.5 - 17.66667)^2}{135.973334}}$$

$$\Rightarrow 15.5113 \pm 2.7198 \Rightarrow (12.7915, 18.2311)$$

We predict that the instructor assigned grade of a Fortran 77 assignment that received an AUTOMARK score of 17.5 will fall in the interval from 12.7915 to 18.2311.

11.53 In Exercise 11.52, it was determined that $y_p - \hat{y}$ is normally distributed with an expected value or mean of 0 and a standard deviation of

$$\sigma\sqrt{1 + \frac{1}{n} + \frac{(x_p - \bar{x})^2}{SS_{xx}}}$$

If we take $y_p - \hat{y}$, subtract its mean and divide by its standard deviation, we will form a normally distributed random variable with a mean of 0 and a standard deviation of 1.

11.55 $P(-t_{\alpha/2} < t < t_{\alpha/2}) = 1 - \alpha$

$$P\left(-t_{\alpha/2} < \frac{y_p - \hat{y}}{s\sqrt{1 + \frac{1}{n} + \frac{(x_p - \bar{x})^2}{SS_{xx}}}} < t_{\alpha/2}\right) = 1 - \alpha$$

$$P\left(-t_{\alpha/2}s\sqrt{1 + \frac{1}{n} + \frac{(x_p - \bar{x})^2}{SS_{xx}}} < y_p - \hat{y} < t_{\alpha/2}s\sqrt{1 + \frac{1}{n} + \frac{(x_p - \bar{x})^2}{SS_{xx}}}\right) = 1 - \alpha$$

$$P\left(\hat{y} - t_{\alpha/2}s\sqrt{1 + \frac{1}{n} + \frac{(x_p - \bar{x})^2}{SS_{xx}}} < y_p < \hat{y} + t_{\alpha/2}s\sqrt{1 + \frac{1}{n} + \frac{(x_p - \bar{x})^2}{SS_{xx}}}\right) = 1 - \alpha$$

The $(1 - \alpha)$ 100% prediction interval for y_p with df $= n - 2$ is:

$$\hat{y} \pm t_{\alpha/2}s\sqrt{1 + \frac{1}{n} + \frac{(x_p - \bar{x})^2}{SS_{xx}}}$$

11.57 From the printout, the least squares prediction equation is $\hat{y} = 1.403 + .101014x$.

To determine if there is a linear relationship between the number of disk I/O's and the number of records, we test:

$H_0:\ \beta_1 = 0$
$H_a:\ \beta_1 \neq 0$

The test statistic is $t = 28.30$ and the p-value is 0.000.

Since the p-value is so small, there is sufficient evidence to indicate there is a linear relationship between the number of disk I/O's and the number of records.

The value of $r^2 = .984$. This indicates that 98.4% of the variability of the number of disk I/O's about their predicted values is explained by the linear relationship between the number of disk I/O's and the number of records. This indicates a very strong linear relationship. Since the value of $\hat{\beta}_1$ is greater than 0, the relationship is positive.

11.59 Some preliminary calculations are:

$$\sum x = 4.951 \qquad\qquad \sum y = 118.76$$

$$\sum x^2 = 2.488995 \qquad\qquad \sum y^2 = 1415.704 \qquad\qquad \sum xy = 59.20694$$

$$SS_{xx} = \sum x^2 - \frac{\left(\sum x\right)^2}{n} = 2.489 - \frac{(4.951)^2}{10} = .0377549$$

$$SS_{yy} = \sum y^2 - \frac{\left(\sum y\right)^2}{n} = 1415.704 - \frac{(118.76)^2}{10} = 5.31024$$

$$SS_{xy} = \sum xy - \frac{\left(\sum x\right)\left(\sum y\right)}{n} = 59.207 - \frac{(4.951)(118.76)}{10} = .408864$$

$$\hat{\beta}_1 = \frac{SS_{xy}}{SS_{xx}} = \frac{.408864}{.0377549} = 10.82942876 \approx 10.829$$

$$\hat{\beta}_0 = \bar{y} - \hat{\beta}_1 \bar{x} = \frac{118.76}{10} - (10.82942876)\left[\frac{4.951}{10}\right] = 6.514349821 \approx 6.514$$

The least squares line is $\hat{y} = 6.514 + 10.829x$.

$$SSE = SS_{yy} - \hat{\beta}_1 SS_{xy} = 5.31024 - 10.82942876(.408864) = .882476438$$

$$s^2 = \frac{SSE}{n-2} = \frac{.882476438}{10-2} = .110309554 \qquad s = \sqrt{.110309554} = .33213$$

To determine if there is a linear relationship between breaking strength and specific gravity, we test:

$H_0:\ \beta_1 = 0$
$H_a:\ \beta_1 \neq 0$

The test statistic is $t = \dfrac{\hat{\beta}_1}{s/\sqrt{SS_{xx}}} = \dfrac{10.829}{.33213/\sqrt{.0377549}} = 6.34$

The rejection region for this small-sample, two-tailed test requires $\alpha/2 = .05/2 = .025$ in each tail of the t distribution with df $= n - 2 = 10 - 2 = 8$. From Table 7, Appendix II, $t_{.025} = 2.306$. The rejection region is $t < -2.306$ or $t < 2.306$.

Since the observed value of the test statistic falls in the rejection region ($t = 6.34 > 2.306$), H_0 is rejected. There is sufficient evidence to indicate there is a linear relationship between breaking strength and specific gravity at $\alpha = .05$.

$$r = \dfrac{SS_{xy}}{\sqrt{SS_{xx}SS_{yy}}} = \dfrac{.408864}{\sqrt{.0377549(5.31024)}} = .9131$$

$$r^2 = .9131^2 = .834$$

The value of $r^2 = .834$. This indicates that 83.4% of the variability of the breaking strengths about their predicted values is explained by the linear relationship between the breaking strength and the specific gravity. This indicates a fairly strong linear relationship. Since the value of $\hat{\beta}_1$ is greater than 0, the relationship is positive.

11.61 Some preliminary calculations are:

$$\sum x = 15 \qquad \sum x^2 = 55 \qquad \sum xy = 16.97$$

$$\sum y = 4.6 \qquad \sum y^2 = 5.2826$$

$$SS_{xy} = \sum xy - \dfrac{\sum x \sum y}{n} = 16.97 - \dfrac{15(4.6)}{5} = 3.17$$

$$SS_{xx} = \sum x^2 - \dfrac{(\sum x)^2}{n} = 55 - \dfrac{15^2}{5} = 10$$

$$SS_{yy} = \sum y^2 - \dfrac{(\sum y)^2}{n} = 5.2826 - \dfrac{4.6^2}{5} = 1.0506$$

$$\bar{x} = \dfrac{\sum x}{n} = \dfrac{15}{5} = 3 \qquad \bar{y} = \dfrac{\sum y}{n} = \dfrac{4.6}{5} = .92$$

$$\hat{\beta}_1 = \dfrac{SS_{xy}}{SS_{xx}} = \dfrac{3.17}{10} = .317$$

$$\hat{\beta}_0 = \bar{y} - \hat{\beta}_1\bar{x} = .92 - .317(3) = -.031$$

The least squares line is $\hat{y} = -.031 + .317x$.

$$SSE = SS_{yy} - \hat{\beta}_1 SS_{xy} = 1.0506 - .317(3.17) = .04571$$

$$s^2 = \frac{SSE}{n-2} = \frac{.04571}{5-2} = .01523667 \qquad s = \sqrt{.01523667} = .12344$$

To determine if there is a linear relationship between terminal response time and number of simultaneous users, we test:

H_0: $\beta_1 = 0$
H_a: $\beta_1 \neq 0$

The test statistic is $t = \dfrac{\hat{\beta}_1}{s/\sqrt{SS_{xx}}} = \dfrac{.317}{.12344/\sqrt{10}} = 8.12$

The rejection region for this small-sample, two-tailed test requires $\alpha/2 = .05/2 = .025$ in each tail of the t distribution with df $= n - 2 = 5 - 2 = 3$. From Table 7, Appendix II, $t_{.025} = 3.182$. The rejection region is $t < -3.182$ or $t > 3.182$.

Since the observed value of the test statistic falls in the rejection region ($t = 8.12 > 3.182$), H_0 is rejected. There is sufficient evidence to indicate there is a linear relationship between breaking terminal response time and number of simultaneous users at $\alpha = .05$.

$$r = \frac{SS_{xy}}{\sqrt{SS_{xx}SS_{yy}}} = \frac{3.17}{\sqrt{10(1.0506)}} = .978$$

$$r^2 = .978^2 = .956$$

The value of $r^2 = .956$. This indicates that 95.6% of the variability of the terminal response times about their predicted values is explained by the linear relationship between the terminal response time and the number of simultaneous users. This indicates a very strong linear relationship. Since the value of $\hat{\beta}_1$ is greater than 0, the relationship is positive.

11.63   a.   Some preliminary calculations are:

$$\sum x = 100 \qquad \sum x^2 = 1332 \qquad \sum xy = 114.49$$

$$\sum y = 8.54 \qquad \sum y^2 = 9.8808$$

$$SS_{xy} = \sum xy - \frac{\sum x \sum y}{n} = 114.49 - \frac{100(8.54)}{8} = 7.74$$

$$SS_{xx} = \sum x^2 - \frac{(\sum x)^2}{n} = 1332 - \frac{100^2}{8} = 82$$

$$SS_{yy} = \sum y^2 - \frac{(\sum y)^2}{n} = 9.8808 - \frac{8.54^2}{8} = .76435$$

$$\bar{x} = \frac{\sum x}{n} = \frac{100}{8} = 12.5 \qquad \bar{y} = \frac{\sum y}{n} = \frac{8.54}{8} = 1.0675$$

$$\hat{\beta}_1 = \frac{SS_{xy}}{SS_{xx}} = \frac{7.74}{82} = .0943902$$

$$\hat{\beta}_0 = \bar{y} - \hat{\beta}_1\bar{x} = 1.0675 - .0943902(12.5) = -.11238$$

The least squares line is $\hat{y} = -.11238 + .09439x$.

b. To determine if the model is useful for predicting flow rate, we test:

$H_0: \beta_1 = 0$
$H_a: \beta_1 \neq 0$

$$SSE = SS_{yy} - \hat{\beta}_1 SS_{xy} = .76435 - .0943902(7.74) = .03376985$$
$$s^2 = \frac{SSE}{n-2} = \frac{.03376985}{8-2} = .005628308 \qquad s = \sqrt{.005628308} = .07502205$$

The test statistic is $t = \dfrac{\hat{\beta}_1}{s/\sqrt{SS_{xx}}} = \dfrac{.0943902}{.07502205/\sqrt{82}} = 11.39$

For a small-sample, two-tailed test, the rejection region requires $\alpha/2 = .05/2 = .025$ in each tail of the t distribution with df $= n - 2 = 8 - 2 = 6$. From Table 7, Appendix II, $t_{.025} = 2.447$.

The rejection region is $t < -2.447$ or $t > 2.447$.

Since the observed value of the test statistic falls in the rejection region ($t = 11.39 > 2.447$), H_0 is rejected. There is sufficient evidence to indicate the model is useful for predicting flow rate at $\alpha = .05$.

c. The 95% prediction interval for the flow rate in a sampling environment in which $x = 11$ is:

$$\hat{y} \pm t_{\alpha/2}s\sqrt{1 + \frac{1}{n} + \frac{(x_p - \bar{x})^2}{SS_{xx}}}$$

$$\hat{y} = -.11238 + .09439(11) = .926$$

For confidence coefficient .95, $\alpha = 1 - .95 = .05$, $\alpha/2 = .05/2 = .025$.

From Table 7, Appendix II, with df $= n - 2 = 8 - 2 = 6$, $t_{.025} = 2.447$.

The prediction interval is:

$$.926 \pm 2.447(.075022)\sqrt{1 + \frac{1}{n} + \frac{(11 - 12.5)^2}{82}} \Rightarrow .926 \pm .197 \Rightarrow (.729, 1.123)$$

11.65 First, we plot y vs. x.

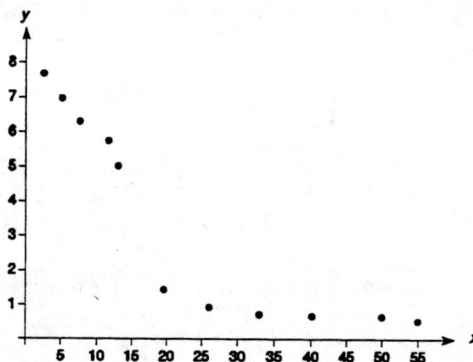

From the plot, there appears to be a negative linear relationship between concentration and corrosion rate.

From the printout, the least squares line is $\hat{y} = 6.735 - .142x$.

To determine if the model provides an adequate fit, we test:

H_0: $\beta_1 = 0$
H_a: $\beta_1 \neq 0$

The test statistic is $t = -5.562$

For a small-sample, two-tailed test, the rejection region requires $\alpha/2 = .05/2 = .025$ in each tail of the t distribution with df $= n - 2 = 11 - 2 = 9$. From Table 7, Appendix II, $t_{.025} = 2.262$.

The rejection region is $t < -2.262$ or $t > 2.262$.

Since the observed value of the test statistic falls in the rejection region ($t = -5.562 < -2.262$), H_0 is rejected. There is sufficient evidence to indicate the model provides an adequate fit at $\alpha = .05$.

$r^2 = .7747$ We know that by using the concentration of $NaPO_4$ to predict measure of corrosion rate with the least squares line, the total sum of squares of deviations of the 11 sample y values about their predicted values has been reduced by 77.47%.

s = Root MSE = 1.48088   The estimated standard deviation of ϵ is 1.48088.

CHAPTER TWELVE

....

Multiple Regression Analysis

12.1   a. $Y = \begin{bmatrix} 1 \\ 2 \\ 2 \\ 3 \\ 5 \\ 5 \end{bmatrix}$ $X = \begin{bmatrix} 1 & 1 \\ 1 & 2 \\ 1 & 3 \\ 1 & 4 \\ 1 & 5 \\ 1 & 6 \end{bmatrix}$

b. $X'X = \begin{bmatrix} 1 & 1 & 1 & 1 & 1 & 1 \\ 1 & 2 & 3 & 4 & 5 & 6 \end{bmatrix} \begin{bmatrix} 1 & 1 \\ 1 & 2 \\ 1 & 3 \\ 1 & 4 \\ 1 & 5 \\ 1 & 6 \end{bmatrix} = \begin{bmatrix} 6 & 21 \\ 21 & 91 \end{bmatrix}$

$X'Y = \begin{bmatrix} 1 & 1 & 1 & 1 & 1 & 1 \\ 1 & 2 & 3 & 4 & 5 & 6 \end{bmatrix} \begin{bmatrix} 1 \\ 2 \\ 2 \\ 3 \\ 5 \\ 5 \end{bmatrix} = \begin{bmatrix} 18 \\ 78 \end{bmatrix}$

c.   To invert $X'X$, we will use the following. Start by placing a 2×2 identity matrix beside $X'X$.

$$X'X = \begin{bmatrix} 6 & 21 \\ 21 & 91 \end{bmatrix} \qquad I = \begin{bmatrix} 1 & 0 \\ 0 & 1 \end{bmatrix}$$

Operation 1: Multiply the second row by 2.

$$\rightarrow \begin{bmatrix} 6 & 21 \\ 42 & 182 \end{bmatrix} \qquad \begin{bmatrix} 1 & 0 \\ 0 & 2 \end{bmatrix}$$

Operation 2: Multiply the first row by 7 and subtract it from row 2.

$$\rightarrow \begin{bmatrix} 6 & 21 \\ 0 & 35 \end{bmatrix} \qquad \begin{bmatrix} 1 & 0 \\ -7 & 2 \end{bmatrix}$$

Operation 3: Divide both rows by 7.

$$\begin{matrix} \rightarrow \\ \rightarrow \end{matrix} \begin{bmatrix} 6/7 & 3 \\ 0 & 5 \end{bmatrix} \qquad \begin{bmatrix} 1/7 & 0 \\ -1 & 2/7 \end{bmatrix}$$

Operation 4: Multiply the first row by 5.

$$\rightarrow \begin{bmatrix} 30/7 & 15 \\ 0 & 5 \end{bmatrix} \qquad \begin{bmatrix} 5/7 & 0 \\ -1 & 2/7 \end{bmatrix}$$

Operation 5: Multiply the second row by 3 and subtract it from the first row.

$$\rightarrow \begin{bmatrix} 30/7 & 0 \\ 0 & 5 \end{bmatrix} \qquad \begin{bmatrix} 26/7 & -6/7 \\ -1 & 2/7 \end{bmatrix}$$

Operation 6: Multiply the first row by 7/30 and divide the second row by 5.

$$\begin{matrix} \rightarrow \\ \rightarrow \end{matrix} \begin{bmatrix} 1 & 0 \\ 0 & 1 \end{bmatrix} \qquad \begin{bmatrix} 26/30 & -1/5 \\ -1/5 & 2/35 \end{bmatrix}$$

Thus, $X'X^{-1} = \begin{bmatrix} 13/15 & -1/5 \\ -1/5 & 2/35 \end{bmatrix}$

d. $\hat{\beta} = (X'X)^{-1}(X'Y) = \begin{bmatrix} 13/15 & -1/5 \\ -1/5 & 2/35 \end{bmatrix} \begin{bmatrix} 18 \\ 78 \end{bmatrix} = \begin{bmatrix} 0 \\ 6/7 \end{bmatrix}$

e. $\hat{y} = 6/7x$

12.3   a. $Y = \begin{bmatrix} 0 \\ 0 \\ 2 \\ 1 \\ 1 \\ 1 \\ 3 \\ 3 \\ 2 \\ 4 \\ 4 \\ 4 \\ 3 \\ 4 \\ 5 \end{bmatrix}$; $X = \begin{bmatrix} 1 & 1 \\ 1 & 2 \\ 1 & 3 \\ 1 & 4 \\ 1 & 5 \\ 1 & 6 \\ 1 & 7 \\ 1 & 8 \\ 1 & 9 \\ 1 & 10 \\ 1 & 11 \\ 1 & 12 \\ 1 & 13 \\ 1 & 14 \\ 1 & 15 \end{bmatrix}$

b. $X'X = \begin{bmatrix} 1 & 1 & 1 & 1 & 1 & 1 & 1 & 1 & 1 & 1 & 1 & 1 & 1 & 1 & 1 \\ 1 & 2 & 3 & 4 & 5 & 6 & 7 & 8 & 9 & 10 & 11 & 12 & 13 & 14 & 15 \end{bmatrix} \begin{bmatrix} 1 & 1 \\ 1 & 2 \\ 1 & 3 \\ 1 & 4 \\ 1 & 5 \\ 1 & 6 \\ 1 & 7 \\ 1 & 8 \\ 1 & 9 \\ 1 & 10 \\ 1 & 11 \\ 1 & 12 \\ 1 & 13 \\ 1 & 14 \\ 1 & 15 \end{bmatrix} = \begin{bmatrix} 15 & 120 \\ 120 & 1240 \end{bmatrix}$

$$X'Y = \begin{bmatrix} 1 & 1 & 1 & 1 & 1 & 1 & 1 & 1 & 1 & 1 & 1 & 1 & 1 & 1 \\ 1 & 2 & 3 & 4 & 5 & 6 & 7 & 8 & 9 & 10 & 11 & 12 & 13 & 14 & 15 \end{bmatrix} \begin{bmatrix} 0 \\ 0 \\ 2 \\ 1 \\ 1 \\ 1 \\ 3 \\ 3 \\ 2 \\ 4 \\ 4 \\ 4 \\ 3 \\ 4 \\ 5 \end{bmatrix} = \begin{bmatrix} 37 \\ 386 \end{bmatrix}$$

c. To invert $X'X$, we will use the following. Start by placing 2×2 identity matrix beside $X'X$.

$$X'X = \begin{bmatrix} 15 & 120 \\ 120 & 1240 \end{bmatrix} \qquad I = \begin{bmatrix} 1 & 0 \\ 0 & 1 \end{bmatrix}$$

Operation 1: Multiply the first row by 8 and subtract from row 2.

$$\rightarrow \begin{bmatrix} 15 & 120 \\ 0 & 280 \end{bmatrix} \qquad \begin{bmatrix} 1 & 0 \\ -8 & 1 \end{bmatrix}$$

Operation 2: Divide the first row by 15 and the second row by 40.

$$\begin{matrix} \rightarrow \\ \rightarrow \end{matrix} \begin{bmatrix} 1 & 8 \\ 0 & 7 \end{bmatrix} \qquad \begin{bmatrix} 1/15 & 0 \\ -1/5 & 1/40 \end{bmatrix}$$

Operation 3: Multiply the first row by 7.

$$\rightarrow \begin{bmatrix} 7 & 56 \\ 0 & 7 \end{bmatrix} \qquad \begin{bmatrix} 7/15 & 0 \\ -1/5 & 1/40 \end{bmatrix}$$

Operation 4: Multiply the second row by 8 and subtract from the first row 5.

$$\rightarrow \begin{bmatrix} 7 & 0 \\ 0 & 7 \end{bmatrix} \qquad \begin{bmatrix} 31/15 & -1/5 \\ -1/5 & 1/40 \end{bmatrix}$$

Operation 5: Divide both rows by 7.

$$\begin{array}{ll} \rightarrow \\ \rightarrow \end{array} \begin{bmatrix} 1 & 0 \\ 0 & 1 \end{bmatrix} \qquad \begin{bmatrix} 31/105 & -1/35 \\ -1/35 & 1/280 \end{bmatrix}$$

Thus, $X'X^{-1} = \begin{bmatrix} 31/105 & -1/35 \\ -1/35 & 1/280 \end{bmatrix}$

$$\hat{\beta} = X'X^{-1}X'Y = \begin{bmatrix} 31/105 & -1/35 \\ -1/35 & 1/280 \end{bmatrix} \begin{bmatrix} 37 \\ 386 \end{bmatrix} = \begin{bmatrix} -.1048 \\ .3214 \end{bmatrix}$$

d.   The prediction equation is $\hat{y} = \hat{\beta}_0 + \hat{\beta}_1 x = -.0148 + .3214x$

12.5   a. $Y = \begin{bmatrix} .231 \\ .107 \\ .053 \\ .129 \\ .069 \\ .030 \\ 1.005 \\ .559 \\ .321 \\ 2.948 \\ 1.633 \\ .934 \end{bmatrix}$ $X = \begin{bmatrix} 1 & 740 & 1.10 \\ 1 & 740 & 0.62 \\ 1 & 740 & 0.31 \\ 1 & 805 & 1.10 \\ 1 & 805 & 0.62 \\ 1 & 805 & 0.31 \\ 1 & 980 & 1.10 \\ 1 & 980 & 0.62 \\ 1 & 980 & 0.31 \\ 1 & 1235 & 1.10 \\ 1 & 1235 & 0.62 \\ 1 & 1235 & 0.31 \end{bmatrix}$

b. $X'X = \begin{bmatrix} 1 & 1 & 1 & 1 & 1 & 1 & 1 & 1 & 1 & 1 & 1 & 1 \\ 740 & 740 & 740 & 805 & 805 & 805 & 980 & 980 & 980 & 1235 & 1235 & 1235 \\ 1.1 & .62 & .31 & 1.1 & .62 & .31 & 1.1 & .62 & .31 & 1.1 & .62 & .31 \end{bmatrix} \begin{bmatrix} 1 & 740 & 1.10 \\ 1 & 740 & 0.62 \\ 1 & 740 & 0.31 \\ 1 & 805 & 1.10 \\ 1 & 805 & 0.62 \\ 1 & 805 & 0.31 \\ 1 & 980 & 1.10 \\ 1 & 980 & 0.62 \\ 1 & 980 & 0.31 \\ 1 & 1235 & 1.10 \\ 1 & 1235 & 0.62 \\ 1 & 1235 & 0.31 \end{bmatrix}$

$$= \begin{bmatrix} 12 & 11,280 & 8.12 \\ 11,280 & 11,043,750 & 7632.8 \\ 8.12 & 7632.8 & 6.762 \end{bmatrix}$$

$$X'Y = \begin{bmatrix} 1 & 1 & 1 & 1 & 1 & 1 & 1 & 1 & 1 & 1 & 1 & 1 \\ 740 & 740 & 740 & 805 & 805 & 805 & 980 & 980 & 980 & 1235 & 1235 & 1235 \\ 1.1 & .62 & .31 & 1.1 & .62 & .31 & 1.1 & .62 & .31 & 1.1 & .62 & .31 \end{bmatrix} \begin{bmatrix} .231 \\ .107 \\ .053 \\ .129 \\ .069 \\ .030 \\ 1.005 \\ .559 \\ .321 \\ 2.948 \\ 1.633 \\ .934 \end{bmatrix}$$

$$= \begin{bmatrix} 8.019 \\ 9,131.205 \\ 6.62724 \end{bmatrix}$$

c.   To find $(X'X)^{-1}$, we use the following steps:

$$X'X = \begin{bmatrix} 12 & 11,280 & 8.12 \\ 11,280 & 11,043,750 & 7632.8 \\ 8.12 & 7632.8 & 6.762 \end{bmatrix} \qquad I = \begin{bmatrix} 1 & 0 & 0 \\ 0 & 1 & 0 \\ 0 & 0 & 1 \end{bmatrix}$$

Multiply row 1 by 2.03 and row 3 by 3. Subtract new row 1 from new row 3.

$$\rightarrow \begin{bmatrix} 12 & 11,280 & 8.12 \\ 11,280 & 11,043,750 & 7632.8 \\ 0 & 0 & 3.8024 \end{bmatrix} \begin{bmatrix} 1 & 0 & 0 \\ 0 & 1 & 0 \\ -2.03 & 0 & 3 \end{bmatrix}$$

Multiply row 1 by 2820 and row 2 by 3. Subtract new row 1 from new row 2.

$$\rightarrow \begin{bmatrix} 12 & 11,280 & 8.12 \\ 0 & 1,321,650 & 0 \\ 0 & 0 & 3.8024 \end{bmatrix} \begin{bmatrix} 1 & 0 & 0 \\ -2820 & 3 & 0 \\ -2.03 & 0 & 3 \end{bmatrix}$$

Multiply row 3 by 2.03 and row 1 by .9506. Subtract new row 3 from new row 1.

$$\rightarrow \begin{bmatrix} 11.4092 & 10,722.768 & 0 \\ 0 & 1,321,650 & 0 \\ 0 & 0 & 3.8024 \end{bmatrix} \begin{bmatrix} 5.0715 & 0 & -6.09 \\ -2820 & 3 & 0 \\ -2.03 & 0 & 3 \end{bmatrix}$$

Multiply row 1 by 82603.125 and row 2 by 670.173. Subtract new row 2 from new row 1.

$$\rightarrow \begin{bmatrix} 942{,}270.3675 & 0 & 0 \\ 0 & 1{,}321{,}650 & 0 \\ 0 & 0 & 3.8024 \end{bmatrix} \begin{bmatrix} 2{,}308{,}809.608 & -2010.519 & -503{,}053.0312 \\ -2820 & 3 & 0 \\ -2.03 & 0 & 3 \end{bmatrix}$$

Divide row 1 by 942,270.3675, row 2 by 1,321.650, and row 3 by 3.8024.

$$\begin{matrix} \rightarrow \\ \rightarrow \\ \rightarrow \end{matrix} \begin{bmatrix} 1 & 0 & 0 \\ 0 & 1 & 0 \\ 0 & 0 & 1 \end{bmatrix} \begin{bmatrix} 2.4502624 & -.0021337 & -.5338733 \\ -.0021337 & .00000227 & 0 \\ -.5338733 & 0 & .7889754 \end{bmatrix}$$

Thus, $(X'X)^{-1} = \begin{bmatrix} 2.450262 & -.00213 & -.53387 \\ -.00213 & .00000227 & 0 \\ -.53387 & 0 & .78897 \end{bmatrix}$

d. $\hat{\beta} = X'X^{-1}X'Y = \begin{bmatrix} 2.4502624 & -.0021337 & -.5338733 \\ -.0021337 & .00000227 & 0 \\ -.5338733 & 0 & .7889754 \end{bmatrix} \begin{bmatrix} 8.019 \\ 9{,}131.205 \\ 6.62724 \end{bmatrix}$

$$= \begin{bmatrix} -3.37270 \\ .00362 \\ .94760 \end{bmatrix}$$

The least squares prediction equation is $\hat{y} = -3.3727 + .00362x_1 + .9476x_2$.

12.7 From Exercise 12.4:

$$Y = \begin{bmatrix} .24 \\ .38 \\ .44 \\ .61 \\ .75 \end{bmatrix} \quad X = \begin{bmatrix} 1 & 1.0 & 1.0 \\ 1 & 3.5 & 12.25 \\ 1 & 6.0 & 36.0 \\ 1 & 8.5 & 72.25 \\ 1 & 11.0 & 121.0 \end{bmatrix} \quad \hat{\beta} = \begin{bmatrix} .213542857 \\ .034914285 \\ .001257142 \end{bmatrix}$$

$$X'X^{-1} = \begin{bmatrix} 1.785828571 & -.611657142 & .042971428 \\ -.611657142 & .279314285 & -.021942857 \\ .042971428 & -.021942857 & .001828571 \end{bmatrix} \quad X'Y = \begin{bmatrix} 2.42 \\ 17.645 \\ 155.5575 \end{bmatrix}$$

Some preliminary calculations are:

$$Y'Y = \begin{bmatrix} .24 & .38 & .44 & .61 & .75 \end{bmatrix} \begin{bmatrix} .24 \\ .38 \\ .44 \\ .61 \\ .75 \end{bmatrix} = 1.3302$$

$$\hat{\beta}'X'Y = \begin{bmatrix} .213542857 & .034914285 & .001257142 \end{bmatrix} \begin{bmatrix} 2.42 \\ 17.645 \\ 155.5575 \end{bmatrix} = 1.32839414$$

$$\text{SSE} = Y'Y - \hat{\beta}'X'Y = 1.3302 - 1.32839414 = .00180586$$

$$s^2 = \frac{\text{SSE}}{n-3} = \frac{.00180586}{5-3} = .00096293$$

$$s = \sqrt{.00090293} = .030048793$$

The test is:

$$H_0: \beta_2 = 0$$
$$H_a: \beta_2 > 0$$

The test statistic is $t = \dfrac{\hat{\beta}_2}{s\sqrt{c_{22}}} = \dfrac{.00126}{.030048793\sqrt{.001828571}} = .98$

The rejection region requires $\alpha = .05$ in the upper tail of the t distribution with df $= n - (k+1) = 5 - (2+1) = 2$. From Table 7, Appendix II, $t_{.05} = 2.920$. The rejection region is $t > 2.920$.

Since the observed value of the test statistic does not fall in the rejection region ($t = .98 \not> -2.920$). H_0 is not rejected. There is insufficient evidence to indicate that upward curvature exists at $\alpha = .05$.

12.9 a.

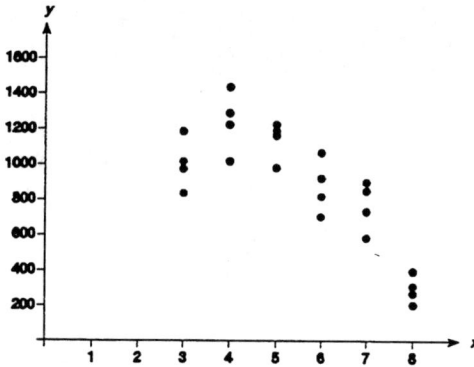

b.   Some preliminary calculations are:

$$\sum x = 132 \qquad \sum x^2 = 796 \qquad \sum xy = 107,099$$

$$\sum y = 21,385 \qquad \sum y^2 = 21,696,775$$

$$SS_{xy} = \sum xy - \frac{\sum x \sum y}{n} = 107,099 - \frac{132(21,385)}{24} = -10518.5$$

$$SS_{xx} = \sum x^2 - \frac{(\sum x)^2}{n} = 796 - \frac{132^2}{24} = 70$$

$$SS_{yy} = \sum y^2 - \frac{(\sum y)^2}{n} = 21,696,775 - \frac{21,385^2}{24} = 2,641,848.959$$

$$\bar{x} = \frac{\sum x}{n} = \frac{132}{24} = 5.5 \qquad \bar{y} = \frac{\sum y}{n} = \frac{21,385}{24} = 891.0416667$$

$$\hat{\beta}_1 = \frac{SS_{xy}}{SS_{xx}} = \frac{-10518.5}{70} = -150.2642857$$

$$\hat{\beta}_0 = \bar{y} - \hat{\beta}_1 \bar{x} = 891.0416667 - (-150.2642857)(5.5) = 1717.4952$$

The least squares line is $\hat{y} = 1717.4952 - 150.2643x$

$$SSE = SS_{yy} - \hat{\beta}_1 SS_{xy} = 2641848.959 - (-150.2642857)(-10518.5)$$
$$= 1061294.07$$

$$s^2 = \frac{SSE}{n-2} = \frac{1061294.07}{24-2} = 48240.63955$$

$$s = \sqrt{48240.63955} = 219.6375185$$

To determine if the model is useful, we test:

$H_0: \beta_1 = 0$
$H_a: \beta_1 \neq 0$

The test statistic is $t = \dfrac{\hat{\beta}_1}{s/\sqrt{SS_{xx}}} = \dfrac{-150.2642857}{219.6375185/\sqrt{70}} = -5.72$

For a small sample, two-tailed test, the rejection region requires $\alpha/2 = .05/2 = .025$ in each tail of the t distribution with df $= n - 2 = 24 - 2 = 22$. From Table 7, Appendix II, $t_{.025} = 2.074$. The rejection region is $t < -2.074$ or $t > 2.074$.

Since the observed value of the test statistic falls in the rejection region ($t = -5.72 < -2.074$), H_0 is rejected. There is sufficient evidence to indicate that the model is useful at $\alpha = .05$.

c. To test if the quadratic term should be included in the model, we test:

H_0: $\beta_2 = 0$
H_a: $\beta_2 \neq 0$

The test statistic is $t = -5.468$.

For a small sample, two-tailed test, the rejection region requires $\alpha/2 = .05/2 = .025$ in each tail of the t distribution, with df $= n - (k + 1) = 24 - (2 + 1) = 21$. From Table 7, Appendix II, $t_{.025} = 2.08$. The rejection region is $t < -2.08$ or $t > 2.08$.

Since the observed value of the test statistic falls in the rejection region ($t = -5.47 < -2.08$), H_0 is rejected. There is sufficient evidence to indicate the quadratic term should be included in the model at $\alpha = .05$.

Note: The p-value is .0001. Since the p-value is less than α (.0001 < .05), H_0 is rejected.

12.11 a.

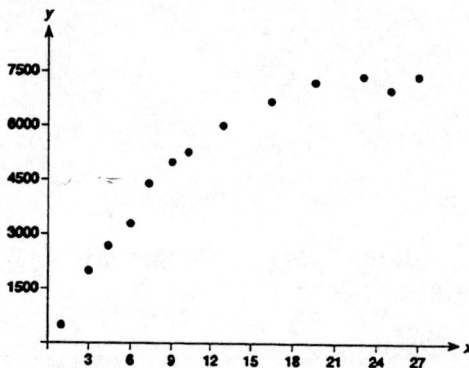

From the scattergram, there appears to be a quadratic relationship between axial strain and deviatoric stress.

b. To determine if deviatoric stress y increases with axial strain x at a decreasing rate, we test:

$$H_0: \beta_2 = 0$$
$$H_a: \beta_2 < 0$$

The test statistic is $t = -15.779$.

For a small sample, one-tailed test, the rejection region requires $\alpha = .05$ in the lower tail of the t distribution, with df $= n - (k + 1) = 13 - (2 + 1) = 10$. From Table 7, Appendix II, $t_{.05} = 1.812$. The rejection region is $t < -1.812$.

Since the observed value of the test statistic falls in the rejection region ($t = -15.779 < -1.812$), H_0 is rejected. There is sufficient evidence to indicate deviatoric stress increases with axial strain at a decreasing rate at $\alpha = .05$.

Note: The p-value is $.0001/2 = .00005$. Since the p-value is less than α ($.00005 < .05$), H_0 is rejected.

c. For a one-tailed test, the p-value $= 1/2$ (PR $> |T|$) $= 1/2(.0001) = .00005$. We would reject H_0 for any values of $\alpha > .00005$.

d. The estimate of $\sigma = s =$ Root MSE $= 188.89198$. This is the estimate of the standard deviation of the random error ϵ that appears in the model.

12.13 From Exercise 12.12, we know that

$$t = \frac{\hat{\beta}_i - \beta_i}{s\sqrt{c_{ii}}} \text{ has a } t \text{ distribution with } [n - (k + 1)] \text{ degrees of freedom.}$$

Thus, $P\left[-t_{\alpha/2} < \dfrac{\hat{\beta}_i - \beta_i}{s\sqrt{c_{ii}}} < t_{\alpha/2}\right] = 1 - \alpha$

$$\Rightarrow P\left(-t_{\alpha/2}s\sqrt{c_{ii}} < \hat{\beta}_i - \beta_i < t_{\alpha/2}s\sqrt{c_{ii}}\right) = 1 - \alpha$$

$$\Rightarrow P\left(\hat{\beta}_i - t_{\alpha/2}s\sqrt{c_{ii}} < \beta_i < \hat{\beta}_i + t_{\alpha/2}s\sqrt{c_{ii}}\right) = 1 - \alpha$$

Thus, the $(1 - \alpha)100\%$ confidence interval for β_i is $\hat{\beta}_i \pm t_{\alpha/2}s\sqrt{c_{ii}}$

12.15 To determine if the model is useful for predicting y, we test:

$$H_0: \beta_1 = \beta_2 = 0$$
$$H_a: \text{At least one of the parameters differs from } 0$$

The test statistic is $F = \dfrac{R^2/k}{(1 - R^2)/[n - (k + 1)]}$

$$= \frac{.78/2}{(1 - .78)/[253 - (2 + 1)]} = \frac{.39}{.00088} = 443.18$$

The rejection region requires $\alpha = .05$ in the upper tail of the F distribution with $\nu_1 = k = 2$ and $\nu_2 = n - (k + 1) = 253 - (2 + 1) = 250$. From Table 10, Appendix II, $F_{.05} \approx 3.00$. The rejection region is $F > 3.00$.

Since the observed value of the test statistic falls in the rejection region ($F = 443.18 > 3.00$), H_0 is rejected. There is sufficient evidence to indicate that the model is useful for predicting the number of defects in modules of the software product at $\alpha = .05$.

12.17   a.   To determine if the model is useful for predicting rate of conversion y, we test:

H_0: $\beta_1 = \beta_2 = \beta_3 = 0$
H_a: At least one of the parameters differs from 0

The test statistic is $F = \dfrac{R^2/k}{(1 - R^2)/[n - (k + 1)]}$

$$= \frac{.899/3}{(1 - .899)/[10 - (3 + 1)]} = \frac{.29966667}{.01683333} = 17.8$$

The rejection region requires $\alpha = .01$ in the upper tail of the F distribution with $\nu_1 = k = 3$ and $\nu_2 = n - (k + 1) = 10 - (3 + 1) = 6$. From Table 12, Appendix II, $F_{.01} = 9.78$. The rejection region is $F > 9.78$.

Since the observed value of the test statistic falls in the rejection region ($F = 17.8 > 9.78$), H_0 is rejected. There is sufficient evidence to indicate that the model is useful for predicting the rate of conversion y at $\alpha = .01$.

b.   To determine whether atom ratio x_1 is a useful predictor of rate of conversion y, we test:

H_0: $\beta_1 = 0$
H_a: $\beta_1 \neq 0$

The test statistic is $t = \dfrac{\hat{\beta}_1}{s_{\hat{\beta}_1}} = \dfrac{-.808}{.231} = -3.50$

For a small sample, two-tailed test, the rejection region requires $\alpha/2 = .05/2 = .025$ in each tail of the t distribution with df $= n - (k + 1) = 10 - (3 + 1) = 6$. From Table 7, Appendix II, $t_{.025} = 2.447$. The rejection region is $t < -2.447$ or $t > 2.447$.

Since the observed value of the test statistic falls in the rejection region ($t = -3.50 < -2.447$), H_0 is rejected. There is sufficient evidence to indicate that atom ratio x_1 is a useful predictor of rate of conversion at $\alpha = .05$.

c. A 95% confidence interval for β_2 is

$$\hat{\beta}_2 \pm t_{\alpha/2} s_{\hat{\beta}_2}$$

For confidence coefficient .95, $\alpha = 1 - .95 = .05$ and $\alpha/2 = .05/2 = .025$. From Table 7, Appendix II, with df $= n - (k + 1) = 10 - (3 + 1) = 6$, $t_{.025} = 2.447$. The 95% confidence interval for β_2 is

$$-6.38 \pm 2.447(1.93) \Rightarrow -6.38 \pm 4.723 \Rightarrow (-11.103, -1.657)$$

We are 95% confident that β_2 is between -11.103 and -1.657.

12.19 a. To determine if the model is useful for predicting the urban travel times of passenger cars, we test:

H_0: $\beta_1 = \beta_2 = 0$
H_a: At least one of the parameters differs from 0

The test statistic is $F = \dfrac{R^2/k}{(1 - R^2)/[n - (k + 1)]}$

$$= \dfrac{.687/2}{(1 - .687)/[567 - (2 + 1)]} = \dfrac{.3435}{.00055496} = 618.96$$

The rejection region requires $\alpha = .05$ in the upper tail of the F distribution with $\nu_1 = k = 2$ and $\nu_2 = n - (k + 1) = 567 - (2 + 1) = 564$. From Table 10, Appendix II, $F_{.05} = 3.00$. The rejection region is $F > 3.00$.

Since the observed value of the test statistic falls in the rejection region ($F = 618.96 > 3.00$), H_0 is rejected. There is sufficient evidence to indicate that the model is useful for predicting the urban travel times of passenger cars at $\alpha = .05$.

b. To determine if the model is useful for predicting the urban travel times of trucks, we test:

H_0: $\beta_1 = \beta_2 = 0$
H_a: At least one of the parameters differs from 0

The test statistic is $F = \dfrac{R^2/k}{(1 - R^2)/[n - (k + 1)]}$

$$= \dfrac{.771/2}{(1 - .771)/[918 - (2 + 1)]} = \dfrac{.3855}{.00025027} = 1540.32$$

The rejection region requires $\alpha = .05$ in the upper tail of the F distribution with $\nu_1 = k = 2$ and $\nu_2 = n - (k + 1) = 918 - (2 + 1) = 915$. From Table 10, Appendix II, $F_{.05} = 3.00$. The rejection region is $F > 3.00$.

Since the observed value of the test statistic falls in the rejection region ($F = 1540.32 >$ 3.00), H_0 is rejected. There is sufficient evidence to indicate that the model is useful for predicting the urban travel times of trucks at $\alpha = .05$.

12.21 To determine if the model is useful for predicting CPU time, we test:

H_0: $\beta_1 = \beta_2 = \beta_3 = \cdots = \beta_{18} = 0$
H_a: At least one of the parameters differs from 0

The test statistic is $F = \dfrac{R^2/k}{(1 - R^2)/[n - (k + 1)]}$

$$= \dfrac{.95/18}{(1 - .95)/[20 - (18 + 1)]} = \dfrac{.052778}{.05} = 1.056$$

The rejection region requires $\alpha = .05$ in the upper tail of the F distribution with $\nu_1 = k = 18$ and $\nu_2 = n - (k + 1) = 20 - (18 + 1) = 1$. From Table 10, Appendix II, $F_{.05} = 248.0$. The rejection region is $F > 248.0$.

Since the observed value of the test statistic does not fall in the rejection region ($F = 1.056 \not> 248$), H_0 is not rejected. There is insufficient evidence to indicate that the model is useful for predicting CPU time at $\alpha = .05$.

12.23 a. From Exercise 12.4, $\hat{y} = .2135 + .0349x + .00126x^2$

$$Y'Y = \begin{bmatrix} .24 & .38 & .44 & .61 & .75 \end{bmatrix} \begin{bmatrix} .24 \\ .38 \\ .44 \\ .61 \\ .75 \end{bmatrix} = 1.3302$$

$$\text{SSE} = Y'Y - \hat{\beta}'X'Y = 1.3302 - \begin{bmatrix} .213542857 & .034914285 & .001257142 \end{bmatrix} \begin{bmatrix} 2.42 \\ 17.645 \\ 155.5575 \end{bmatrix}$$

$$= 1.3302 - 1.32839414 = .00180586$$

$$s^2 = \dfrac{\text{SSE}}{[n - (k + 1)]} = \dfrac{.00180586}{[5 - (2 + 1)]} = .00090293$$

$$s = \sqrt{.00090293} = .030048793$$

For $x = 7$, $\hat{y} = .2135 + .0349(7) + .00126(7^2) = .51954$

$$a = \begin{bmatrix} 1 \\ 7 \\ 49 \end{bmatrix}$$

A confidence interval for the mean value of y is

$$\ell \pm t_{\alpha/2}s\sqrt{a'(X'X)^{-1}a} \quad \text{where } \ell = \hat{y}$$

For confidence coefficient .90, $\alpha = 1 - .90 = .10$ and $\alpha/2 = .10/2 = .05$. From Table 7, Appendix II, with df $= n - (k + 1) = 5 - (2 + 1) = 2$, $t_{.05} = 2.920$. The confidence interval is:

$$.5195 \pm 2.920(.03005)\sqrt{\begin{bmatrix}1 & 7 & 49\end{bmatrix}\begin{bmatrix}1.7785828571 & -.611657142 & .042971428 \\ -.611657142 & .279314285 & -.021942857 \\ .042971428 & -.021942857 & .001828571\end{bmatrix}\begin{bmatrix}1 \\ 7 \\ 49\end{bmatrix}}$$

where $(X'X)^{-1}$ is from Exercise 12.4.

$$\Rightarrow .5195 \pm .087746\sqrt{\begin{bmatrix}-.390171451 & .26834286 & 0.021028592\end{bmatrix}\begin{bmatrix}1 \\ 7 \\ 49\end{bmatrix}}$$

$$\Rightarrow .5195 \pm .087746\sqrt{.457827561} \Rightarrow .5195 \pm .0594 \Rightarrow (.4601, .5789)$$

We are 90% confident the mean grain size of tensile specimens 7 centimeters from the ingot chill face is between .4601 and .5789.

b. The prediction interval for y is

$$\ell \pm t_{\alpha/2}s\sqrt{1 + a'(X'X)^{-1}a}$$

From part a, $\ell = \hat{y} = .5195$, $t_{.05} = 2.920$, $a'(X'X)^{-1}a = .457827561$ and $s = .03005$.

The 90% prediction interval is

$$.5195 \pm 2.920(.03005)\sqrt{1 + .457827561} \Rightarrow .5195 \pm .087746(1.207405301)$$
$$\Rightarrow .5195 \pm .1059 \Rightarrow (.4136, .6254)$$

c. The width of the interval in a is 2(.0594) = .1188 and the width of the interval in b is 2(.1059) = .2118. The prediction interval in b is wider than the confidence interval in a. The predictor for the actual value of y has 2 variances associated with it—the variance for locating the mean value of y plus the variance of y once the mean has been located. Thus, the variance for the predictor of the actual value of y is larger than the variance for the predictor of the mean value of y.

12.25 **a.** The 95% confidence interval for $E(y)$ has the form:

$$\hat{y} \pm t_{\alpha/2}s\sqrt{a'(X'X)^{-1}a}$$

From Exercise 12.10, $\hat{y} = 22.925 - 3.525x - .375x_2$. Thus,

$$\hat{y} = 22.925 - 3.525(-1) - .375(1) = 26.075$$

From Exercise 12.8, $(X'X)^{-1} = \begin{bmatrix} .25 & 0 & 0 \\ 0 & .25 & 0 \\ 0 & 0 & .25 \end{bmatrix}$, and $s = 4.65$.

$$a'(X'X)^{-1}a = \begin{bmatrix} 1 & -1 & 1 \end{bmatrix} \begin{bmatrix} .25 & 0 & 0 \\ 0 & .25 & 0 \\ 0 & 0 & .25 \end{bmatrix} \begin{bmatrix} 1 \\ -1 \\ 1 \end{bmatrix} = .75$$

For confidence coefficient .95, $\alpha = .05$ and $\alpha/2 = .05/2 = .025$. From Table 7, Appendix II, with df $= n - (k + 1) = 4 - (2 + 1) = 1$. $t_{.025} = 12.706$. The 95% confidence interval is:

$$26.075 \pm 12.706(4.65)\sqrt{.75} \Rightarrow 26.075 \pm 51.167 \Rightarrow (-25.092, 77.242)$$

We are 95% confident the mean yield for observations with temperature $= 50°$ and pressure $= 20$ lbs/inch^2 will fall in the interval -25.092 to 77.242.

b. The 95% prediction interval has the form:

$$\hat{y} \pm t_{\alpha/2}s\sqrt{1 + a'(X'X)^{-1}a}$$

From part a, we get:

$$26.075 \pm 12.706(4.65)\sqrt{1 + .75} \Rightarrow 26.075 \pm 78.159 \Rightarrow (-52.084, 104.234)$$

We are 95% confident the mean yield for a particular observation with temperature $= 50°$ and pressure $= 20$ lbs/inch^2 will fall in the interval -52.084 to 104.234.

c. Part b is wider because it is the confidence interval for a single observation, while part a is the confidence interval for a mean value.

12.27 Show $t = \dfrac{\ell - E(\ell)}{s\sqrt{a'(X'X)^{-1}a}}$ has a Student's t distribution with $[n - (k + 1)]$ degrees of freedom.

From Theorem 12.3, the sampling distribution of ℓ is normal with

$$E(\ell) = a_0\beta_0 + a_1\beta_1 + \cdots + a_k\beta_k,$$
$$V(\ell) = [a'(X'X)^{-1}a]\sigma^2$$

Thus, $\dfrac{\ell - E(\ell)}{\sqrt{\sigma^2 a'(X'X)^{-1}a}}$ has a standard normal distribution.

From Theorem 12.2, $\chi^2 = \dfrac{SSE}{\sigma^2} = \dfrac{[n - (k + 1)]s^2}{\sigma^2}$ has a χ^2 distribution with $\nu = [n - (k + 1)]$ degrees of freedom.

By definition, a t random variable is the ratio of a standard normal random variable and the square root of an independent chi-squared random variable divided by its degrees of freedom.

Thus, $t = \dfrac{\dfrac{\ell - E(\ell)}{\sqrt{\sigma^2 a'(X'X)^{-1}a}}}{\sqrt{\dfrac{\dfrac{[n - (k + 1)]s^2}{\sigma^2}}{[n - (k + 1)]}}} = \dfrac{\dfrac{\ell - E(\ell)}{\sqrt{\sigma^2 a'(X'X)^{-1}a}}}{\sqrt{\dfrac{s^2}{\sigma^2}}} = \dfrac{\ell - E(\ell)}{s\sqrt{a'(X'X)^{-1}a}}$ with df $= [n - (k + 1)]$.

12.29   a.   By the assumptions of Multiple Regression Analysis, the probability distribution of ϵ is normal. Thus,

$$y = \beta_0 + \beta_1 x_1 + \cdots + \beta_k x_k + \epsilon$$

has a normal distribution for a particular setting of $x_1, x_2, \ldots, x_k$ because ϵ is the only variable.

$y = \hat{\beta}_0 + \hat{\beta}_1 x_1 + \cdots + \hat{\beta}_k x_k$ is a linear combination of normal random variables, because each $\hat{\beta}_i$ is normally distributed by Theorem 12.1. Thus, $\hat{y}$ is also normally distributed.

Thus, $(\hat{y} - y)$ is also a linear combination of normal random variables, and therefore is also normal.

b. $E(\hat{y} - y) = E[\hat{\beta}_0 + \hat{\beta}_1 x_1 + \cdots + \hat{\beta}_k x_k - (\beta_0 + \beta_1 x_1 + \cdots + \beta_k + \epsilon)]$
$\qquad\qquad = \beta_0 + \beta_1 x_1 + \cdots + \beta_k x_k - (\beta_0 + \beta_1 x_1 + \cdots + \beta_k x_k + 0) = 0$

$V(\hat{y} - y) = V(\hat{y}) + V(y) - 2\,\text{Cov}(\hat{y}, y) = V(\hat{y}) + V(y)$
$\qquad\qquad = [a'(X'X)^{-1}a]\sigma^2 + \sigma^2 \qquad$ (by Theorem 12.3)
$\qquad\qquad = [1 + a'(X'X)^{-1}a]\sigma^2$

12.31
$$P\left[-t_{\alpha/2} < t < t_{\alpha/2}\right] = 1 - \alpha$$

$$P\left[-t_{\alpha/2} < \frac{\hat{y} - y}{s\sqrt{1 + a'(X'X)^{-1}a}} < t_{\alpha/2}\right] = 1 - \alpha$$

$$P\left[-t_{\alpha/2}s\sqrt{1 + a'(X'X)^{-1}a} < \hat{y} - y < t_{\alpha/2}s\sqrt{1 + a'(X'X)^{-1}a}\right] = 1 - \alpha$$

Therefore, a $(1 - \alpha)$ 100 prediction interval for y is

$$\hat{y} \pm t_{\alpha/2}s\sqrt{1 + a'(X'X)^{-1}a} \quad \text{with df} = [n - (k + 1)].$$

12.33 **a.** To determine if the model is adequate, we test:

$$H_0: \; \beta_1 = 0$$
$$H_a: \; \beta_1 \neq 0$$

The test statistic is $t = 4.666$ and the p-value is .0001.

Since the p-value is so small, there is strong evidence to indicate the model is adequate for predicting y.

b. Using SAS, the residual plot is:

Plot of Residual *PCB84.   Legend: A = 1 obs, B = 2 obs, etc.

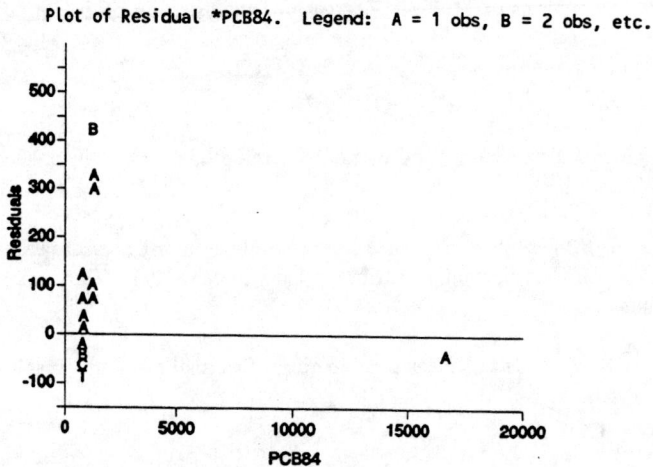

In order for an observation to be considered an outlier, it needs to be more than 3 standard deviations from 0. From the printout, $s = 145.7252$. Three standard deviations is $3(145.7252) = 437.1756$. The largest residual has a value of 429.6 which is less than 437.1756. Thus, there is no evidence of any outliers. However, observation #4 looks to be quite unusual with its extremely large value of PCB84 or x.

c. Using SAS, the output from fitting the 36 observations is:

Dependent Variable: PCB85

Analysis of Variance

Source	DF	Sum of Squares	Mean Square	F Value	Prob>F
Model	1	717106.87280	717106.87280	297.197	0.0001
Error	34	82038.74470	2412.90426		
C Total	35	799145.61750			

Root MSE	49.12132	R-square	0.8973
Dep Mean	89.66500	Adj R-sq	0.8943
C.V.	54.78316		

Parameter Estimates

Variable	DF	Parameter Estimate	Standard Error	T for H0: Parameter=0	Prob > |T|
INTERCEP	1	4.007395	9.57670301	0.418	0.6782
PCB84	1	0.986340	0.05721432	17.239	0.0001

Obs	PCB84	Dep Var PCB85	Predict Value	Residual
1	95.28	77.5500	97.9859	-20.4359
2	52.97	29.2300	56.2538	-27.0238
3	533.58	403.1	530.3	-127.2
4	308.46	192.2	308.3	-116.1
5	159.96	220.6	161.8	58.8177
6	10	8.6200	13.8708	-5.2508
7	234.43	174.3	235.2	-60.9251
8	443.89	529.3	441.8	87.4461
9	2.5	130.7	6.4732	124.2
10	51	39.7400	54.3107	-14.5707
11	0	0	4.0074	-4.0074
12	9.1	8.4300	12.9831	-4.5531
13	0	0	4.0074	-4.0074
14	140	120.0	142.1	-22.0550
15	0	0	4.0074	-4.0074
16	12	11.9300	15.8435	-3.9135
17	0	0	4.0074	-4.0074
18	0	0	4.0074	-4.0074
19	34	30.1400	37.5430	-7.4030
20	0	0	4.0074	-4.0074
21	0	0	4.0074	-4.0074
22	0	0	4.0074	-4.0074
23	422.1	531.7	420.3	111.3
24	6.74	9.3000	10.6553	-1.3553
25	7.06	5.7400	10.9710	-5.2310
26	46.71	46.4700	50.0793	-3.6093
27	159.56	176.9	161.4	15.5122
28	14	13.6900	17.8162	-4.1262
29	4.18	4.8900	8.1303	-3.2403
30	3.19	6.6000	7.1538	-0.5538
31	8.77	6.7300	12.6576	-5.9276
32	4.23	4.2800	8.1796	-3.8996
33	20.6	20.500	24.3260	-3.8260
34	329.97	414.5	329.5	85.0300
35	5.5	5.8000	9.4323	-3.6323
36	6.6	5.0800	10.5172	-5.4372

Sum of Residuals	0
Sum of Squared Residuals	82038.7447
Predicted Resid SS (Press)	116035.2035

To determine if the model is adequate, we test:

H_0: $\beta_1 = 0$
H_a: $\beta_1 \neq 0$

The test statistic is $t = 17.239$ and the p-value is .0001.

Since the p-value is so small, there is strong evidence to indicate the model is adequate for predicting y.

However $r^2 = .8973$ for the model with the one observation removed, compared with $r^2 = .3835$ for the model with all the observations. This implies a better fit with the one observation removed. Also, Root MSE for the model with the one observation removed is 49.12132, compared with Root MSE = 145.7252 for the model with all the observations. Again, this implies a better fit with the one observation removed.

d. To determine if the model is adequate, we test:

H_0: $\beta_1 = 0$
H_a: $\beta_1 \neq 0$

The test statistic is $t = 15.848$ and the p-value is .0001.

Since the p-value is so small, there is strong evidence to indicate the model is adequate for predicting natural $\log(y + 1)$.

In order for an observation to be considered an outlier, it needs to be more than 3 standard deviations from 0. From the printout, $s = 0.76132$. Three standard deviations is $3(0.76132) = 2.28396$. The largest residual has a value of 3.3893 which is larger than 2.28396. Thus, it appears that there is one outlier.

Also, the residual for Boston Harbor is -2.1157. The z-score for this residual is

$z = \dfrac{-2.1157 - 0}{.76132} = -2.779$. Although this value is less than 3 in magnitude, it is still

fairly close to 3. This point may be a suspect outlier.

12.35 First, we look at the plot of the residuals versus $\hat{y}$. From this plot, there is no evidence of a curvilinear trend among the residuals, so it appears that the model is not misspecified. Also from this plot, there is no evidence of a cone-shaped distribution or a football-shaped distribution. Thus, it appears that the error terms have equal variances.

To check for normality, we look at the stem-and-leaf display. The plot appears to be slightly skewed to the left, but fairly mound-shaped. Thus, we would conclude that the error terms are probably normally distributed. We could also look at the normal probability plot. It looks like a fairly straight line, indicating the error terms are normally distributed.

From the printout, $s = .37592$. An observation is considered an outlier if it is more than 3 standard deviations from 0. Three standard deviations is $3(.37592) = 1.2776$. All residuals are less than 1.12776 in absolute magnitude. Thus, there is no indication of any outliers

The data were not collected sequentially, so there is no need to check for independence of error terms.

From the above analysis, it appears that all the assumptions are met.

12.37 a. There appears to be a problem with the equal variance assumption. As the values of $\hat{y}$ get larger, the spread of the residuals increases. This indicates that as $\hat{y}$ increases, the variance increases.

      b. Since the data appear to be Poisson in nature, we would suggest the variance stabilizing transformation:

$$y^* = \sqrt{y}$$

12.39 We only know what the relationship between y and the independent variables looks like within the observed range. We have no information about what the relationship looks like outside the observed range.

Therefore, it is dangerous to predict values of y for values of the independent variable outside the experimental region.

12.41 Since the correlation between Importance and Support is fairly high (.6991), we should drop one of the two variables from the regression analysis. Since the correlation between Support and Replace ($-.0531$) is smaller than the correlation between Importance and Replace (.2682), we would recommend deleting the variable Importance and keeping the variables Support and Replace.

12.43 To determine if the model is adequate, we test:

H_0: $\beta_1 = \beta_2 = 0$
H_a: At least one of the parameters differ from 0

We are unable to test for model adequacy since there are no degrees of freedom for estimating σ^2 (df $= n - (k + 1) = 3 - (2 + 1) = 0$).

12.45 a. Some preliminary calculations are:

$$Y = \begin{bmatrix} 13.6 \\ 16.6 \\ 23.5 \\ \cdot \\ \cdot \\ \cdot \\ 14.9 \end{bmatrix} \qquad X_1 = \begin{bmatrix} 1 & 14.1 \\ 1 & 16.0 \\ 1 & 29.8 \\ \cdot & \cdot \\ \cdot & \cdot \\ \cdot & \cdot \\ 1 & 12.0 \end{bmatrix}$$

$$X_1'X_1 = \begin{bmatrix} 25 & 305.4 \\ 305.4 & 4501.2 \end{bmatrix} \qquad X_1'Y = \begin{bmatrix} 313.2 \\ 4443.15 \end{bmatrix}$$

$$(X_1'X_1)^{-1} = \begin{bmatrix} .233696972 & -.015856006 \\ -.015856006 & .00129797 \end{bmatrix}$$

$$\hat{\beta}_1 = (X_1'X_1)^{-1}X_1'Y = \begin{bmatrix} 2.7433 \\ .800974 \end{bmatrix}$$

The least squares line is $\hat{y} = 2.7433 + .800974x_1$

$$\text{SSE} = Y'Y - \hat{\beta}_1 X_1'Y = 4462.92 - 4418.0492 = 44.8708$$

$$s^2 = \frac{\text{SSE}}{n-2} = \frac{44.8708}{25-2} = 1.9509$$

$$s = \sqrt{1.9509} = 1.3967$$

$$SS_{x_1x_1} = \sum x_1^2 - \frac{\left(\sum x_1\right)^2}{n} = 4501.2 - \frac{305.4^2}{25} = 770.4336$$

To determine if the tar content (x_1) is useful for predicting carbon monoxide content (y), we test:

H_0: $\beta_1 = 0$
H_a: $\beta_1 \neq 0$

The test statistic is $t = \dfrac{\hat{\beta}_1}{s/\sqrt{SS_{x_1x_1}}} = \dfrac{.800974}{1.3967/\sqrt{770.4336}} = 15.92$

For a small sample, two-tailed test, the rejection region requires $\alpha/2 = .05/2 = .025$ in each tail of the t distribution with df $= n - 2 = 25 - 2 = 23$. From Table 7, Appendix II, $t_{.025} = 2.069$. The rejection region is $t < -2.069$ or $t > 2.069$.

Since the observed value of the test statistic falls in the rejection region ($t = 15.92 > 2.069$), H_0 is rejected. There is sufficient evidence to indicate that the tar content is useful for predicting carbon monoxide content at $\alpha = .05$.

b. Some preliminary calculations are:

$$Y = \begin{bmatrix} 13.6 \\ 16.6 \\ 23.5 \\ \cdot \\ \cdot \\ \cdot \\ 14.9 \end{bmatrix} \qquad X_2 = \begin{bmatrix} 1 & .86 \\ 1 & 1.06 \\ 1 & 2.03 \\ \cdot & \cdot \\ \cdot & \cdot \\ \cdot & \cdot \\ 1 & .82 \end{bmatrix}$$

$$X_2'X_2 = \begin{bmatrix} 25 & 21.91 \\ 21.91 & 22.2105 \end{bmatrix} \qquad X_2'Y = \begin{bmatrix} 313.2 \\ 311.781 \end{bmatrix}$$

$$(X_2'X_2)^{-1} = \begin{bmatrix} .295295847 & -.291300602 \\ -.291300602 & .33238316 \end{bmatrix}$$

$$\hat{\beta}_2 = (X_2'X_2)^{-1}X_2'Y = \begin{bmatrix} 1.664666288 \\ 12.39540546 \end{bmatrix}$$

The least squares line is $\hat{y} = 1.6647 + 12.3954x_2$.

$$\text{SSE} = Y'Y - \hat{\beta}_2 X_2'Y = 4462.92 - 4386.025391 = 76.894609$$

$$s^2 = \frac{\text{SSE}}{n-2} = \frac{76.894609}{25-2} = 3.3432$$

$$s = \sqrt{3.3432} = 1.8285$$

$$\text{SS}_{x_2 x_2} = \sum x_2^2 - \frac{\left(\sum x_2\right)^2}{n} = 22.2105 - \frac{21.91^2}{25} = 3.008576$$

To determine if the nicotine content (x_2) is useful for predicting carbon monoxide content (y), we test:

$$H_0: \ \beta_2 = 0$$
$$H_a: \ \beta_2 \neq 0$$

The test statistic is $t = \dfrac{\hat{\beta}_2}{s/\sqrt{\text{SS}_{x_2 x_2}}} = \dfrac{12.3954}{1.8285/\sqrt{3.008576}} = 11.76$

For a small sample, two-tailed test, the rejection region requires $\alpha/2 = .05/2 = .025$ in each tail of the t distribution with df $= n - 2 = 25 - 2 = 23$. From Table 7, Appendix II, $t_{.025} = 2.069$. The rejection region is $t < -2.069$ or $t > 2.069$.

Since the observed value of the test statistic falls in the rejection region ($t = 11.76 > 2.069$), H_0 is rejected. There is sufficient evidence to indicate that the nicotine content is useful for predicting carbon monoxide content at $\alpha = .05$.

c. Some preliminary calculations are:

$$Y = \begin{bmatrix} 13.6 \\ 16.6 \\ 23.5 \\ \cdot \\ \cdot \\ \cdot \\ 14.9 \end{bmatrix} \qquad X_3 = \begin{bmatrix} 1 & .9853 \\ 1 & 1.0938 \\ 1 & 1.1650 \\ \cdot & \cdot \\ \cdot & \cdot \\ \cdot & \cdot \\ 1 & 1.1184 \end{bmatrix}$$

$$X_3'X_3 = \begin{bmatrix} 25 & 24.2571 \\ 24.2571 & 23.7209575 \end{bmatrix} \qquad X_3'Y = \begin{bmatrix} 313.2 \\ 308.52258 \end{bmatrix}$$

$$(X_3'X_3)^{-1} = \begin{bmatrix} 5.13770131 & -5.25382394 \\ -5.25382394 & 5.4147280 \end{bmatrix}$$

$$\hat{\beta}_3 = (X_3'X_3)^{-1}X_3'Y = \begin{bmatrix} -11.7953 \\ 25.0682 \end{bmatrix}$$

The least squares line is $\hat{y} = -11.7953 + 25.0682x_3$.

$$\text{SSE} = Y'Y - \hat{\beta}_3 X_3'Y = 4462.92 - 4039.8161 = 423.1039$$

$$s^2 = \frac{\text{SSE}}{n-2} = \frac{423.1039}{25-2} = 18.3958$$

$$s = \sqrt{18.3958} = 4.2890$$

$$\text{SS}_{x_3 x_3} = \sum x_3 - \frac{\left(\sum x_3\right)^2}{n} = 23.7209575 - \frac{24.2571^2}{25} = .1846815$$

To determine if the weight (x_3) is useful for predicting carbon monoxide content (y), we test:

$$H_0: \beta_3 = 0$$
$$H_a: \beta_3 \neq 0$$

The test statistic is $t = \dfrac{\hat{\beta}_3}{s/\sqrt{\text{SS}_{x_3 x_3}}} = \dfrac{25.0682}{4.2890/\sqrt{.1846815}} = 2.51$

For a small sample, two-tailed test, the rejection region requires $\alpha/2 = .05/2 = .025$ in each tail of the t distribution with df $= n - 2 = 25 - 2 = 23$. From Table 7, Appendix II, $t_{.025} = 2.069$. The rejection region is $t < -2.069$ or $t > 2.069$.

Since the observed value of the test statistic falls in the rejection region ($t = 2.51 > 2.069$), H_0 is rejected. There is sufficient evidence to indicate that the weight is useful for predicting carbon monoxide content at $\alpha = .05$.

    d.   In Example 12.13,

$\hat{\beta}_1$ is positive

$\hat{\beta}_2$ is negative

$\hat{\beta}_3$ is negative

In part a, we found $\hat{\beta}_1$ is positive as in Example 12.13.

In part b, we found $\hat{\beta}_2$ is positive, but in Example 12.13 it was negative.

In part c, we found $\hat{\beta}_2$ is positive, but in Example 12.13 it was negative.

12.47   a.   From the printout, $\hat{y} = .13202 - 9.307122x_1 + 1.557563x_2$

    b.   To test the overall model utility, we test:

H_0: $\beta_1 = \beta_2 = 0$
H_a: At least one $\beta_i \neq 0$

The test statistic is $F = 35.843$.

The rejection region requires $\alpha = .05$ in the upper tail of the F distribution with $\nu_1 = k = 2$ and $\nu_2 = n - (k + 1) = 9 - (2 + 1) = 6$. From Table 10, Appendix II, $F_{.05} = 5.14$. The rejection region is $F > 5.14$.

Since the observed value of the test statistic falls in the rejection region ($F = 35.843 > 5.14$), H_0 is rejected. There is sufficient evidence to indicate at least one of the independent variables (porosity and/or estimated slope coefficient) contributes information for the prediction of the coefficient of permeability at $\alpha = .05$.

    c.   To determine if porosity is a useful predictor, we test:

H_0: $\beta_1 = 0$
H_a: $\beta_1 \neq 0$

The test statistic is $t = -1.840$

The rejection region for this two-tailed test requires $\alpha/2 = .05/2 = .025$ in each tail of the t distribution with df $= n - (k + 1) = 9 - (2 + 1) = 6$. From Table 7, Appendix II, $t_{.025} = 2.447$. The rejection region is $t > 2.447$ or $t < -2.447$.

Since the observed value of the test statistic does not fall in the rejection region ($t = -1.840 \not< -2.447$), H_0 is not rejected. There is insufficient evidence to indicate porosity is a useful predictor of coefficient of permeability at $\alpha = .05$.

d. To determine if the estimated water outflow-time slope is a useful predictor, we test:

H_0: $\beta_2 = 0$
H_a: $\beta_2 \neq 0$

The test statistic is $t = 8.467$.

The rejection region is $t > 2.447$ or $t < -2.447$ (from part c).

Since the observed value of the test statistic falls in the rejection region ($t = 8.467 > 2.447$), H_0 is rejected. There is sufficient evidence to indicate the estimated water outflow-time slope is a useful predictor of coefficient of permeability at $\alpha = .05$.

e. $R^2 = $ R-square $= .9228$. We know that by using the model containing both porosity and estimated water outflow-time slope coefficient to predict the coefficient of permeability, the total sum of squares of deviations of the nine sample values about their predicted values has been reduced by 92.28%.

f. The estimate of σ is $s = $ Root MSE $= .15214$. Most of the observed values of coefficient of permeability will fall within $2s$ or $2(.15214)$ or $.30428$ units of their respective least squares predicted values.

12.49 a. To determine if the model is adequate for predicting porosity, we test:

H_0: $\beta_1 = \beta_2 = 0$
H_a: At least one $\beta_i \neq 0$

The test statistic is $F = 12.27$.

The rejection region requires $\alpha = .05$ in the upper tail of the F distribution with $\nu_1 = k = 2$ and $\nu_2 = n - (k + 1) = 5 - (2 + 1) = 2$. From Table 10, Appendix II, $F_{.05} = 19.00$. The rejection region is $F > 19.00$.

Since the observed value of the test statistic does not fall in the rejection region ($F = 12.27 \not> 19.00$), H_0 is not rejected. There is insufficient evidence to indicate the model is adequate for predicting porosity at $\alpha = .05$.

$R^2 = $ R-square $= 92.5\%$. We know that by using the model containing both thickness and thickness squared to predict porosity, the total sum of squares of deviations of the five sample values about their predicted values has been reduced by 92.5%.

$s = 6.801$. Most of the observed porosity values will fall within $2s$ or $2(6.801)$ or 13.602 units of their respective least squares predicted values.

b. The interval is $(-7.98, 40.05)$. We are 95% confident the mean porosity will be between -7.98 and 40.05 when the thickness of the chromium deposit is .3. The interval is so wide because (1) the model is not adequate for predicting porosity and (2) the degrees of freedom for error is so small (df $= 2$).

12.51 From the printout, $R^2 = .9043$. This implies that 90.43% of the variability in the percentage improvement at the end of the decade measurements (y) can be explained by the regression model including $x =$ cost of modifying fleet and x^2.

$s = \sqrt{\text{MSE}} = 4.5485$. This is an estimate of the standard deviation of the random error ϵ that appears in the model.

To test for overall model adequacy, we test:

H_0: $\beta_1 = \beta_2 = 0$
H_a: At least one parameter differs from 0

The test statistic is $F = 33.079$.

The rejection region requires $\alpha = .05$ in the upper tail of the F distribution with $\nu_1 = k = 2$ and $\nu_2 = n - (k + 1) = 10 - (2 + 1) = 7$. From Table 10, Appendix II, $F_{.05} = 4.74$. The rejection region is $F > 4.74$.

Since the observed value of the test statistic falls in the rejection region ($F = 33.079 > 4.74$), H_0 is rejected. There is sufficient evidence to indicate the overall model is useful at $\alpha = .05$.

The p-value from the printout is .0003. Since the p-value is less than $\alpha = .05$, H_0 is rejected.

To determine if y increases more quickly for more costly fleet modifications than for less costly fleet modifications, we test:

H_0: $\beta_2 = 0$
H_a: $\beta_2 > 0$

The test statistic is $t = 2.131$.

For a small sample, one-tailed test, the rejection regions requires $\alpha = .05$ in the upper tail of the t distribution with df $= n - (k + 1) = 10 - (2 + 1) = 7$. From Table 7, Appendix II, $t_{.05} = 1.895$. The rejection region is $t > 1.895$.

Since the observed value of the test statistic falls in the rejection region ($t = 2.131 > 1.895$), H_0 is rejected. There is sufficient evidence to indicate that the percentage improvement y increases more quickly for more costly fleet modifications than for less costly ones at $\alpha = .05$.

For a one-tailed test, p-value $= 1/2(\text{PR} > |\text{T}|) = 1/2(.0706) = .0353$.

Since the p-value is less than $\alpha = .05$, H_0 is rejected.

12.53 a.   Some preliminary calculations are:

$$\sum x = 20 \qquad \sum x^2 = 60$$

$$\sum y = 5.12 \qquad \sum xy = 5.96$$

$$SS_{xy} = \sum xy - \frac{\sum x \sum y}{n} = 5.96 - \frac{20(5.12)}{10} = -4.28$$

$$SS_{xx} = \sum x^2 - \frac{(\sum x)^2}{n} = 60 - \frac{20^2}{10} = 20$$

$$\bar{x} = \frac{\sum x}{n} = \frac{20}{10} = 2 \qquad \bar{y} = \frac{\sum y}{n} = \frac{5.12}{10} = .512$$

$$\hat{\beta}_1 = \frac{SS_{xy}}{SS_{xx}} = \frac{-4.28}{20} = -.214$$

$$\hat{\beta}_0 = \bar{y} - \hat{\beta}_1 \bar{x} = .512 - (-.214)2 = .94$$

The fitted model is $\hat{y} = .94 - .214x$.

b.   For $x = 0$, $\hat{y} = .94 - .214(0) = .94$
   The residual is $y - \hat{y} = .94 - .94 = 0$

$x = 0$, $\hat{y} = .94$
   The residual is $.96 - .94 = .02$

$x = 1$, $\hat{y} = .94 - .214(1) = .726$
   The residual is $.7 - .726 = -.026$

$x = 1$, $\hat{y} = .726$
   The residual is $.76 - .726 = .034$

$x = 2$, $\hat{y} = .94 - .214(2) = .512$
   The residual is $.6 - .512 = .088$

$x = 2$, $\hat{y} = .512$
   The residual is $.4 - .512 = -.112$

$x = 3$, $\hat{y} = .94 - .214(3) = .298$
   The residual is $.24 - .298 = -.058$

$x = 3$, $\hat{y} = .298$
   The residual is $.3 - .298 = .002$

$x = 4$, $\hat{y} = .94 - .214(4) = .084$
   The residual is $.12 - .084 = .036$

$x = 4$, $\hat{y} = .084$
   The residual is $.1 - .084 = .016$

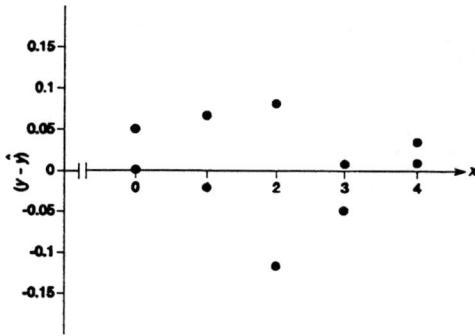

c.   The plot has a football shape. This implies the variances are not equal.

d.   In the problem, y is a proportion. To stabilize the variance, we will transform the data using

$$y^* = \sin^{-1}\sqrt{y} \text{ and fit } y^* = \beta_0 + \beta_1 x_1 + \epsilon$$

e. $y^* = \sin^{-1}\sqrt{y}$ For $y = .94$, $y^* = \sin^{-1}\sqrt{.94} = 75.821°$

To convert degrees to radians, multiply the degrees by π and divide by 180°. Thus,

$$75.821° \Rightarrow \frac{75.821(3.14159)}{180} = 1.3233 \text{ rad.}$$

For $y = .96$, $y^* = \dfrac{\left(\sin^{-1}\sqrt{.96}\right)\pi}{180} = 1.3694$

For $y = .7$, $y^* = \dfrac{\left(\sin^{-1}\sqrt{.7}\right)\pi}{180} = .9912$

For $y = .76$, $y^* = \dfrac{\left(\sin^{-1}\sqrt{.76}\right)\pi}{180} = 1.0588$

For $y = .6$, $y^* = \dfrac{\left(\sin^{-1}\sqrt{.6}\right)\pi}{180} = .8861$

For $y = .4$, $y^* = \dfrac{\left(\sin^{-1}\sqrt{.4}\right)\pi}{180} = .6847$

For $y = .24$, $y^* = \dfrac{\left(\sin^{-1}\sqrt{.24}\right)\pi}{180} = .5120$

For $y = .3$, $y* = \dfrac{\left(\sin^{-1}\sqrt{.3}\,\right)\pi}{180} = .5796$

For $y = .12$, $y* = \dfrac{\left(\sin^{-1}\sqrt{.12}\,\right)\pi}{180} = .3537$

For $y = .1$, $y* = \dfrac{\left(\sin^{-1}\sqrt{.1}\,\right)\pi}{180} = .3218$

$\sum y* = 8.0806 \qquad \sum xy* = 11.1684$

$\text{SS}_{xy*} = \sum xy* - \dfrac{\sum x \sum y*}{n} = 11.1684 - \dfrac{20(8.0806)}{10} = -4.9928$

$\bar{y}* = \dfrac{\sum y*}{n} = \dfrac{8.0806}{10} = .80806$

$\hat{\beta}_1^* = \dfrac{\text{SS}_{xy}*}{\text{SS}_{xx}} = \dfrac{-4.9928}{20} = -.24964$

$\hat{\beta}_0^* = \bar{y}* - \beta_1^* \bar{x} = .80806 - (-.24964)(2) = 1.30734$

The fitted model is $\hat{y}* = 1.307 - .2496x$.

For $x = 0$, $\hat{y}* = 1.307 - .2496(0) = 1.307$

The residual is $y* - \hat{y}* = 1.3233 - 1.307 = .0163$

$x = 0, \hat{y}* = 1.307 \quad y* - \hat{y}* = 1.3694 - 1.307 = .0624$
$x = 1, \hat{y}* = 1.0574 \quad y* - \hat{y}* = .9912 - 1.0574 = -.0662$
$x = 1, \hat{y}* = 1.0574 \quad y* - \hat{y}* = 1.0588 - 1.0574 = .0014$
$x = 2, \hat{y}* = .8078 \quad y* - \hat{y}* = .8861 - .8078 = .0783$
$x = 2, \hat{y}* = .8078 \quad y* - \hat{y}* = .6847 - .8078 = -.1231$
$x = 3, \hat{y}* = .5582 \quad y* - \hat{y}* = .5120 - .5582 = -.0462$
$x = 3, \hat{y}* = .5582 \quad y* - \hat{y}* = .5796 - .5582 = .0214$
$x = 4, \hat{y}* = .3086 \quad y* - \hat{y}* = .3537 - .3086 = .0451$
$x = 4, \hat{y}* = .3086 \quad y* - \hat{y}* = .3218 - .3086 = .0132$

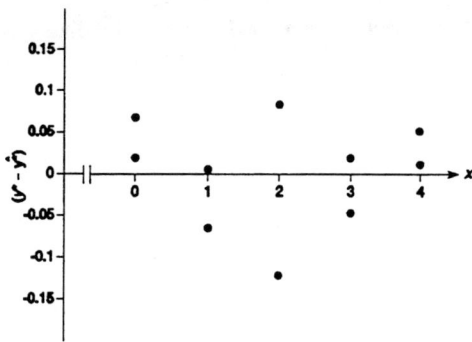

The assumption of equal variances could be satisfied. The variances still may be different (football shape).

12.55 a.

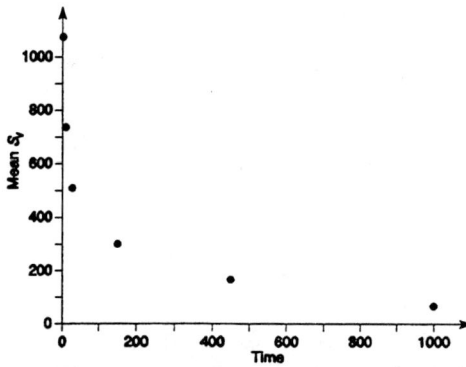

Since these means plot a curve, we hypothesize:

$$E(S_v) = \beta_0 + \beta_1 x + \beta_2 x^2$$

b.

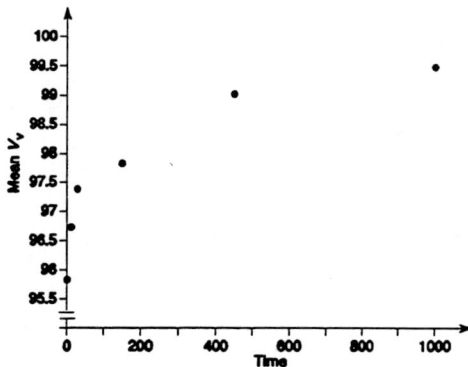

Since the means plot a curve, we hypothesize:

$$E(V_v) = \beta_0 + \beta_1 x + \beta_2 x^2$$

c.   If a linear model was fit to the data when a quadratic model is indicated, the model will be misspecified.

12.57   a.

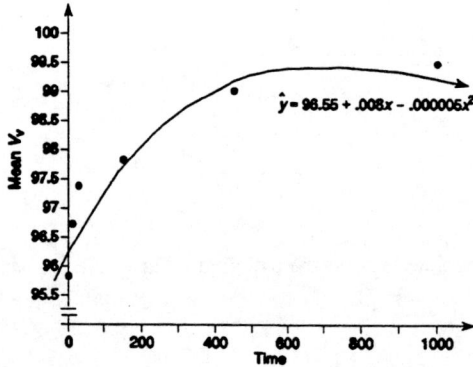

b.   To test if the quadratic term is significant, we test:

$$H_0: \beta_2 = 0$$
$$H_a: \beta_2 \neq 0$$

The test statistic is $t = -2.105$.

The p-value is .1260.

Since the p-value is greater than α (.1260 > .05), H_0 is not rejected. There is insufficient evidence to indicate that the quadratic term is significant.

c.   The interval is (95.7, 99.7). We are 95% confident that the value of a particular V_v, with sintering time = 150 minutes, will fall in the interval 95.7 to 99.7.

12.59   a. $Y = \begin{bmatrix} 1.1 \\ .5 \\ 1.8 \\ 2.0 \\ 2.0 \\ 2.9 \\ 3.8 \\ 3.4 \\ 4.1 \\ 5.0 \\ 5.0 \\ 5.8 \end{bmatrix}$ $X = \begin{bmatrix} 1 & 1 \\ 1 & 1 \\ 1 & 2 \\ 1 & 2 \\ 1 & 3 \\ 1 & 3 \\ 1 & 4 \\ 1 & 4 \\ 1 & 5 \\ 1 & 5 \\ 1 & 6 \\ 1 & 6 \end{bmatrix}$

b. $X'X = \begin{bmatrix} 1 & 1 & 1 & 1 & 1 & 1 & 1 & 1 & 1 & 1 & 1 & 1 \\ 1 & 1 & 2 & 2 & 3 & 3 & 4 & 4 & 5 & 5 & 6 & 6 \end{bmatrix} \begin{bmatrix} 1 & 1 \\ 1 & 1 \\ 1 & 2 \\ 1 & 2 \\ 1 & 3 \\ 1 & 3 \\ 1 & 4 \\ 1 & 4 \\ 1 & 5 \\ 1 & 5 \\ 1 & 6 \\ 1 & 6 \end{bmatrix} = \begin{bmatrix} 12 & 42 \\ 42 & 182 \end{bmatrix}$

$X'Y = \begin{bmatrix} 1 & 1 & 1 & 1 & 1 & 1 & 1 & 1 & 1 & 1 & 1 & 1 \\ 1 & 1 & 2 & 2 & 3 & 3 & 4 & 4 & 5 & 5 & 6 & 6 \end{bmatrix} \begin{bmatrix} 1.1 \\ .5 \\ 1.8 \\ 2.0 \\ 2.0 \\ 2.9 \\ 3.8 \\ 3.4 \\ 4.1 \\ 5.0 \\ 5.0 \\ 5.8 \end{bmatrix} = \begin{bmatrix} 37.4 \\ 163.0 \end{bmatrix}$

c. In Exercise 12.1, $X'X = \begin{bmatrix} 6 & 21 \\ 21 & 91 \end{bmatrix}$

The elements of the $X'X$ matrix in Exercise 12.1 are doubled in the $X'X$ matrix of Exercise 12.59.

d. In Exercise 12.1, $(X'X)^{-1} = \begin{bmatrix} 13/15 & -1/5 \\ -1/5 & 2/35 \end{bmatrix}$

For Exercise 12.59, the $(X'X)^{-1}$ contains elements that are half those in the $(X'X)^{-1}$ of Exercise 12.1. Thus,

$$(X'X)^{-1} = \begin{bmatrix} 13/30 & -1/10 \\ -1/10 & 1/35 \end{bmatrix}$$

To verify that this is the correct inverse matrix,

$$X'X(X'X)^{-1} = \begin{bmatrix} 12 & 42 \\ 42 & 182 \end{bmatrix} \begin{bmatrix} 13/30 & -1/10 \\ -1/10 & 1/35 \end{bmatrix} = \begin{bmatrix} 1 & 0 \\ 0 & 1 \end{bmatrix}$$

e. $\hat{\beta} = (X'X)^{-1}X'Y = \begin{bmatrix} 13/30 & -1/10 \\ -1/10 & 1/35 \end{bmatrix} \begin{bmatrix} 37.4 \\ 163.0 \end{bmatrix} = \begin{bmatrix} -.093333 \\ .917143 \end{bmatrix}$

The prediction equation is $\hat{y} = -.093 + .917x$.

CHAPTER THIRTEEN

. . . .

Model Building

13.1 a.   Number of preincident psychological symptoms is quantitative. This number could range from 0 to about 20.

b.   Years of experience is quantitative. Years of experience could range from 0 to about 50.

c.   Cigarette smoking behavior is qualitative. Levels of this might be "smokes" or "does not smoke."

d.   Level of social support is qualitative. Levels of this variable might be "gets no social support," "gets some social support," or "gets much social support."

e.   Marital status is qualitative. Levels of this variable might be "single," "married," "divorced," "widowed," or "other."

f.   Age is quantitative. Age could range from 18 to about 70.

g.   Ethnic status is qualitative. Levels of this variable might be "White," "American Indian," "Black," "Asian," or "Hispanic."

h.   Exposure to a chemical fire is qualitative. Levels of this variable might be "exposed to chemical fire" or "not exposed to chemical fire."

i.   Educational level is qualitative. Levels might be "Junior High School," "High School graduate," "College graduate," or "Advanced College degree."

j.   Distance lived from site of incident is quantitative. The distance could range from 0 to about 100 miles.

k.   Gender is qualitative. Levels of gender could be "male" or "female."

13.3 a.   The independent variable for CPU time is quantitative. We would expect to observe levels ranging from 0 to 20 minutes.

b.   The independent variable for the software system is qualitative. We would expect such levels as Fortran, Cobol, SAS, SPSSX, or BMDP.

c.   The independent variable for the number of lines of output is quantitative. We would expect to observe levels ranging from 0 to 20,000 lines.

d.   The independent variable for job cost is quantitative. We would expect to observe levels ranging from $0 to $100.

. . . .

e.	The independent variable for the date of submission is qualitative. We would expect to observe levels such as June 17, 1994; July 5, 1994, etc.

13.5	a.	First-order since x is to the first power.

b.	Since this is a first-order model, the graph is a straight line.

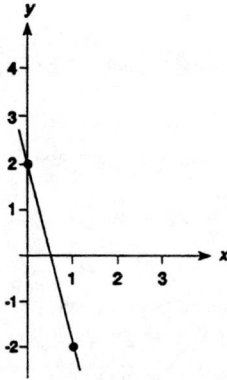

c.	If the coefficient of x was positive, the line would have a positive or upward slope (go from lower left to upper right.

13.7	a.	i.	First-order
ii.	Third-order
iii.	First-order
iv.	Second-order

b.	i.	$E(y) = \beta_0 + \beta_1 x$
ii.	$E(y) = \beta_0 + \beta_1 x + \beta_2 x^2 + \beta_3 x^3$
iii.	$E(y) = \beta_0 + \beta_1 x$
iv.	$E(y) = \beta_0 + \beta_1 x + \beta_2 x^2$

c.	i.	$\beta_1 > 0$ since the slope of the line is positive.
ii.	$\beta_3 > 0$ refer to Figure 13.4.
iii.	$\beta_1 < 0$ since the slope of the line is negative.
iv.	$\beta_2 < 0$ since the parabola opens downward.

13.9	a.	For a curvilinear relationship, the model would be:

$$E(y) = \beta_0 + \beta_1 x_2 + \beta_2 x_2^2$$

A sketch of what this relationship might look like is:

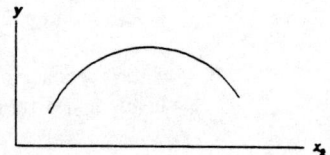

b.	For a third-order relationship, the model would be:

$$E(y) = \beta_0 + \beta_1 x_1 + \beta_2 x_1^2 + \beta_3 x_1^3$$

A sketch of what this relationship might look like is:

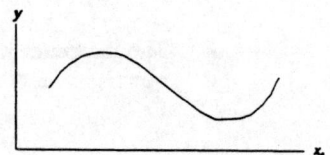

13.11 a.

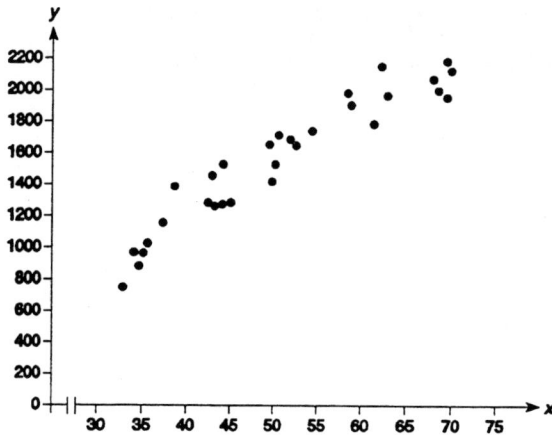

b. There appears to be a slight curve to the scattergram. Thus, the model would be

$$E(y) = \beta_0 + \beta_1 x + \beta_2 x^2$$

13.13 Since the maximum amount of pollution that can be emitted increases as the plant's output increases, the model would be:

$$E(y) = \beta_0 + \beta_1 x$$

where y = maximum amount of pollution permitted (in parts per million)
   x = plant's output (in megawatts)

13.15 To determine if the rate of increase in output per unit increase of input decreases as the input increases, we test:

$H_0: \beta_2 = 0$
$H_a: \beta_2 < 0$

The test statistic is $t = -6.60$ (from printout)

The rejection region for this small-sample, one-tailed test requires $\alpha = .05$ in the lower tail of the t distribution with df $= n - (k + 1) = 25 - (2 + 1) = 22$. From Table 7, Appendix II, $t_{.05} = 1.717$. The rejection region is $t < -1.717$.

Since the observed value of the test statistic falls in the rejection region ($t = -6.60 < -1.717$), H_0 is rejected. There is sufficient evidence to indicate that the rate of increase in output per unit increase of input decreases as the input increases at $\alpha = .05$.

13.17 $E(y) = 4 - x_1 + 2x_2 + x_1 x_2$

a. The resulting surface is three-dimensional. It would appear as a twisted plane similar to the one in Figure 13.9.

Model Building

b. For $x_1 = 2$, $E(y) = 4 - 2 + 2x_2 + 2x_2 = 2 + 4x_2$
    $x_1 = 3$, $E(y) = 4 - 3 + 2x_2 + 3x_2 = 1 + 5x_2$
    $x_1 = 4$, $E(y) = 4 - 4 + 2x_2 + 4x_2 = 6x_2$

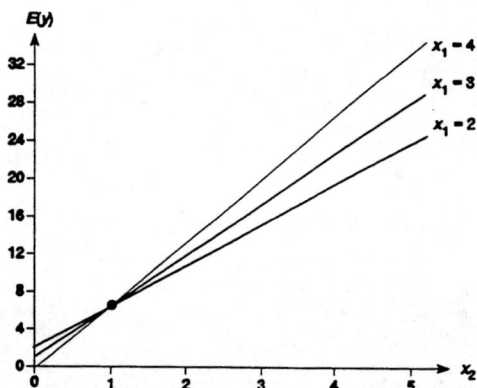

c. For $x_2 = 2$, $E(y) = 4 - x_1 + 4 + 2x_1 = 8 + x_1$
    $x_2 = 3$, $E(y) = 4 - x_1 + 6 + 3x_1 = 10 + 2x_1$
    $x_2 = 4$, $E(y) = 4 - x_1 + 8 + 4x_1 = 12 + 3x_1$

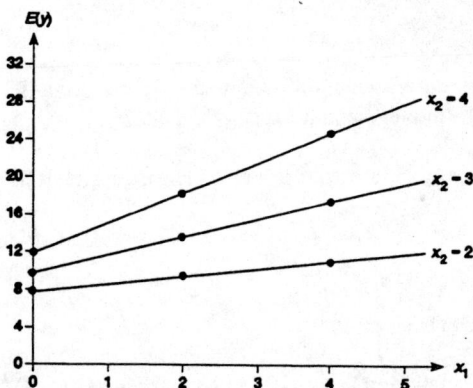

d. From the plot in part b, as x_2 increases from 0 to 5, $E(y)$ also increases. However, the rate of increase depends on the value of x_1. The rate of increase in $E(y)$ is smallest when $x_1 = 2$ and largest when $x_1 = 4$.

From the plot in part c, as x_1 increases from 0 to 5, $E(y)$ also increases. However, the rate of increase depends on the value of x_2. The rate of increase in $E(y)$ is smallest when $x_2 = 2$ and largest when $x_2 = 4$.

e. From the graph in b, when x_1 is 4 and x_2 is 1, $E(y) = 4 - 4 + 2(1) + 4(1) = 6$. When x_1 is 2 and x_2 is 2, $E(y) = 4 - 2 + 2(2) + 2(2) = 10$. So $E(y)$ changes from 6 to 10, an increase of 4 units.

In Exercise 13.16, when no interaction term was present, $E(y)$ increased as x_2 increased, but decreased as x_1 increased. With the interaction term present, within the given ranges of x_1 and x_2, $E(y)$ increases as x_1 increases and also increases as x_2 increases.

13.19   a.   Both independent variables are quantitative.

    b.   The first-order model for $E(y)$ is:

$$E(y) = \beta_0 + \beta_1 x_1 + \beta_2 x_2$$

    c.   The complete second-order model is:

$$E(y) = \beta_0 + \beta_1 x_1 + \beta_2 x_2 + \beta_3 x_1 x_2 + \beta_4 x_1^2 + \beta_5 x_2^2$$

    d.   The null and alternative hypotheses would be:

$$H_0:\ \beta_3 = 0$$
$$H_a:\ \beta_3 \neq 0$$

13.19   a.   Both independent variables are quantitative.

    b.   The first-order model for $E(y)$ is:

$$E(y) = \beta_0 + \beta_1 x_1 + \beta_2 x_2$$

    c.   The model is:

$$E(y) = \beta_0 + \beta_1 x_1 + \beta_2 x_2 + \beta_3 x_1 x_2$$

A typical graph might look like the following:

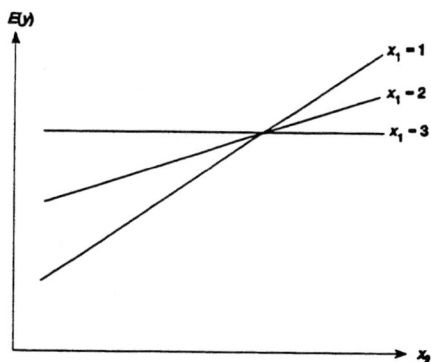

    d.   The complete second-order model is:

$$E(y) = \beta_0 + \beta_1 x_1 + \beta_2 x_2 + \beta_3 x_1 x_2 + \beta_4 x_1^2 + \beta_5 x_2^2$$

13.23   a. $E(y) = \beta_0 + \beta_1 x_1 + \beta_2 x_2 + \beta_3 x_1 x_2 + \beta_4 x_1^2 + \beta_5 x_2^2$

    b. $E(y) = \beta_0 + \beta_1 x_1 + \beta_2 x_2$

c. $E(y) = \beta_0 + \beta_1 x_1 + \beta_2 x_2 + \beta_3 x_1 x_2$

d. $\beta_1 + \beta_3 x_2$

e. $\beta_2 + \beta_3 x_1$

13.25 a. Some preliminary calculations are:

$$\sum x = 132 \qquad \sum x^2 = 796$$

$$\bar{x} = \frac{132}{24} = 5.5$$

$$s_x^2 = \frac{\sum x^2 - \dfrac{(\sum x)^2}{n}}{n - 1} = \frac{796 - \dfrac{132^2}{24}}{24 - 1} = 3.043478261$$

$$s_x = \sqrt{s_x^2} = 1.744556752$$

$$u = \frac{x - \bar{x}}{s_x} = \frac{x - 5.5}{1.745}$$

b. When
$$\begin{aligned}
x &= 3, u = -1.433 \\
x &= 4, u = -.860 \\
x &= 5, u = -.287 \\
x &= 6, u = .287 \\
x &= 7, u = .860 \\
x &= 8, u = 1.433
\end{aligned}$$

c. Some preliminary calculations are:

$$\sum x = 132 \qquad \sum x \cdot x^2 = \sum x^3 = 5148$$

$$\sum x^2 = 796 \qquad \sum (x^2)^2 = \sum x^4 = 35{,}020$$

$$SS_{xx} = \sum x^2 - \frac{(\sum x)^2}{n} = 796 - \frac{132^2}{24} = 70$$

$$SS_{xx^2} = \sum x \cdot x^2 - \frac{\sum x \sum x^2}{n} = 5148 - \frac{132(796)}{24} = 770$$

$$SS_{x^2x^2} = \sum (x^2)^2 - \frac{(\sum x^2)^2}{n} = 35{,}020 - \frac{796^2}{24} = 8619.333334$$

The coefficient of correlation, $r = \dfrac{SS_{xx^2}}{\sqrt{SS_{xx}SS_{x^2x^2}}} = \dfrac{770}{\sqrt{70(8619.333334)}} = .9913$

d. Some preliminary calculations are:

$$\sum u = 0 \qquad\qquad \sum u \cdot u^2 = \sum u^3 = 0$$

$$\sum u^2 = 23.003664 \qquad \sum \left(u^2\right)^2 = \sum u^4 = 38.16487908$$

$$SS_{uu} = \sum u^2 - \frac{\left(\sum u\right)^2}{n} = 23.003664 - \frac{0^2}{24} = 23.003664$$

$$SS_{uu^2} = \sum u \cdot u^2 - \frac{\sum u \sum u^2}{n} = 0 - \frac{0(23.003664)}{24} = 0$$

$$SS_{u^2 u^2} = \sum \left(u^2\right)^2 - \frac{\left(\sum u^2\right)^2}{n} = 38.16487908 - \frac{23.003664^2}{24} = 16.11618919$$

The coefficient of correlation, $r = \dfrac{SS_{uu^2}}{\sqrt{SS_{uu} SS_{u^2 u^2}}} = \dfrac{0}{\sqrt{23.003664(16.11618919)}} = 0$

The coefficient of correlation between x and x^2 (part c) is quite large (.9913). The coefficient of correlation between u and u^2 found above, is 0. Therefore, u and u^2 are not correlated.

e. $\hat{y} = 1079.55 - 262.13u - 196.67u^2$

13.27 To determine if the profitability of any major airline is related only to overall industry conditions but not to any unchanging features of that airline, we test:

H_0: $\beta_3 = \beta_4 = \cdots = \beta_{30} = 0$
H_a: At least one $\beta_i \neq 0$ ($i = 3, 4, \ldots, 30$)

The test statistic is $F = 3.59$ and the p-value is .0001.

Since the p-value is so small, H_0 is rejected for any value of $\alpha > .0001$. There is sufficient evidence to indicate that profitability is related to at least one unchanging feature of one of the airlines. Thus, the profitability hypothesis is not supported.

13.29 To determine if the second-order model contributes more information than the first-order model, we test:

H_0: $\beta_3 = \beta_4 = \beta_5 = 0$
H_a: At least one β_i is not 0 ($i = 3, 4, 5$)

The test statistic is $F = \dfrac{(SSE_R - SSE_C)/(k - g)}{SSE_C/[n - (k + 1)]} = \dfrac{(2094.4 - 159.94)/(5 - 2)}{159.94/[12 - (5 + 1)]} = 24.19$

The rejection region requires $\alpha = .05$ in the upper tail of the F distribution with $\nu_1 = k - g = 5 - 2 = 3$ and $\nu_2 = n - (k + 1) = 12 - (5 + 1) = 6$. From Table 10, Appendix II, $F_{.05} = 4.76$. The rejection region is $F > 4.76$.

Since the observed value of the test statistic falls in the rejection region ($F = 24.19 > 4.76$), H_0 is rejected. The conclusion is yes, there is sufficient evidence to indicate that the second-order model does contribute more information for the prediction of the signal-to-noise ratio, y, than the first-order model at $\alpha = .05$.

13.31 a. To compare the models 1 and 2, we test:

$$H_0: \beta_4 = \beta_5 = \beta_{18} = \beta_{19} = \beta_{20} = \beta_{21} = \beta_{22} = \beta_{23} = 0$$
$$H_a: \text{At least one of the } \beta_i \neq 0 \text{ for } i = 4, 5, 18, 19, 20, 21, 22, 23$$

b. The test statistic is $F = \dfrac{(SSE_R - SSE_C)/(k - g)}{SSE_C/[n - (k + 1)]} = \dfrac{(227,520 - 203,570)/(23 - 15)}{203,570/[132 - (23 + 1)]}$

$$= 1.59$$

The rejection region requires $\alpha = .05$ in the upper tail of the F distribution with numerator df $= k - g = 23 - 15 = 8$ and denominator df $= n - (k + 1) = 132 - (23 + 1) = 108$. From Table 10, Appendix II, $F_{.05} \approx 2.08$. The rejection region is $F > 2.08$.

Since the observed value of the test statistic does not fall in the rejection region ($F = 1.59 \not> 2.08$), H_0 is not rejected. There is insufficient evidence to indicate the quadratic terms add to the model at $\alpha = .05$.

c. No. there is no indication that they contribute information to the model.

d. To compare the models 2 and 3, we test:

$$H_0: \beta_7 = \beta_8 = \beta_9 = \beta_{10} = \beta_{11} = \beta_{12} = \beta_{13} = \beta_{14} = \beta_{15} = 0$$
$$H_a: \text{At least one of the } \beta_i \neq 0 \text{ for } i = 7, 8, 9, \ldots, 15$$

e. The test statistic is $F = \dfrac{(SSE_R - SSE_C)/(k - g)}{SSE_C/[n - (k + 1)]} = \dfrac{(395,165 - 227,520)/(15 - 6)}{227,520/[132 - (15 + 1)]}$

$$= 9.50$$

The rejection region requires $\alpha = .05$ in the upper tail of the F distribution with numerator df $= k - g = 15 - 6 = 9$ and denominator df $= n - (k + 1) = 132 - (15 + 1) = 116$. From Table 10, Appendix II, $F_{.05} \approx 1.96$. The rejection region is $F > 1.96$.

Since the observed value of the test statistic falls in the rejection region ($F = 9.50 > 1.96$), H_0 is rejected. There is sufficient evidence to indicate at least one of the interaction terms adds to the model at $\alpha = .05$.

f. Based on the results of the above two tests, we would recommend using model 2.

g. We would use model 2 with no terms involving x_3. The reduced model would be:

$$E(y) = \beta_0 + \beta_1 x_1 + \beta_2 x_2 + \beta_3 x_1 x_2 + \beta_4 x_4 + \beta_5 x_1 x_4 + \beta_6 x_2 x_4 + \beta_7 x_1 x_2 x_4$$

13.33 **a.** If the qualitative variable is at level 1, then $x_1 = x_2 = x_3 = 0$. The least squares prediction line is $\hat{y} = 10.2$.

If the qualitative variable is at level 2, then $x_1 = 1$ and $x_2 = x_3 = 0$. The least squares prediction line is $y = 10.2 - 4(1) = 6.2$.

If the qualitative variable is at level 3, then $x_2 = 1$ and $x_1 = x_3 = 0$. The least squares prediction line is $\hat{y} = 10.2 + 12(1) = 22.2$.

If the qualitative variable is at level 4, then $x_3 = 1$ and $x_1 = x_2 = 0$. The least squares prediction line is $\hat{y} = 10.2 + 2(1) = 12.2$.

b. The null and alternative hypotheses would be:

$$H_0: \beta_1 = \beta_2 = \beta_3 = 0$$
$$H_a: \text{At least one } \beta_i \neq 0, \; i = 1, 2, 3$$

13.35 **a.** Group is the qualitative independent variable. It must be coded into 2 dummy variables since it has 3 levels.

b. $E(y) = \beta_0 + \beta_1 x_1 + \beta_2 x_2$

where $\quad x_1 = \begin{cases} 1 & \text{if group 2} \\ 0 & \text{otherwise} \end{cases} \qquad x_2 = \begin{cases} 1 & \text{if group 3} \\ 0 & \text{otherwise} \end{cases}$

c. $\beta_0 = \mu_1 \qquad$ mean milk production of cows in group 1 (man-made shade structure)

$\beta_1 = \mu_2 - \mu_1 \quad$ difference in mean milk production between cows in group 2 and group 1 (tree shade—man-made shade structure)

$\beta_2 = \mu_3 - \mu_1 \quad$ difference in mean milk production between cows in group 3 and group 1 (no shade—man-made shade structure)

13.37 **a.** The first-order, main effects model for $E(y)$ is

$$E(y) = \beta_0 + \beta_1 x_1 + \beta_2 x_3 + \beta_3 x_4 + \beta_4 x_5 + \beta_5 x_6$$

where $\quad x_3 = \begin{cases} 1 & \text{if benezene} \\ 0 & \text{otherwise} \end{cases} \qquad x_4 = \begin{cases} 1 & \text{if toluene} \\ 0 & \text{otherwise} \end{cases}$

$\qquad\qquad x_5 = \begin{cases} 1 & \text{if chloroform} \\ 0 & \text{otherwise} \end{cases} \qquad x_6 = \begin{cases} 1 & \text{if methanol} \\ 0 & \text{otherwise} \end{cases}$

A sketch of the model might look like:

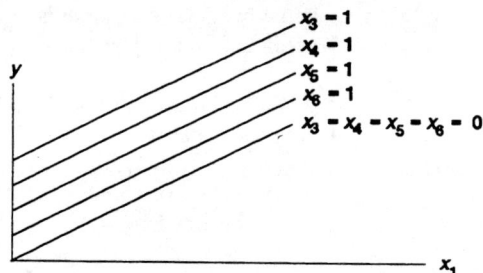

b. β_0 = y-intercept for benzene.

β_1 = change in mean retention coefficient for each unit change in temperature, holding organic compound constant

β_2 = difference in mean retention coefficient between benzene and anisole for a fixed temperature.

β_3 = difference in mean retention coefficient between toluene and anisole for a fixed temperature.

β_4 = difference in mean retention coefficient between chloroform and anisole for a fixed temperature.

β_5 = difference in mean retention coefficient between methanol and anisole for a fixed temperature.

c.   The model for $E(y)$ is

$$E(y) = \beta_0 + \beta_1 x_2 + \beta_2 x_3 + \beta_3 x_4 + \beta_4 x_5 + \beta_5 x_6 + \beta_6 x_2 x_3 + \beta_7 x_2 x_4 + \beta_8 x_2 x_5 + \beta_9 x_2 x_6$$

where x_3, x_4, x_5 and x_6 are defined above.

A sketch of the model might look like:

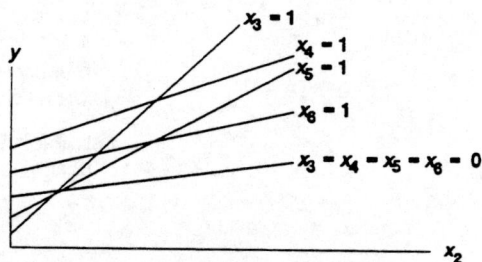

d. | **Organic Compound** | **Slope** |
|---|---|
| benzene | $\beta_1 + \beta_6$ |
| toluene | $\beta_1 + \beta_7$ |
| chloroform | $\beta_1 + \beta_8$ |
| methanol | $\beta_1 + \beta_9$ |
| anisole | β_1 |

13.39 a. Let $x_2 = \begin{cases} 1 & \text{if method} = G \\ 0 & \text{otherwise} \end{cases}$ Let $x_3 = \begin{cases} 1 & \text{if method } R_1 \\ 0 & \text{otherwise} \end{cases}$

The first-order, main effects model would be:

$$E(y) = \beta_0 + \beta_1 x_1 + \beta_2 x_2 + \beta_3 x_3$$

b. β_0 = y-intercept for method R_2

β_1 = change in mean estimated shelf life for each unit change in drug potency, with method held constant

β_2 = change in estimated shelf life between methods G and R_2

β_3 = difference in estimated shelf life between methods R_1 and R_2

c. The model allowing the slopes to differ is:

$$E(y) = \beta_0 + \beta_1 x_1 + \beta_2 x_2 + \beta_3 x_3 + \beta_4 x_1 x_2 + \beta_5 x_1 x_3$$

d. For method G, the slope of the line is $\beta_1 + \beta_4$
For method R_1, the slope of the line is $\beta_1 + \beta_5$
For method R_2, the slope of the line is β_1

13.41 a. If $x_2 = 0$, $x_3 = 0 \Rightarrow E(y) = \beta_0 + \beta_1 x_1$
If $x_2 = 1$, $x_3 = 0 \Rightarrow E(y) = \beta_0 + \beta_1 x_1 + \beta_2$
If $x_2 = 0$, $x_3 = 1 \Rightarrow E(y) = \beta_0 + \beta_1 x_1 + \beta_3$

To determine if the mean compressive strength $E(y)$ differs for the 3 cement mixes, we test:

H_0: $\beta_2 = \beta_3 = 0$
H_a: At least one β (β_2 or β_3) is not 0

b. The test statistic is $F = \dfrac{(SSE_R - SSE_C)/(k - g)}{SSE_C/[n - (k + 1)]} = \dfrac{(183.2 - 140.5)/(3 - 1)}{140.5/[50 - (3 + 1)]} = 6.99$

The rejection region requires $\alpha = .05$ in the upper tail of the F distribution with $\nu_1 = n - g = 3 - 1 = 2$ and $\nu_2 = n - (k + 1) = 50 - (3 + 1) = 46$. From Table 10, Appendix II, $F_{.05} \approx 3.23$. The rejection region is $F > 3.23$.

Since the observed value of the test statistic falls in the rejection region ($F = 6.99 >$ 3.23), H_0 is rejected. There is sufficient evidence to indicate the mean compressive strength, y, differs for the 3 cement mixes at $\alpha = .05$.

c. To test if the slope of the linear relationship between mean compressive strength $E(y)$ and hardening time x_1 varies according to type of cement mix, fit the model stated earlier with the interaction terms added for hardening time and type of cement mix and test if the interaction terms can be dropped from the model. Therefore, fit the model $E(y) = \beta_0 + \beta_1 x_1 + \beta_2 x_2 + \beta_3 x_3 + \beta_4 x_1 x_2 + \beta_5 x_1 x_3$

If $x_2 = 0$, $x_3 = 0 \Rightarrow E(y) = \beta_0 + \beta_1 x_1$
If $x_2 = 1$, $x_3 = 0 \Rightarrow E(y) = \beta_0 + (\beta_1 + \beta_4)x_1 + \beta_2$
If $x_2 = 0$, $x_3 = 1 \Rightarrow E(y) = \beta_0 + (\beta_1 + \beta_5)x_1 + \beta_3$

To determine if the slope varies according to type of cement mix, we test:

H_0: $\beta_4 = \beta_5 = 0$
H_a: At least one β (β_4 or β_5) is not 0

13.43 a. $E(y) = \beta_0 + \beta_1 x_1 + \beta_2 x_2 + \beta_3 x_3$ (no interaction)

b. $E(y) = \beta_0 + \beta_1 x_1 + \beta_2 x_2 + \beta_3 x_3 + \beta_4 x_1 x_2 + \beta_5 x_1 x_3$

c. TDS-3A waste: $\beta_1 + \beta_4$
FW waste: $\beta_1 + \beta_5$
AL waste: β_1

d. To test for the presence of temperature-waste type interaction, test if the interaction terms are needed in the model.

H_0: $\beta_4 = \beta_5 = 0$
H_a: At least one β (β_4 or β_5) is not 0

To perform this test, we would first fit the complete model (model given in b) and then fit the reduced model (model give in part a). By comparing the two models, we could decide whether to reject H_0 or not.

13.45 a. To determine whether the rate of increase of emotional distress with experience is different for the two groups of firefighters, we test:

H_0: $\beta_4 = \beta_5 = 0$
H_a: At least one $\beta_i \neq 0$, $i = 4, 5$

b. To test if there are differences in mean emotional distress levels that are attributable to exposure group, we test:

H_0: $\beta_3 = \beta_4 = \beta_5 = 0$
H_a: At least one $\beta_i \neq 0$, $i = 3, 4, 5$

To test the hypothesis in part **b**, the test statistic is

$$F = \frac{(SSE_R - SSE_C)/(k - g)}{SSE_C/[n - (k + 1)]} = \frac{(795.23 - 783.90/(5 - 2)}{783.90/[200 - (5 + 1)]} = .935$$

The rejection region requires $\alpha = .05$ in the upper tail of the F distribution with numerator df $= k - g = 5 - 2 = 3$ and denominator df $= n - (k + 1) = 200 - (5 + 1) = 194$. From Table 10, Appendix II, $F_{.05} \approx 2.68$. The rejection region is $F > 2.68$.

Since the observed value of the test statistic does not fall in the rejection region ($F = .935 \not> 2.68$), H_0 cannot be rejected. There is insufficient evidence to indicate a difference exists in mean emotional distress levels for the exposure groups at $\alpha = .05$.

13.47   a. $E(y) = \beta_0 + \beta_1 x_1 + \beta_2 x_1^2 + \beta_3 x_2 + \beta_4 x_3 + \beta_5 x_1 x_2 + \beta_6 x_1 x_3 + \beta_7 x_1^2 x_2 + \beta_8 x_1^2 x_3$

       b.   To determine if the three response curves have the same shape but different y-intercepts, test if the interaction terms can be dropped from the model:

           H_0: $\beta_5 = \beta_6 = \beta_7 = \beta_8 = 0$
           H_a: At least one β_i is not 0, $i = 5, 6, 7, 8$

13.49   To test if the second-order (x_1^2) term is necessary, we test:

           H_0: $\beta_2 = 0$
           H_a: $\beta_2 \neq 0$

The test statistic is $F = \dfrac{(SSE_R - SSE_C)/(k - g)}{SSE_C/[n - (k + 1)]} = \dfrac{(182 - 128)/(3 - 2)}{128/[20 - (3 + 1)]} = 6.75$

The rejection region requires $\alpha = .05$ in the upper tail of the F distribution with $\nu_1 = k - g = 3 - 2 = 1$ and $\nu_2 = n - (n + 1) = 20 - (3 + 1) = 16$. From Table 10, Appendix II, $F_{.05} = 4.49$. The rejection region is $F > 4.49$.

Since the observed value of the test statistic falls in the rejection region ($F = 6.71 > 4.49$), H_0 is rejected. The conclusion is yes, there is sufficient evidence to conclude that the second-order term is necessary in the model at $\alpha = .05$.

13.51   a.   According to the SAS printout, the independent variables x_4, x_5, and x_6 should be used in the model. When x_3 is entered in step 4, the F value is not significant, so it is taken back out of the model in step 5. Therefore, only x_4, x_5, and x_6 should be included.

       b.   No, there may be other important variables that are not specified. We would be unable to assume all the important independent variables are specified.

       c. $E(y) = \beta_0 + \beta_1 x_4 + \beta_2 x_5 + \beta_3 x_6 + \beta_4 x_4 x_5 + \beta_5 x_4 x_6 + \beta_6 x_5 x_6$

d. To determine if the model in part c is better than the first-order model, we test:

H_0: $\beta_4 = \beta_5 = \beta_6 = 0$
H_a: At least one β_i is not 0, $i = 4, 5, 6$

If we fail to reject H_0, there is insufficient evidence to conclude the model in c is better than the first-order model.

e. The biologist might want to see if he can find any more variables that might be important in this model. There might be a variable he has not yet considered. That would increase the fit of the model. Also, squared terms or interaction terms involving current variables could be added to the model.

13.53 If only 3 data points are used to fit a second-order model, SSE = 0 and df for error = $n - (k + 1) = 3 - (2 + 1) = 0$. Thus, there is no estimate for σ^2. With no estimate for σ^2, no tests can be performed.

13.55 a. To test if the quadratic terms are useful, we test:

H_0: $\beta_2 = \beta_3 = 0$
H_a: At least one β (β_2 or β_3) is not 0

b. To determine if there is a difference between the mean delivery times by rail and truck, we test:

H_0: $\beta_3 = \beta_4 = \beta_5 = 0$
H_a: At least one β (β_3, β_4, or β_5) is not 0

c. To determine if the mean delivery time differs for rail and truck deliveries, we test:

H_0: $\beta_3 = \beta_4 = \beta_5 = 0$
H_a: At least one β_i is not 0, $i = 3, 4, 5$

The test statistic is $F = \dfrac{(SSE_R - SSE_C)/(k - g)}{SSE_C/[n - (k + 1)]} = \dfrac{(259.34 - 226.12)/(5 - 2)}{226.12/[50 - (5 + 1)]}$
$$= 2.15$$

The rejection region requires $\alpha = .05$ in the upper tail of the F distribution with $\nu_1 = k - g = 5 - 2 = 3$ and $\nu_2 = n - (k + 1) = 50 - (5 + 1) = 44$. From Table 10, Appendix II, $F_{.05} \approx 2.84$. The rejection region is $F > 2.84$.

Since the observed value of the test statistic does not fall in the rejection region ($F = 2.15 \not> 2.84$), H_0 is not rejected. There is insufficient evidence to indicate the mean delivery times differ for rail and truck deliveries at $\alpha = .05$.

13.57 a. $E(y) = \beta_0 + \beta_1 x_1 + \beta_2 x_1^2 + \beta_3 x_2 + \beta_4 x_1 x_2 + \beta_5 x_1^2 x_2$

b.

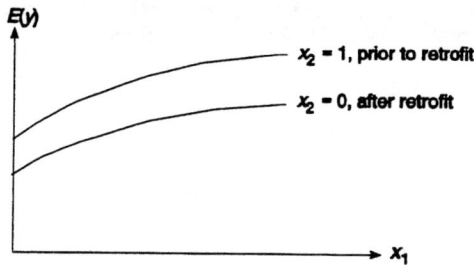

If $x_2 = 1$, $E(y) = \beta_0 + \beta_1 x_1 + \beta_2 x_1^2 + \beta_3 + \beta_4 x_1 + \beta_5 x_1^2$

$\qquad\qquad = (\beta_0 + \beta_3) + (\beta_1 + \beta_4)x_1 + (\beta_2 + \beta_5)x_1^2$

If $x_2 = 0$, $E(y) = \beta_0 + \beta_1 x_1 + \beta_2 x_1^2$

The two response curves have different shapes and different y-intercepts.

c. $E(y) = \beta_0 + \beta_1 x_1 + \beta_2 x_2$

d.

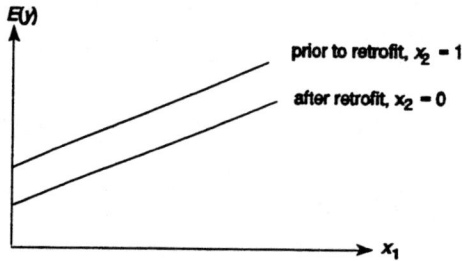

If $x_2 = 1$, $E(y) = \beta_0 + \beta_1 x_1 + \beta_2 = (\beta_0 + \beta_2) + \beta_1 x_1$

If $x_2 = 0$, $E(y) = \beta_0 + \beta_1 x_1$

The slopes of the 2 contour lines are the same, but they have different y-intercepts.

13.59 a. $E(y) = \beta_0 + \beta_1 x_1 + \beta_2 x_2 + \beta_3 x_3$

where $x_2 = \begin{cases} 1 & \text{if program B} \\ 0 & \text{otherwise} \end{cases}$ $\qquad$ $x_3 = \begin{cases} 1 & \text{if program C} \\ 0 & \text{otherwise} \end{cases}$

b. If $x_2 = 0$, $x_3 = 0 \Rightarrow E(y) = \beta_0 + \beta_1 x_1$
If $x_2 = 1$, $x_3 = 0 \Rightarrow E(y) = \beta_0 + \beta_1 x_1 + \beta_2$
If $x_2 = 0$, $x_3 = 1 \Rightarrow E(y) = \beta_0 + \beta_1 x_1 + \beta_3$

To determine whether mean work-hours differ for the 3 safety programs, we test:

H_0: $\beta_2 = \beta_3 = 0$
H_a: At least one β (β_2 or β_3) is not 0

13.61 a. Using SAS, the stepwise regression output is given below. Since no α value was given, $\alpha = .15$ was used for both entry into the model and to stay in the model.

STEPWISE REGRESSION PROCEDURE FOR DEPENDENT VARIABLE Y

NOTE: SLENTRY AND SLSTAY HAVE BEEN SET TO
      .15 FOR THE STEPWISE TECHNIQUE.

STEP 1   VARIABLE X5 ENTERED   R SQUARE = 0.34494362
                                         C(P) =   18.77496123

	DF	SUM OF SQUARES	MEAN SQUARE	F	PROB>F
REGRESSION	1	4247.12255742	4247.1225574	26.33	0.0001
ERROR	50	8065.38975028	161.3077950		
TOTAL	51	12312.51230769			

	B VALUE	STD ERROR	TYPE II SS	F	PROB>F
INTERCEPT	88.19303012				
X5	0.04914227	0.00957713	4247.1225574	26.33	0.0001

BOUNDS ON CONDITION NUMBER: 1,   1

STEP 2   VARIABLE X2 ENTERED   R SQUARE = 0.43626225
                                         C(P) =   11.46614765

	DF	SUM OF SQUARES	MEAN SQUARE	F	PROB>F
REGRESSION	2	5371.48426800	2685.7421340	18.96	0.0001
ERROR	49	6941.02803970	141.6536335		
TOTAL	51	12312.51230769			

	B VALUE	STD ERROR	TYPE II SS	F	PROB>F
INTERCEPT	75.96639260				
X2	0.13176422	0.04676903	1124.3617106	7.94	0.0070
X5	0.04954409	0.00897587	4315.7712404	30.47	0.0001

BOUNDS ON CONDITION NUMBER: 1.000253,   4.00101

STEP 3   VARIABLE X4 ENTERED   R SQUARE = 0.48058428
                                         C(P) =   8.94806006

	DF	SUM OF SQUARES	MEAN SQUARE	F	PROB>F
REGRESSION	3	5917.19985762	1972.3999525	14.80	0.0001
ERROR	48	6395.31245007	133.2356760		
TOTAL	51	12312.51230769			

	B VALUE	STD ERROR	TYPE II SS	F	PROB>F
INTERCEPT	77.72563951				
X2	0.13626447	0.04541256	1199.5926508	9.00	0.0043
X4	-0.03468898	0.01714031	545.7155896	4.10	0.0486
X5	0.05826780	0.00971385	4793.9636264	35.98	0.0001

BOUNDS ON CONDITION NUMBER: 1.246889,   10.48516

```
STEP 4     VARIABLE X6 ENTERED          R SQUARE = 0.51820132
                                        C(P) =    7.11346468

                DF    SUM OF SQUARES   MEAN SQUARE      F    PROB>F

REGRESSION      4     6380.36008577   1595.0900214   12.64   0.0001
ERROR          47     5932.15222192    126.2160047
TOTAL          51    12312.51230769

                B VALUE     STD ERROR    TYPE II SS      F    PROB>F

INTERCEPT   70.44909662
X2           0.10212354   0.04765800    579.5547117    4.59   0.0373
X4          -0.03397824   0.01668679    523.3238499    4.15   0.0474
X5           0.05074647   0.01023736   3101.3436075   24.57   0.0001
X6           0.25225652   0.13168429    463.1602281    3.67   0.0615

BOUNDS ON CONDITION NUMBER:     1.460313,    20.94023
----------------------------------------------------------------

STEP 5     VARIABLE X3 ENTERED          R SQUARE = 0.54497598
                                        C(P) =    6.38411606

                DF    SUM OF SQUARES   MEAN SQUARE      F    PROB>F

REGRESSION      5     6710.02348002   1342.0046960   11.02   0.0001
ERROR          46     5602.48882767    121.7932354
TOTAL          51    12312.51230769

                B VALUE     STD ERROR    TYPE II SS      F    PROB>F

INTERCEPT   68.27443064
X2           0.08308635   0.04822437    361.5342621    2.97   0.0916
X3           0.01386446   0.00842712    329.6633943    2.71   0.1067
X4          -0.04344508   0.01737245    761.6968264    6.25   0.0160
X5           0.04471142   0.01070453   2124.8325090   17.45   0.0001
X6           0.22909462   0.13012037    377.5394993    3.10   0.0849

BOUNDS ON CONDITION NUMBER:     1.674835,    36.72629
----------------------------------------------------------------

NO OTHER VARIABLES MET THE 0.1500 SIGNIFICANCE LEVEL FOR ENTRY

SUMMARY OF STEPWISE REGRESSION PROCEDURE FOR DEPENDENT VARIABLE Y

              VARIABLE         NUMBER   PARTIAL   MODEL
STEP    ENTERED    REMOVED       IN      R**2     R**2      C(P)

  1     X5                        1     0.3449   0.3449   18.7750
  2     X2                        2     0.0913   0.4363   11.4661
  3     X4                        3     0.0443   0.4806    8.9481
  4     X6                        4     0.0376   0.5182    7.1135
  5     X3                        5     0.0268   0.5450    6.3841

                   VARIABLE
        STEP    ENTERED    REMOVED         F        PROB>F

          1     X5                      26.3293     0.0001
          2     X2                       7.9374     0.0070
          3     X4                       4.0959     0.0486
          4     X6                       3.6696     0.0615
          5     X3                       2.7067     0.1067
```

From the output, the variables included in the model are x_2, x_3, x_4, x_5, and x_6.

b. $\hat{\beta}_0 = 68.274$ This is the y-intercept.

$\hat{\beta}_2 = 0.0831$ We estimate the mean number of hours worked per day will increase by .0831 hours for each additional money order or gift certificate sold, all other variables held constant.

$\hat{\beta}_3 = 0.0139$ We estimate the mean number of hours worked per day will increase by .0139 hours for each additional window payment transacted, all other variables held constant.

$\hat{\beta}_4 = -0.0434$ We estimate the mean number of hours worked per day will decrease by .0434 hours for each additional change order transaction processed, all other variables held constant.

$\hat{\beta}_5 = 0.0447$ We estimate the mean number of hours worked per day will increase by .0447 hours for each additional check cashed, all other variables held constant.

$\hat{\beta}_6 = 0.2291$ We estimate the mean number of hours worked per day will increase by .2291 hours for each additional piece of miscellaneous mail processed on "as available" basis, all other variables held constant.

c. The dangers of drawing inferences from this stepwise model are that this may not be the best model. An extremely large number of single parameter t tests have been conducted, and the probability is quite high that one or more errors have been made in including or excluding variables. That is, we may have either included some unimportant variables, or excluded some important variables. This is a screening process, and gives us a start. Also, we have not included any higher order terms or interaction terms. These variables should be considered in the next step of the process.

13.63 a. The model for this problem would be:

$$E(y) = \beta_0 + \beta_1 x_1 + \beta_2 x_2 + \beta_3 x_3 + \beta_4 x_4$$

where $x_1 = \begin{cases} 1 & \text{if variety B} \\ 0 & \text{otherwise} \end{cases}$ $x_2 = \begin{cases} 1 & \text{if variety C} \\ 0 & \text{otherwise} \end{cases}$

$x_3 = \begin{cases} 1 & \text{if variety D} \\ 0 & \text{otherwise} \end{cases}$ $x_4 = \begin{cases} 1 & \text{if variety E} \\ 0 & \text{otherwise} \end{cases}$

$\beta_0 = \mu_A$
$\beta_1 = \mu_B - \mu_A$
$\beta_2 = \mu_C - \mu_A$
$\beta_3 = \mu_D - \mu_A$
$\beta_4 = \mu_E - \mu_A$

where $\mu_A, \mu_B, \mu_C, \mu_D, \mu_E$ represent the mean yield in bushels of peas for variety i, i = A, B, C, D, or E.

b. The output from fitting the model using SAS is:

GENERAL LINEAR MODELS PROCEDURE

DEPENDENT VARIABLE: Y

SOURCE	DF	SUM OF SQUARES	MEAN SQUARE
MODEL	4	342.04000000	85.51000000
ERROR	15	53.52000000	3.56800000
CORRECTED TOTAL	19	395.56000000	

MODEL F =	23.97		PR > F = 0.0001

R-SQUARE	C.V.	ROOT MSE	Y MEAN
0.864698	7.3499	1.88891503	25.70000000

SOURCE	DF	TYPE I SS	F VALUE	PR > F
X1	1	52.81250000	14.80	0.0016
X2	1	243.90083333	68.36	0.0001
X3	1	0.20166667	0.06	0.8153
X4	1	45.12500000	12.65	0.0029

SOURCE	DF	TYPE III SS	F VALUE	PR > F
X1	1	29.64500000	8.31	0.0114
X2	1	85.80500000	24.05	0.0002
X3	1	14.04500000	3.94	0.0659
X4	1	45.12500000	12.65	0.0029

PARAMETER	ESTIMATE	T FOR H0: PARAMETER=0	PR > |T|	STD ERROR OF ESTIMATE
INTERCEPT	25.10000000	26.58	0.0001	0.94445752
X1	3.85000000	2.88	0.0114	1.33566463
X2	6.55000000	4.90	0.0002	1.33566463
X3	-2.65000000	-1.98	0.0659	1.33566463
X4	-4.75000000	-3.56	0.0029	1.33566463

The fitted model is:

$$\hat{y} = 25.1 + 3.85x_1 + 6.55x_2 - 2.65x_3 - 4.75x_4$$

c. To determine if the model is useful for predicting harvest yield, we test:

H_0: $\beta_1 = \beta_2 = \beta_3 = \beta_4 = 0$
H_a: At least one $\beta_i \neq 0$, $i = 1, 2, 3, 4$

The test statistic is $F = 23.97$ from printout.

The rejection region requires $\alpha = .05$ in the upper tail of the F distribution with $\nu_1 = k = 4$ and $\nu_2 = n - (k + 1) = 20 - (4 + 1) = 15$. From Table 10, Appendix II, $F_{.05} = 3.06$. The rejection region is $F > 3.06$.

Since the observed value of the test statistic falls in the rejection region $(F = 23.97 > 3.06)$, H_0 is rejected. There is sufficient evidence to indicate the model is useful for predicting harvest yield at $\alpha = .05$.

13.65 A stepwise regression was run using SAS and the following terms:

$$x_1 = \begin{cases} 1 & \text{if Bulb surface = D} \\ 0 & \text{otherwise} \end{cases} \qquad x_2 = \text{length of operation}$$

and the higher order terms x_2^2, x_1x_2, and $x_1x_2^2$

The results of the stepwise regression follow:

```
STEPWISE REGRESSION PROCEDURE FOR DEPENDENT VARIABLE Y

NOTE: SLENTRY AND SLSTAY HAVE BEEN SET TO
      .15 FOR THE STEPWISE TECHNIQUE.

STEP 1     VARIABLE X1 ENTERED          R SQUARE = 0.47398750
                                        C(P) =    453.12589074

               DF      SUM OF SQUARES    MEAN SQUARE       F   PROB>F

REGRESSION      1      1045.78571429    1045.7857143    10.81   0.0065
ERROR          12      1160.57142857      96.7142857
TOTAL          13      2206.35714286

               B VALUE     STD ERROR     TYPE II SS       F   PROB>F

INTERCEPT   24.42857143
X1         -17.28571429    5.25667700   1045.7857143    10.81   0.0065

BOUNDS ON CONDITION NUMBER:                1,            1
----------------------------------------------------------------

STEP 2     VARIABLE X2 ENTERED          R SQUARE = 0.85510214
                                        C(P) =    119.57482185

               DF      SUM OF SQUARES    MEAN SQUARE       F   PROB>F

REGRESSION      2      1886.66071429    943.33035714    32.46   0.0001
ERROR          11       319.69642857     29.06331169
TOTAL          13      2206.35714286

               B VALUE     STD ERROR     TYPE II SS       F   PROB>F

INTERCEPT   12.80357143
X1         -17.28571429    2.88163206   1045.7857143    35.98   0.0001
X2           0.00968750    0.00180102    840.8750000    28.93   0.0002

BOUNDS ON CONDITION NUMBER:                1,            4
----------------------------------------------------------------

STEP 3     VARIABLE X1X2 ENTERED        R SQUARE = 0.95482210
                                        C(P) =     33.77672209

               DF      SUM OF SQUARES    MEAN SQUARE       F   PROB>F

REGRESSION      3      2106.67857143    702.22619048    70.45   0.0001
ERROR          10        99.67857143      9.96785714
TOTAL          13      2206.35714286

               B VALUE     STD ERROR     TYPE II SS       F   PROB>F

INTERCEPT    6.85714286
X1          -5.39285714    3.04234570     31.32005495     3.14   0.1067
X2           0.01464286    0.00149163    960.57142857    96.37   0.0001
X1X2        -0.00991071    0.00210949    220.01785714    22.07   0.0008

BOUNDS ON CONDITION NUMBER:              4.25,          28.5
----------------------------------------------------------------
```

```
STEP 4      VARIABLE X2SQ ENTERED          R SQUARE = 0.98020601
                                           C(P) =    13.42755344

                DF      SUM OF SQUARES    MEAN SQUARE      F    PROB>F

REGRESSION      4       2162.68452381    540.67113095   111.42  0.0001
ERROR           9         43.67261905      4.85251323
TOTAL          13       2206.35714286

                B VALUE      STD ERROR     TYPE II SS      F    PROB>F

INTERCEPT     3.97023810
X1           -5.39285714     2.12271174    31.32005495    6.45  0.0317
X2            0.02330357     0.00275355   347.55612245   71.62  0.0001
X1X2         -0.00991071     0.00147184   220.01785714   45.34  0.0001
X2SQ         -0.00000361     0.00000106    56.00595238   11.54  0.0079

BOUNDS ON CONDITION NUMBER:                  14,       138
-----------------------------------------------------------------

STEP 5      VARIABLE X1X2SQ ENTERED        R SQUARE = 0.99091370
                                           C(P) =     6.00000000

                DF      SUM OF SQUARES    MEAN SQUARE      F    PROB>F

REGRESSION      5       2186.30952381    437.26190476   174.49  0.0001
ERROR           8         20.04761905      2.50595238
TOTAL          13       2206.35714286

                B VALUE      STD ERROR     TYPE II SS      F    PROB>F

INTERCEPT     2.09523810
X1           -1.64285714     1.95412234     1.77120536    0.71  0.4249
X2            0.02892857     0.00269662   288.39560440   115.08  0.0001
X1X2         -0.02116071     0.00381359    77.15521978   30.79  0.0005
X2SQ         -0.00000595     0.00000108    76.19047619   30.40  0.0006
X1X2SQ        0.00000469     0.00000153    23.62500000    9.43  0.0153

BOUNDS ON CONDITION NUMBER:                55.25,    763.3333
-----------------------------------------------------------------

STEP 6      VARIABLE X1 REMOVED            R SQUARE = 0.99011093
                                           C(P) =     4.70679929

                DF      SUM OF SQUARES    MEAN SQUARE      F    PROB>F

REGRESSION      4       2184.53831845    546.13457961   225.27  0.0001
ERROR           9         21.81882440      2.42431382
TOTAL          13       2206.35714286

                B VALUE      STD ERROR     TYPE II SS      F    PROB>F

INTERCEPT     1.27380952
X2            0.03017997     0.00221165   451.43178267   186.21  0.0001
X1X2         -0.02366350     0.00234435   247.00330014   101.89  0.0001
X2SQ         -0.00000635     0.00000095   107.87326991    44.50  0.0001
X1X2SQ        0.00000549     0.00000117    53.17384959    21.93  0.0011

BOUNDS ON CONDITION NUMBER:             24.42578,    340.0312
-----------------------------------------------------------------
```

NO OTHER VARIABLES MET THE 0.1500 SIGNIFICANCE LEVEL FOR ENTRY

SUMMARY OF STEPWISE REGRESSION PROCEDURE FOR DEPENDENT VARIABLE Y

STEP	VARIABLE ENTERED	REMOVED	NUMBER IN	PARTIAL R**2	MODEL R**2	C(P)
1	X1		1	0.4740	0.4740	453.126
2	X2		2	0.3811	0.8551	119.575
3	X1X2		3	0.0997	0.9548	33.777
4	X2SQ		4	0.0254	0.9802	13.428
5	X1X2SQ		5	0.0107	0.9909	6.000
6		X1	4	0.0008	0.9901	4.707

STEP	VARIABLE ENTERED	REMOVED	F	PROB>F
1	X1		10.8131	0.0065
2	X2		28.9325	0.0002
3	X1X2		22.0727	0.0079
4	X2SQ		11.5416	0.0079
5	X1X2SQ		9.4276	0.0153
6		X1	0.7068	0.4249

From the output, all terms are significant at the .05 level except x_1. However, x_1 is involved in higher order interaction terms, and thus must be included in the model as a location variable. The fitted model is:

$$\hat{y} = 2.0952 - 1.6429x_1 + .0289x_2 - .0212x_1x_2 - .00000595x_2^2 + .00000469x_1x_2^2$$

CHAPTER FOURTEEN

....

Analysis of Variance for Designed Experiments

14.1 The two factors that affect the quantity of information in an experiment are noise (variability) and volume (n).

14.3 **a.** This experiment involves a single factor, work scheduling, at three levels: flextime, staggered starting hours, and fixed hours. Since work scheduling is the only factor, these levels represent the treatments.

b. To assign treatments in a completely random manner, with equal numbers of workers in each treatment, number the 60 workers from 1 to 60. Use Table 6 in Appendix II to select two-digit numbers, discarding those that are larger than 60 or are identical, until there is a total of 40 two-digit numbers. The workers who have been assigned the first 20 numbers in the sequence are assigned to flextime, the second group of 20 workers are assigned to staggered starting times, and the remaining workers are assigned to fixed hours.

c. The linear model is $E(y) = \beta_0 + \beta_1 x_1 + \beta_2 x_2$

where $x_1 = \begin{cases} 1 & \text{if flextime} \\ 0 & \text{otherwise} \end{cases}$

and $x_2 = \begin{cases} 1 & \text{if staggered} \\ 0 & \text{otherwise} \end{cases}$

14.5 **a.** The model for the first observation for engineer B ($x_1 = 0$, $x_2 = 1$, $x_3 = 0$) is:

$$y_{B1} = \beta_0 + \beta_2 + \beta_4 + \epsilon_{B1}$$

The models for the rest of the observations for engineer B are:

$$y_{B2} = \beta_0 + \beta_2 + \beta_5 + \epsilon_{B2}$$
$$y_{B3} = \beta_0 + \beta_2 + \beta_6 + \epsilon_{B3}$$
$$y_{B4} = \beta_0 + \beta_2 + \beta_7 + \epsilon_{B4}$$
$$y_{B5} = \beta_0 + \beta_2 + \beta_8 + \epsilon_{B5}$$
$$y_{B6} = \beta_0 + \beta_2 + \beta_9 + \epsilon_{B6}$$
$$y_{B7} = \beta_0 + \beta_2 + \beta_{10} + \epsilon_{B7}$$
$$y_{B8} = \beta_0 + \beta_2 + \beta_{11} + \epsilon_{B8}$$
$$y_{B9} = \beta_0 + \beta_2 + \beta_{12} + \epsilon_{B9}$$
$$y_{B10} = \beta_0 + \beta_2 + \epsilon_{B10}$$

The average of the 10 observations for engineer B is:

$$\bar{y}_B = \beta_0 + \beta_2 + \frac{\beta_4 + \beta_5 + \cdots + \beta_{12}}{10} + \bar{\epsilon}_B$$

b.   The models for the observations for engineer D $(x_1 = 0, x_2 = 1, x_3 = 0)$ are:

$$y_{D1} = \beta_0 + \beta_4 + \epsilon_{D1}$$
$$y_{D2} = \beta_0 + \beta_5 + \epsilon_{D2}$$
$$y_{D3} = \beta_0 + \beta_6 + \epsilon_{D3}$$
$$y_{D4} = \beta_0 + \beta_7 + \epsilon_{D4}$$
$$y_{D5} = \beta_0 + \beta_8 + \epsilon_{D5}$$
$$y_{D6} = \beta_0 + \beta_9 + \epsilon_{D6}$$
$$y_{D7} = \beta_0 + \beta_{10} + \epsilon_{D7}$$
$$y_{D8} = \beta_0 + \beta_{11} + \epsilon_{D8}$$
$$y_{D9} = \beta_0 + \beta_{12} + \epsilon_{D9}$$
$$y_{D10} = \beta_0 + \epsilon_{D10}$$

The average of the 10 observations for engineer D is:

$$\bar{y}_D = \beta_0 + \frac{\beta_4 + \beta_5 + \cdots + \beta_{12}}{10} + \bar{\epsilon}_D$$

c. $$(\bar{y}_B - \bar{y}_D) = \left[\beta_0 + \beta_2 + \frac{\beta_4 + \cdots + \beta_{12}}{10} + \bar{\epsilon}_B \right] - \left[\beta_0 + \frac{\beta_4 + \cdots + \beta_{12}}{10} + \bar{\epsilon}_D \right]$$

$$= \beta_2 + (\bar{\epsilon}_B - \bar{\epsilon}_D)$$

14.7 a.   Since y will be observed for all factor-level combinations, this is a complete 3×3 factorial experiment.

b.   The factors are pay rate (quantitative) and length of workday (quantitative).

c.   The treatments will include all $3 \times 3 = 9$ factor-level combinations:

$$P_1L_1, P_1L_2, P_1L_3, P_2L_1, P_2L_2, P_2L_3, P_3L_1, P_3L_2, P_3L_3$$

14.9 a.   This is not a complete factorial experiment since the treatments do not include all $3 \times 3 = 9$ factor-level combinations.

b.   Interaction between two factors implies the effect of one factor on the dependent variable depends on the level of the second factor. There are 3 levels of factor A at B_1. However, there is only 1 level of A at B_2 and only 1 level of A at B_3. Thus, we cannot measure the effect of A at levels B_2 and B_3. Therefore, we cannot determine if the effect of A differs at different levels of B.

14.11 Let x_1 be the quantitative factor A,

$$x_2 = \begin{cases} 1 & \text{if factor } B \text{ at level 1} \\ 0 & \text{otherwise} \end{cases} \qquad x_3 = \begin{cases} 1 & \text{if factor } B \text{ at level 2} \\ 0 & \text{otherwise} \end{cases}$$

$$x_4 = \begin{cases} 1 & \text{if factor } C \text{ at level 1} \\ 0 & \text{otherwise} \end{cases} \qquad x_5 = \begin{cases} 1 & \text{if factor } C \text{ at level 2} \\ 0 & \text{otherwise} \end{cases}$$

The model is:

$$\begin{aligned} E(y) &= \beta_0 + \beta_1 x_1 + \beta_2 x_2 + \beta_3 x_3 + \beta_4 x_4 + \beta_5 x_5 + \beta_6 x_1 x_2 + \beta_7 x_1 x_3 + \beta_8 x_1 x_4 \\ &+ \beta_9 x_1 x_5 + \beta_{10} x_2 x_4 + \beta_{11} x_2 x_5 + \beta_{12} x_3 x_4 + \beta_{13} x_3 x_5 + \beta_{14} x_1 x_2 x_4 \\ &+ \beta_{15} x_1 x_2 x_5 + \beta_{16} x_1 x_3 x_4 + \beta_{17} x_1 x_3 x_5 \end{aligned}$$

14.13 If a randomized block design is used to investigate the effects of two qualitative factors, there will be only one observation per treatment. There will not be enough degrees of freedom (or observations) to test for interaction between the two factors.

14.15 To solve for the number of replicates, r, we want to solve the equation

$$t_{\alpha/2} \frac{s_{\hat{\beta}_3}}{\sqrt{r}} = B$$

We know the estimate of $s_{\hat{\beta}_3}$ is 3 and $B = 2$. For $\alpha = .10$, $\alpha/2 = .05$.

To determine the value of $t_{.05}$, we need to know the degrees of freedom. We know the degrees of freedom for t will be df (Error) $= n - 4 = 4r - 4 = 4(r - 1)$. At minimum, we require 2 replicates, so we will start with $r = 2$. Thus, with df $= 4$, $t_{.05} = 2.132$. Substituting into the formula, we get

$$2.132 \frac{3}{\sqrt{r}} = 2 \Rightarrow r = \frac{2.132^2(3^2)}{2^2} = 10.23$$

Since this number is quite a bit larger than the 2 replicates that we used to get the original t value, we will redo the problem using the $t_{.05}$ with df $= 4(r - 1) = 4(10 - 1) = 36$. This value is $t_{.05} \approx 1.69$.

$$1.690 \frac{3}{\sqrt{r}} = 2 \Rightarrow r = \frac{1.690^2(3^2)}{2^2} = 6.43 \approx 7$$

Using 7 as the number of replicates, df $= 4(r - 1) = 24$ and $t_{.05} = 1.711$.

$$1.711 \frac{3}{\sqrt{r}} = 2 \Rightarrow r = \frac{1.711^2(3^2)}{2^2} = 6.59 \approx 7$$

Thus, we should use either 6 or 7 replicates. To be conservative, we should use 7 replicates.

14.17 **a.** Since the intervals for box types B and D do not overlap, there is evidence that these two box types are significantly different.

    b. No. Since many of the intervals overlap, there is no evidence that all five box types are significantly different. The intervals for box types A, B, and E are entirely included in the interval for box type C. The interval for box type D and box type C also overlap.

14.19 **a.** This is a completely randomized design. The 45 subjects were randomly divided into three groups.

    b. The regression model is

$$E(y) = \beta_0 + \beta_1 x_1 + \beta_2 x_2$$

where $x_1 = \begin{cases} 1 & \text{if touch-tone} \\ 0 & \text{if not} \end{cases}$ $\quad x_2 = \begin{cases} 1 & \text{if human operator} \\ 0 & \text{if not} \end{cases}$

    c. To determine if the means of the three groups differ, we test:

H_0: $\mu_T = \mu_H = \mu_S$
H_a: At least two means differ

    d. To determine if the means of the three groups differ, we test:

H_0: $\beta_1 = \beta_2 = 0$
H_a: At least one $\beta_i \neq 0$, $i = 1, 2$

    e. In order to test the above hypotheses, we compare the sample variance between the groups to the sample variance within the groups. Even though the sample means are different, the sample variance within the groups is large compared to the sample variance between the groups. Thus, there is no evidence to indicate the population means of the three groups are different.

14.21 **a.** The linear model is:

$$E(y) = \beta_0 + \beta_1 x \qquad \text{where } x = \begin{cases} 1 & \text{if Treatment 1} \\ 0 & \text{otherwise} \end{cases}$$

    b. Some preliminary calculations are:

$$n = 13 \quad \sum x = 7 \quad \sum y = 128 \quad \sum xy = 64$$
$$\sum x^2 = 7 \quad \sum y^2 = 1294$$

$$SS_{xx} = 7 - \frac{7^2}{13} = 3.230769231$$

$$SS_{xy} = 64 - \frac{7(128)}{13} = -4.92307692$$

$$SS_{yy} = 1294 - \frac{128^2}{13} = 33.692308$$

$$\hat{\beta}_1 = \frac{SS_{xy}}{SS_{xx}} = \frac{-4.92307692}{3.230769231} = -1.523809523 \approx -1.524$$

$$\hat{\beta}_0 = \bar{y} - \hat{\beta}_1\bar{x} = \frac{128}{13} - (-1.523809523)\left[\frac{7}{13}\right] = 10.666666667 \approx 10.667$$

The fitted model is $\hat{y} = 10.667 - 1.524x$.

$$SSE = SS_{yy} - \hat{\beta}_1 SS_{xy} = 33.692308 - (-1.523809523)(-4.92307692) = 26.19047651$$

$$s^2 = \frac{SSE}{n-2} = \frac{26.19047651}{13-2} = 2.38095241 \qquad s = \sqrt{2.38095241} = 1.543$$

$$H_0: \beta_1 = 0$$
$$H_a: \beta_1 \neq 0$$

Test statistic: $t = \dfrac{\hat{\beta}_1}{s_{\hat{\beta}_1}} = \dfrac{-1.524}{\dfrac{1.543}{\sqrt{3.23077}}} = -1.775$

The rejection region requires $\alpha/2 = .05/2 = .025$ in each tail of the t distribution with df $= n - 2 = 11$. From Table 7, Appendix, II, $t_{.025} = 2.201$. The rejection region is $t < -2.201$ or $t > 2.201$.

Since the observed value of the test statistic does not fall in the rejection region ($t = -1.775 \not< -2.201$, H_0 is not rejected. There is insufficient evidence to indicate a difference in the treatment means at $\alpha = .05$.

Note: We could also test the hypothesis using the F test. To calculate the F statistic, we must first calculate r^2.

$$r^2 = 1 - \frac{SSE}{SS_{yy}} = 1 - \frac{26.1905}{33.6923} = .2227$$

Test statistic: $F = \dfrac{r^2/k}{(1-r^2)/[n-(k+1)]} = \dfrac{.2227/1}{(1-.2227)/[13-(1+1)]} = 3.15$

The rejection region requires $\alpha = .05$ in the upper tail of the F distribution with $\nu_1 = k = 1$, $\nu_2 = n - (k+1) = 11$. From Table 10, Appendix II, $F_{.05} = 4.84$. The rejection region is $F > 4.84$.

The conclusion is the same as using the t test.

. . . .
Analysis of Variance for Designed Experiments

271

14.23 **a.** Some preliminary calculations:

$$\bar{y}_1 = \frac{64}{7} = 9.143 \qquad\qquad \bar{y}_2 = \frac{64}{6} = 10.667$$

$$s_1^2 = \frac{596 - \dfrac{64^2}{7}}{7 - 1} = 1.8095 \qquad s_2^2 = \frac{698 - \dfrac{64^2}{6}}{6 - 1} = 3.0667$$

$$s_p^2 = \frac{(n_1 - 1)s_1^2 + (n_2 - 1)s_2^2}{n_1 + n_2 - 2} = \frac{(7 - 1)1.8095 + (6 - 1)3.0667}{7 + 6 - 2} = 2.3810$$

$H_0:\ \mu_1 = \mu_2$
$H_a:\ \mu_1 \neq \mu_2$

The test statistic is $t = \dfrac{\bar{y}_1 - \bar{y}_2}{\sqrt{s_p^2\left[\dfrac{1}{n_1} + \dfrac{1}{n_2}\right]}} = \dfrac{9.143 - 10.667}{\sqrt{2.381\left[\dfrac{1}{7} + \dfrac{1}{6}\right]}} = -1.775$

The rejection region requires $\alpha/2 = .05/2 = .025$ in each tail of the t distribution with df $(n_1 + n_2 - 2) = 11$. From Table 7, Appendix II, $t_{.025} = 2.201$. The rejection region is $t < -2.201$ or $t > 2.201$.

Since the observed value of the test statistic does not fall in the rejection region ($t = -1.775 \not< -2.201$), H_0 is not rejected. There is insufficient evidence to indicate a difference in the treatment means at $\alpha = .05$.

b. $t^2 = (-1.775)^2 = 3.15 = F$

c. The analysis of variance F test for comparing two population means is always a two-tailed test with the alternative hypothesis $H_a:\ \mu_1 \neq \mu_2$

14.25 **a.** To determine if the mean RoP differs for at least two of the three bits, we test:

$H_0:\ \mu_1 = \mu_2 = \mu_3$
$H_a:$ At least two of the treatment means differ

The test statistic is $F = 9.50$.

The rejection region requires $\alpha = .05$ in the upper tail of the F distribution with numerator df $= \nu_1 = p - 1 = 3 - 1 = 2$ and denominator df $= \nu_2 = n - p = 12 - 3 = 9$. From Table 10, Appendix II, $F_{.05} = 4.26$. The rejection region is $F > 4.26$.

Since the observed value of the test statistic falls in the rejection region ($F = 9.50 > 4.26$), H_0 is rejected. There is sufficient evidence to indicate the mean RoP differs for at least two of the three drill bits at $\alpha = .05$.

Also, since the p-value is less than α ($.006 < .05$), H_0 is rejected.

14.27 a. Some preliminary calculations are:

$$CM = \frac{\left(\sum y\right)^2}{n} = \frac{2.39^2}{24} = .2380$$

$$SS(\text{Total}) = \sum y^2 - CM = .2577 - .2380 = .0197$$

$$SST = \frac{T_1^2}{n_1} + \frac{T_2^2}{n_2} + \frac{T_3^2}{n_3} + \frac{T_4^2}{n_4} - CM$$

$$= \frac{.49^2}{6} + \frac{.66^2}{6} + \frac{.73^2}{6} + \frac{.51^2}{6} - .2380 = .24478 - .23800 = .00678$$

$$SSE = SS(\text{Total}) - SST = .01970 - .00678 = .01292$$

$$MST = \frac{SST}{p-1} = \frac{.00678}{4-1} = .00226 \quad MSE = \frac{SSE}{n-p} = \frac{.01292}{24-4} = .00065$$

$$F = \frac{MST}{MSE} = \frac{.00226}{.00065} = 3.48$$

Source	df	SS	MS	F
Location	3	.00678	.00226	3.48
Error	20	.01292	.00065	
Total	23	.01970		

b. To determine if differences in the mean ozone content among the four locations exist, we test:

H_0: $\mu_1 = \mu_2 = \mu_3 = \mu_4$
H_a: At least two of the treatment means differ

The test statistic is $F = 3.48$.

The rejection region requires $\alpha = .05$ in the upper tail of the F distribution with numerator df $= \nu_1 = p - 1 = 4 - 1 = 3$ and denominator df $= \nu_2 = n - p = 24 - 4 = 20$. From Table 10, Appendix II, $F_{.05} = 3.10$. The rejection region is $F > 3.10$.

Since the observed value of the test statistic falls in the rejection region ($F = 3.48 > 3.10$), H_0 is rejected. There is sufficient evidence to indicate that the mean ozone contents of at least two locations differ $\alpha = .05$.

c. Recall the formula for computing the sample sizes to estimate the difference between two means is:

$$n_1 = n_2 = \left[\frac{z_{\alpha/2}}{H}\right]^2 \left(\sigma_1^2 + \sigma_2^2\right)$$

The variances, σ_1^2 and σ_2^2, can be estimated with $s^2 = MSE = .00065$ and $z_{\alpha/2}$ necessary to give a confidence coefficient of .95 is 1.96. The necessary sample sizes are:

$$n_1 = n_2 = \left[\frac{1.96}{.01}\right]^2 (.00065 + .00065) = 49.9408 \approx 50$$

14.29 a. To find the totals for each sample, we multiply the sample mean by the sample size.

Species	T_i	n_i
Northern white cedar	102,480	28
Western red cedar	2,147,850	387
Pacific silver fir	558,260	103
Coastal Douglas fir	988,840	118
Southern pine	1,303,890	147

$$\sum y_i = 5,101,320 \qquad \sum n_i = n = 783$$

$$CM = \frac{\left(\sum y_i\right)^2}{n} = \frac{(5,101,320)^2}{783} = 33,235,588,432$$

$$SST = \frac{T_1^2}{n_1} + \frac{T_2^2}{n_2} + \cdots + \frac{T_5^2}{n_5} - CM$$

$$= \frac{(102,480)^2}{28} + \frac{(2,147,850)^2}{387} + \cdots + \frac{(1,303,890)^2}{147} - 33,235,588,432$$

$$= 1,937,808,567.8$$

b. $SSE = (n_1 - 1)s_1^2 + (n_2 - 1)s_2^2 + \cdots + (n_5 - 1)s_5^2$
$= (28 - 1)(203.33)^2 + (387 - 1)(298.39)^2 + \cdots + (147 - 1)(611.72)^2$
$= 148,746,738.4$

c. $SS(Total) = SSE + SST = 148,746,738.4 + 1,937,808,567.8 = 2,086,555,306.2$

d.

Source	df	SS	MS	F
Treatments	4	1,937,808,567.8	484,452,141.95	2,533.86
Error	778	148,746,738.4	191,191.18	
Total	782	2,086,555,306.2		

e. To test if there are differences in the treatment means, we test:

H_0: $\mu_1 = \mu_2 = \mu_3 = \mu_4 = \mu_5$
H_a: At least two means differ

The test statistic is $F = 2,533.86$.

The rejection region requires $\alpha = .05$ in the upper tail of the F distribution with numerator df $= (p - 1) = 5 - 1 = 4$ and denominator df $= (n - p) = 783 - 5 = 778$. From Table 10, Appendix II, $F_{.05} \approx 2.37$. The rejection region is $F > 2.37$.

Since the observed value of the test statistic falls in the rejection region ($F = 2,533.86 > 2.37$), H_0 can be rejected. There is sufficient evidence to indicate the means differ at $\alpha = .05$.

f. The form of the confidence interval is:

$$\bar{T}_1 \pm t_{\alpha/2} \left[\frac{s}{\sqrt{n_i}} \right]$$

$$\bar{T}_1 = \frac{T_1}{n_1} = \frac{104,480}{28} = 3,660$$

$$s = \sqrt{MSE} = \sqrt{191,191.18} = 437.254$$

For confidence coefficient .90, $\alpha = .10$ and $\alpha/2 = .10/2 = .05$. From Table 7, Appendix II, $t_{.05} \approx 1.645$ with df $= n - p = 783 - 5 = 778$. The 90% confidence interval is

$$3,660 \pm 1.645 \frac{(437.254)}{\sqrt{28}} \Rightarrow 3,660 \pm 135.94 \Rightarrow (3524.06, 3795.94)$$

g. The form of the confidence interval is:

$$\left(\bar{T}_1 - \bar{T}_5 \right) \pm t_{\alpha/2} s \sqrt{\frac{1}{n_i} + \frac{1}{n_j}}$$

$$\bar{T}_1 = 3,660 \quad \text{(from part f)}$$

$$\bar{T}_5 = \frac{1,303,890}{147} = 8,870$$

The 90% confidence interval is

$$(3,660 - 8,870) \pm 1.645(437.254) \sqrt{\frac{1}{28} + \frac{1}{147}}$$

$$\Rightarrow -5,210 \pm 148.31 \Rightarrow (-5358.31, -5061.69)$$

14.31 a. Since we wish to compare the mean medial rotation measurements of the three positions, the treatments in this experiment are the three positions (beginning, direction change, ending).

b. The blocks in this experiment are the ten college students.

c.   The response variable is the medial rotation measurement.

d.   In this experiment, only one factor (the three positions) is really of interest. However, we recognize that different students will, inherently, cause variation in the medial rotation measurements. A randomized block design allows us to account for the variation caused by the students.

e.   The appropriate model is

$$E(y) = \beta_0 + \beta_1 x_1 + \beta_2 x_2 + \beta_3 x_3 + \cdots + \beta_{11} x_{11}$$

where $x_1 = \begin{cases} 1 & \text{if beginning position} \\ 0 & \text{otherwise} \end{cases}$ $x_2 = \begin{cases} 1 & \text{if direction change} \\ 0 & \text{otherwise} \end{cases}$

$x_3 = \begin{cases} 1 & \text{if student \#1} \\ 0 & \text{otherwise} \end{cases}$ $x_4 = \begin{cases} 1 & \text{if student \#2} \\ 0 & \text{otherwise} \end{cases}$ $\cdots$ $x_{11} = \begin{cases} 1 & \text{if student \#9} \\ 0 & \text{otherwise} \end{cases}$

14.33   To determine whether the mean crack widths differ for the four time periods, we test:

$H_0: \mu_1 = \mu_2 = \mu_3 = \mu_4$
H_a: At least two treatment means differ

The test statistic is $F = \dfrac{MS(\text{Period})}{MSE} = 57.99$ (from printout)

The rejection region requires $\alpha = .05$ in the upper tail of the F distribution with $\nu_1 = p - 1 = 4 - 1 = 3$ and $\nu_2 = n - p - b + 1 = 48 - 4 - 12 = 33$. From Table 10, Appendix II, $F_{.05} \approx 2.92$. The rejection region is $F > 2.92$.

Since the observed value of the test statistic falls in the rejection region ($F = 57.99 > 2.92$), H_0 is rejected. There is sufficient evidence to indicate that at least two of the mean crack widths differ among the four time periods at $\alpha = .05$.

14.35 a.   This experiment employed a randomized complete block design. The two treatments (methods) are randomly assigned to experimental units within each of 16 blocks (increments), one unit per treatment.

b.   Some preliminary calculations are:

$$CM = \frac{(\sum y)^2}{n} = \frac{2016.28^2}{32} = 127{,}043.2825$$

$$SS(\text{Total}) = \sum y^2 - CM = 127{,}060.3546 - 127{,}043.2825 = 17.0721$$

$$SST = \frac{T_1^2}{n_1} + \frac{T_2^2}{n_2} - CM = \frac{1006.13^2}{16} + \frac{1010.15^2}{16} - 127{,}043.2825$$

$$= 127{,}043.7875 - 127{,}043.2825 = .5050$$

$$SSB = \frac{B_1^2}{n_1} + \frac{B_2^2}{n_2} + \cdots + \frac{B_{16}^2}{n_{16}} - CM$$

$$= \frac{126.58^2}{2} + \frac{126.51^2}{2} + \frac{126.86^2}{2} + \frac{126.28^2}{2} + \frac{125.04^2}{2} + \frac{128.04^2}{2}$$

$$+ \frac{126.09^2}{2} + \frac{127.77^2}{2} + \frac{123.78^2}{2} + \frac{125.32^2}{2} + \frac{127.42^2}{2} + \frac{125.52^2}{2}$$

$$+ \frac{126.72^2}{2} + \frac{125.81^2}{2} + \frac{124.76^2}{2} + \frac{123.78^2}{2} - 127{,}043.2825$$

$$= 127{,}055.4284 - 127{,}043.2825 = 12.1459$$

$$SSE = SS(Total) - SST - SSB = 17.0721 - .505 - 12.1459 = 4.4212$$

$$MST = \frac{SST}{p-1} = \frac{.505}{2-1} = .5050 \qquad MSB = \frac{SSB}{b-1} \quad \frac{12.1459}{16-1} = .8097$$

$$MSE = \frac{SSE}{n-p-b+1} \quad \frac{4.4212}{15} = .2947$$

$$F(\text{treatments}) = \frac{MST}{MSE} = \frac{.5050}{.2947} = 1.71$$

$$F(\text{blocks}) = \frac{MSB}{MSE} = \frac{.8097}{.2947} = 2.75$$

Source	df	SS	MS	F
Methods	1	.505	.5050	1.71
Increments	15	12.1459	.8097	2.75
Error	15	4.4212	.2947	
Total	31	17.0721		

c. To see if there is a difference in the two method means, we test:

$H_0: \mu_1 = \mu_2$
$H_a: \mu_1 \neq \mu_2$

The test statistic is $F = \dfrac{MST}{MSE} = 1.71$

The rejection region requires $\alpha = .05$ in the upper tail of the F distribution with $\nu_1 = p - 1 = 2 - 1 = 1$ and $\nu_2 = n - p - b + 1 = 32 - 2 - 16 = 15$. From Table 10, Appendix II, $F_{.05} = 4.54$. The rejection region is $F > 4.54$.

Since the observed value of the test statistic does not fall in the rejection region ($F = 1.71 \not> 4.54$), H_0 is not rejected. There is insufficient evidence to indicate a difference in the means of the two methods at $\alpha = .05$.

d. Some preliminary calculations are:

Increment	d	Increment	d
1	−1.26	9	−.28
2	−.77	10	.98
3	−.42	11	−1.26
4	−.26	12	.92
5	−.84	13	−.28
6	−1.18	14	.35
7	.35	15	.98
8	−.63	16	−.42

$$\bar{d} = \frac{\sum d}{n} = \frac{-4.02}{16} = -.251$$

$$s_d^2 = \frac{\sum d^2 - \frac{\left(\sum d\right)^2}{n}}{n-1} = \frac{9.8524 - \frac{16.1604^2}{16}}{16 - 1} = .5895$$

$$s_d = \sqrt{.5895} = .768$$

The test statistic is $t = \dfrac{\bar{d} - 0}{s_d/\sqrt{n}} = \dfrac{-.251 - 0}{.768/\sqrt{16}} = -1.309$

The rejection region for a small sample, two-tailed test requires $\alpha/2 = .05/2 = .025$ in the each tail of the t distribution with df $= n - 1 = 16 - 1 = 15$. From Table 7, Appendix II, $t_{.025} = 2.131$. The rejection region is $t < -2.131$ or $t > 2.131$.

Since the observed value of the test statistic does not fall in the rejection region ($t = -1.309 \not< -2.131$), H_0 is not rejected. This is the same result that was found in part c.

e. $F = t^2 = (-1.309)^2 = 1.71$.

14.37 a. Since we want to find which bonding agent is best, the three agents (nickel, iron, and copper) are the treatments.

b. The ingots cause variation so we will use them as blocks in the experiment.

c. To test for a difference in pressure required to separate the components among the three bonding agents, we test:

H_0: $\mu_1 = \mu_2 = \mu_3$
H_a: At least two means differ

The test statistic is $F = \dfrac{MST}{MSE} = \dfrac{SST/(p-1)}{SSE/(n-b-p+1)} = \dfrac{131.90/2}{124.46/12} = 6.36$

The rejection region requires $\alpha = .05$ in the upper tail of the F distribution with numerator df $= p - 1 = 3 - 1 = 2$ and denominator df $= n - b - p + 1 = 21 - 7 - 3 + 1 = 12$. From Table 10, Appendix II, $F_{.05} = 3.89$. The rejection region is $F > 3.89$.

Since the observed value of the test statistic does fall in the rejection region ($F = 6.36 > 3.89$), H_0 is rejected. There is sufficient evidence to indicate the means differ at $\alpha = .05$.

14.39 In a two factor factorial experiment, the first thing we test for is interaction between the two factors.

To determine if interaction exists between aid and order, we test:

H_0: No interaction between aid and order
H_a: Aid and order interact to affect travel time

The test statistic is $F = 1.29$.

The p-value is greater than .10. Since the p-value is so large, H_0 is not rejected. There is insufficient evidence to indicate aid and order interact to affect travel time.

Since interaction is not significant, tests on the main effects are performed.

To determine if there is a difference in the mean travel times among the three different aids, we test:

H_0: $\mu_S = \mu_M = \mu_N$
H_a: At least two of the means are different

The test statistic is $F = 76.67$.

The p-value is less than .0001. Since the p-value is so small, H_0 is rejected. There is sufficient evidence to indicate a difference in mean travel time among the three different levels of aid.

To determine if there is a difference in the mean travel times among the two different orders, we test:

H_0: $\mu_{c/w} = \mu_{w/c}$
H_a: $\mu_{c/w} \neq \mu_{w/c}$

The test statistic is $F = 1.95$.

The p-value is greater than .10. Since the p-value is so large, H_0 is not rejected. There is insufficient evidence to indicate a difference in mean travel time between the two different levels of order.

. . . .

Analysis of Variance for Designed Experiments

279

14.41 **a.** From the printout, the ANOVA table is:

Source	df	SS	MS	F
Amount (A)	3	104.19	34.73	20.12
Method (B)	3	28.63	9.54	5.53
AB	9	25.13	2.79	1.62
Error	32	55.25	1.73	
Total	47	213.20		

b. To determine if interaction exists, we test:

H_0: No interaction between amount and method exists
H_a: Amount and method interact

The test statistic is $F = \dfrac{MS(AB)}{MSE} = 1.62$

The rejection region requires $\alpha = .01$ in the upper tail of the F distribution with $\nu_1 = (a-1)(b-1) = (4-1)(4-1) = 9$ and $\nu_2 = ab(r-1) = 4(4)(3-1) = 32$. From Table 12, Appendix II, $F_{.01} \approx 3.07$. The rejection region is $F > 3.07$.

Since the observed value of the test statistic does not fall in the rejection region ($F = 1.62 \ngtr 3.07$), H_0 is not rejected. There is insufficient evidence to indicate interaction between amount and method exists at $\alpha = .01$.

c. Because interaction is not present, the effects of one independent variable (amount) on the dependent variable (shear strength) is the same at each level of the second independent variable (method).

d. To test for the main effects of factor A, we test:

H_0: There are no differences among the mean shear strengths for the different amounts of antimony
H_a: At least two of the means differ

The test statistic is $F = \dfrac{MS(A)}{MSE} = 20.12$

The rejection region requires $\alpha = .01$ in the upper tail of the F distribution with $\nu_1 = a - 1 = 3$ and $\nu_2 = ab(r-1) = 32$. From Table 12, Appendix II, $F_{.01} \approx 4.51$. The rejection region is $F > 4.51$.

Since the observed value of the test statistic falls in the rejection region ($F = 20.12 > 4.51$), H_0 is rejected. There is sufficient evidence to indicate differences in the mean shear strengths among the different amounts of antimony at $\alpha = .01$.

To test for the main effects of factor B, we test:

H_0: There are no differences among the mean shear strengths for the different cooling methods
H_a: At least two of the means differ

The test statistic is $F = \dfrac{MS(B)}{MSE} = 5.53$

The rejection region requires $\alpha = .01$ in the upper tail of the F distribution with $\nu_1 = b - 1 = 3$ and $\nu_2 = ab(r - 1) = 32$. From Table 12, Appendix II, $F_{.01} \approx 4.51$. The rejection region is $F > 4.51$.

Since the observed value of the test statistic falls in the rejection region ($F = 5.53 > 4.51$), H_0 is rejected. There is sufficient evidence to indicate differences in the mean shear strengths among the different cooling methods at $\alpha = .01$.

14.43   To determine if interaction exists between inspection levels and burn-in hours, we test:

H_0: No interaction between inspection level and burn-in
H_a: Inspection level and burn-in interact

The test statistic is $F = \dfrac{MS(AB)}{MSE} = 95.27$ (from printout)

The p-value is .0001. Since the p-value is so small, H_0 is rejected. There is sufficient evidence to indicate that inspection levels and burn-in hours interact to affect early part failure.

Since interaction exists, no main effect tests are run. The next step is to find which treatment combination gives the optimal detection of early part failure.

14.45   a.   To use the traditional analysis of variance, we need repetitions for each factor-level combinations. In this problem, we have only one observation per factor-level combination.

b. $E(y) = \beta_0 + \beta_1 x_1 + \beta_2 x_2 + \beta_3 x_1 x_2 + \beta_4 x_1^2 + \beta_5 x_2^2$ where x_1 = temperature and x_2 = mole fraction.

c.   SSE = .00015, s^2 = MSE = .00001, SS(Total) = 24.85834

d.   From the printout, $R^2 = 1.0000$

$$R^2 = 1 - \frac{SSE}{SS(Total)} = 1 - \frac{.00015}{24.85834} = 1 - .000006 = .999994$$

R^2 gives the proportion of the variability of the dependent variable that can be explained by the model. In this case, 99.9994% of the variability in the rates of combustion can be explained by the complete second-order model.

e.   When no interaction is present, the relationship between the dependent variable (rate of combustion) and one independent variable (say temperature) is independent of the second independent variable (mole fraction) or the effect of temperature on rate of combustion does not depend on the value of mole fraction.

Analysis of Variance for Designed Experiments

f. To determine if interaction between mole fraction and temperature is present, we test:

H_0: $\beta_3 = 0$ (no interaction)
H_a: $\beta_3 \neq 0$ (interaction)

The test statistic from the printout is $t = 12.41$.

From the SAS printout, the p-value is .0001. Since the p-value is less than $\alpha = .05$, H_0 is rejected. There is sufficient evidence to indicate interaction between temperature and mole fraction is present.

g. From the printout,

$$\hat{y} = -.280150 + .001000x_1 - .285549x_2 + .000733x_1x_2 + .000000959x_1^2$$
$$+ .551414x_2^2$$

h. $\hat{y} = -.280150 + .001000(1300) - .285549(.017) + .000733(1300)(.017)$
$$+ .000000959(1300)^2 + .551414(.017)^2 = 2.6521$$

i. From the SAS printout, the confidence interval for observation 11 is (2.6494, 2.6538). (The 11th observation has temperature = 1300 and mole fraction = .017.)

14.47 a. To compute the total time measurements for each of the categories, we multiply the mean by $n = 8$. The totals are:

$AB_{11} = n(\bar{y}_{11}) = 8(18.30) = 146.4$
$AB_{12} = n(\bar{y}_{12}) = 8(14.50) = 116.0$
$AB_{21} = n(\bar{y}_{21}) = 8(13.00) = 104.0$
$AB_{22} = n(\bar{y}_{22}) = 8(12.25) = 98.0$

b. The sum of all 32 observations is $\sum y = 146.4 + 116 + 104 + 98 = 464.4$

$$CM = \frac{(\sum y)^2}{n} = \frac{464.4^2}{32} = 6739.605$$

c. Let Factor A = Sex and Factor B = Weight

$$SS(A) = \frac{\sum A_i^2}{br} - CM = \frac{262.4^2 + 202^2}{2(8)} - 6739.605 = 6853.61 - 6739.605$$
$$= 114.005$$

$$SS(B) = \frac{\sum B_i^2}{ar} - CM = \frac{250.4^2 + 98^2}{2(8)} - 6739.605 = 6781.01 - 6739.605$$
$$= 41.405$$

$$SS(AB) = \frac{\sum AB_{ij}^2}{r} - SS(A) - SS(B) - CM$$

$$= \frac{146.4^2 + 116^2 + 104^2 + 98^2}{8} - 114.005 - 41.405 - 6739.605$$

$$= 6913.62 - 114.005 - 41.405 - 6739.605 = 18.605$$

d. $s_{11}^2 = 6.81^2 = 46.3761$ $s_{12}^2 = 2.93^2 = 8.5849$

$s_{21}^2 = 5.04^2 = 25.4016$ $s_{22}^2 = 5.70^2 = 32.49$

The sum of squares within each sample is found by multiplying the sample variance by $n_{ij} - 1$. The four sum of squares within each sample are:

$$\sum (y - \bar{y}_{11})^2 = (8 - 1)(46.3761) = 324.6327$$
$$\sum (y - \bar{y}_{12})^2 = (8 - 1)(8.5849) = 60.0943$$
$$\sum (y - \bar{y}_{21})^2 = (8 - 1)(25.4016) = 177.8112$$
$$\sum (y - \bar{y}_{22})^2 = (8 - 1)(32.4900) = 227.4300$$

e. SSE = sum of the four sum of squares within each sample
$$= 324.6327 + 60.0943 + 177.8112 + 227.4300 = 789.9682$$

f. SS(Total) = SS(A) + SS(B) + SS(AB) + SSE
$$= 114.005 + 41.405 + 18.605 + 789.9682 = 963.9832$$

g. The ANOVA table is:

Source	df	SS	MS	F
Sex (A)	1	114.005	114.005	4.04
Weight (B)	1	41.405	41.405	1.47
Sex $\times$ Weight (AB)	1	18.605	18.605	0.66
Error	28	789.9682	28.2132	
Total	31	963.9832		

h. If sex and weight interact, the effect of sex on the ability to perform laborious tasks that require strength will depend on the weight of the person.

To determine if sex and weight interact, we test:

H_0: Sex and weight do not interact
H_a: Sex and weight interact

The test statistic is $F = \dfrac{MS(AB)}{MSE} = 0.66$

The rejection region requires $\alpha = .05$ in the upper tail of the F distribution with $\nu_1 = (a - 1)(b - 1) = (2 - 1)(2 - 1) = 1$ and $\nu_2 = ab(r - 1) = 2(2)(8 - 1) = 28$. From Table 10, Appendix II, $F_{.05} = 3.34$. The rejection region is $F > 3.34$.

Since the observed value of the test statistic does not fall in the rejection region ($F = 0.66 \not> 3.34$), H_0 is not rejected. There is insufficient evidence to indicate that sex and weight interact at $\alpha = .05$.

14.49 **a.** To determine if any of the interaction terms are significant, we test:

$$H_0: \beta_3 = \beta_5 = \beta_6 = \beta_7 = 0$$
$$H_a: \text{At least one } \beta_i \neq 0 \quad (i = 3, 5, 6, 7)$$

The test statistic is $F = \dfrac{(\text{SSE}_R - \text{SSE}_C)/7}{\text{MSE}_C} = \dfrac{(360.4945833 - 8.145)/7}{.67875} = 74.16$

The rejection region requires $\alpha = .05$ in the upper tail of the F distribution with $\nu_1 = 7$ and $\nu_2 = 12$. From Table 10, Appendix II, $F_{.05} = 2.91$. The rejection region is $F > 2.91$.

Since the observed value of the test statistic falls in the rejection region ($F = 74.16 > 2.91$), H_0 is rejected. There is sufficient evidence to indicate that at least one interaction term is significant at $\alpha = .05$.

b. From the printout, Alloy*Matcond has an F value of 500.56 with a p-value of .0001. Also, Alloy*Time has an F value of 5.88 with a p-value of .0166. Both of these interactions would be significant at $\alpha = .05$.

14.51 To determine if differences among the second-order models relating $E(y)$ to x_1 for the four categories of alloy type and material condition, we test:

$$H_0: \beta_3 = \beta_4 = \cdots = \beta_{11} = 0$$
$$H_a: \text{At least one of the } \beta_i \text{ parameters is not 0, } i = 3, 4, \ldots, 11$$

The test statistic is $F = \dfrac{\text{MS(Drop)}}{\text{MSE(Complete)}}$

$\text{MS(Drop)} = \dfrac{\text{SSE}_R - \text{SSE}_C}{\text{df}} = \dfrac{1868.83875 - 8.145}{9} = 206.74$

$F = \dfrac{206.74}{.67875} = 304.59$

The rejection region requires $\alpha = .05$ in the upper tail of the F distribution with $\nu_1 = $ df(Drop) $= 21 - 12 = 9$ and $\nu_2 = $ df(Error, complete) $= 12$. From Table 10, Appendix II, $F_{.05} = 2.80$. The rejection region is $F > 2.80$.

Since the observed value of the test statistic falls in the rejection region ($F = 304.59 > 2.80$), H_0 is rejected. There is sufficient evidence to indicate differences among the second-order models relating $E(y)$ to x_1 for the four categories of alloy type and material condition at $\alpha = .05$.

14.53 **a.** By using the weeks to partition out a source of variation in this experiment, the design being used is a randomized block design.

b.

Source	df
Temperature (T)	1
Pressure (P)	1
$T \times P$	1
Week	2
Error	6
Total	11

14.55 a.   There are 3 first-stage observations in this sample: Site 1, Site 2, and Site 3.

b.   For each first-stage unit, there are 5 second-stage units selected.

c.   The total number of observations is $3 \times 5 = 15$.

d.   The probabilistic model is $y_{ij} = \mu + \alpha_i$ and ϵ_{ij}. $(i = 1, 2\ 3; j = 1, 2, 3, 4, 5)$

e.   From the printout, $\hat{\sigma}^2 = .0558$.

From Table 14.19, $E(MS(A)) = \sigma^2 + n_2\sigma_\alpha^2$

Thus, $\hat{\sigma}_\alpha^2 = \dfrac{MS(A) - MS(B \text{ in } A)}{n_2} = \dfrac{.00129 - .0558}{5} = -.0109$

Since this number is less than 0, our estimate of σ_α^2 is 0.

f.   To determine if the variation in specimen densities between sites exceeds the variation within sites, we test:

H_0: $\sigma_\alpha^2 = 0$
H_a: $\sigma_\alpha^2 > 0$

The test statistic is $F = \dfrac{MS(A)}{MS(B \text{ in } A)} = .02$

The rejection region requires $\alpha = .10$ in the upper tail of the F distribution with numerator df $= \nu_1 = n_1 - 1 = 3 - 1 = 2$ and denominator df $= \nu_2 = n_1(n_2 - 1) = 3(5 - 1) = 12$. From Table 9, Appendix II, $F_{.10} = 2.81$. The rejection region is $F > 2.81$.

Since the observed value of the test statistic does not fall in the rejection region ($F = .02$ ⊁ 2.81), H_0 is not rejected. There is insufficient evidence to indicate the variance between the sites exceeds the variation within the sties at $\alpha = .10$.

14.57

Source	df
Production Lot (A)	$n_1 - 1 = 10 - 1 = 9$
Batch within Lot $(B$ in $A)$	$n_1(n_2 - 1) = 10(5 - 1) = 40$
Shipping Lot within Batch $(C$ in $B)$	$n_1 n_2(n_3 - 1) = 10(5)(19) = 950$
Total	999

14.59 a. The estimate of σ_w^2 is $\hat{\sigma}_w^2 = MSE = MS(B$ in $A) = .057464$.

The estimate of σ_B^2 is $\hat{\sigma}_B^2 = \dfrac{MS(A) - MS(B \text{ in } A)}{n_2} = \dfrac{.364125 - .057464}{8} = .038333$

b. To determine if between lots variation exceeds within lots variation, we test:

$$H_0: \sigma_B^2 = 0$$
$$H_a: \sigma_B^2 > 0$$

The test statistic is $F = \dfrac{MS(A)}{MS(B \text{ in } A)} = \dfrac{.3641}{.0575} = 6.34$

The rejection region requires $\alpha = .05$ in the upper tail of the F distribution with $\nu_1 = n_1 - 1 = 5 - 1 = 4$ and $\nu_2 = n_1(n_2 - 1) = 5(8 - 1) = 35$. From Table 10, Appendix II, $F_{.05} = 2.65$. The rejection region is $F > 2.65$.

Since the observed value of the test statistic falls in the rejection region ($F = 6.34 > 2.65$), H_0 is rejected. There is sufficient evidence to indicate the variation between the lots exceeds the variation within the lots at $\alpha = .05$.

14.61 From Exercise 14.20, $\bar{x}_O = 1.33$, $\bar{x}_C = 3.16$, $\bar{x}_B = .41$ and $\bar{x}_H = .146$

For this experiment, $p = 4$, $q = p(p - 1)/2 = 4(3)/2 = 6$, $\alpha^* = \alpha/g = .06/6 = .01$, $s = \sqrt{MSE} = .6027$, $\nu = 10$, $n_1 = n_2 = n_3 = 3$, and $n_4 = 5$, and $t_{.01} = 2.764$ from Table 7, Appendix II.

There are two critical values for Bonferroni's comparison procedure:

$$B_{12} = B_{13} = B_{23} = t_{.01} s \sqrt{\dfrac{1}{n_1} + \dfrac{1}{n_2}} = 2.764(.6027)\sqrt{\dfrac{1}{3} + \dfrac{1}{3}} = 1.360$$

$$B_{14} = B_{23} = B_{34} = t_{.01} s \sqrt{\dfrac{1}{n_1} + \dfrac{1}{n_4}} = 2.764(.6027)\sqrt{\dfrac{1}{3} + \dfrac{1}{5}} = 1.217$$

The means arranged in order are:

$\bar{x}_H$	$\bar{x}_B$	$\bar{x}_O$	$\bar{x}_C$
.146	.41	1.33	3.16

C − H = 3.16 − .146 = 3.014 > 1.217 significantly different
C − B = 3.16 − .41 = 2.75 > 1.360   significantly different
C − O = 3.16 − 1.33 = 1.83 > 1.36   significantly different
O − H = 1.33 − .146 = 1.184 > 1.217 not significantly different

Thus, μ_C is significantly greater than μ_H, μ_B, and μ_O. There are no other significant differences.

14.63   First, compute the mean rates of penetration for the drill bits:

Mean rates of penetration

PD-1	IADC 1-2-6	IADC 5-1-7
34.3	28.05	20.775

From Exercise 14.25, $p = 3$, $n_t = 4$, $\alpha = .05$ and $s = \sqrt{19.3} = 4.393$. Since MSE is based on $\nu = 9$ df, from Table 13, Appendix II, $q_{.05}(3, 9) = 3.95$. The critical difference is:

$$\omega = q_{.05}(3, 9) \cdot \frac{s}{\sqrt{n_t}} = 3.95 \frac{(4.393)}{\sqrt{4}} = 8.68$$

Sample means that differ by more than 8.68 indicate a difference in the corresponding population means. After ranking the sample means from smallest to largest, we compare the differences to $\omega = 8.68$.

20.775   28.05   34.3

The only difference we can detect is between the means for the PD-1 and the IADC 5-1-7 drill bits.

14.65   a.   With a p-value of .0003, the Accuracy × Vocabulary interaction is significant. This means that the effect of Accuracy on mean task completion time depends on the level of Vocabulary.

b.   Yes. Since the interaction is significant, the tests for the main effects are meaningless. Thus, we must compare the mean task completion times for the three levels of Accuracy at each level of Vocabulary.

c.   For 75% Vocabulary. Since there are no lines connecting any of the means, all three accuracy levels are significantly different.

For 87.5% Vocabulary. Since there are no lines connecting any of the means, all three accuracy levels are significantly different.

For 100% Vocabulary. Since there is a line connecting the means for 99% and 95% Accuracy, there is no significant difference between the mean task completion times for 99% and 95% Accuracy. The mean task completion time for 90% Accuracy is significantly larger than for 99% and 95% Accuracy.

14.67 For a completely randomized design, the residuals are found by subtracting the treatment sample mean from each of the observations in each treatment. From Exercise 14.61, the treatment sample means are:

$$\bar{x}_O = 1.33, \ \bar{x}_C = 3.16, \ \bar{x}_B = .41 \text{ and } \bar{x}_H = .146$$

The residuals for each treatment are:

Oyster Tissue	Citrus Leaves	Bovine Liver	Human Serum
1.02	−.84	−.02	−.046
−.03	−.09	.13	.024
−.99	.93	−.11	−.006
			.014
			.014

To check for normality, we will construct stem-and-leaf displays of the residuals for each of the treatments using MINITAB. The stem-and-leaf displays are:

```
Stem-and-Leaf of Oyster    N = 3      Stem-and-Leaf of Citrus    N = 3
Leaf Unit = 0.10                      Leaf Unit = 0.10

   1     -0 9                            1     -0 8
  (1)    -0 0                           (1)    -0 0
   1      0                              1      0
   1      0                              1      0 9
   1      1 0

Stem-and-Leaf of Bovine    N = 3      Stem-and-Leaf of Human     N = 5
Leaf Unit = 0.10                      Leaf Unit = 0.10

   1     -1 1                            1     -0 4
  (1)    -0 2                            1     -0
   1      0                              2     -0 0
   1      1 3                           (2)     0 11
                                        1      0 2
```

Since there are so few observations for each of the treatments, it is very difficult to determine if the residuals are normally distributed. From the stem-and-leaf displays, there is no strong evidence that the residuals are not normally distributed.

To check for unequal variances, we will construct residual frequency plots for each of the treatments. The frequency plots are:

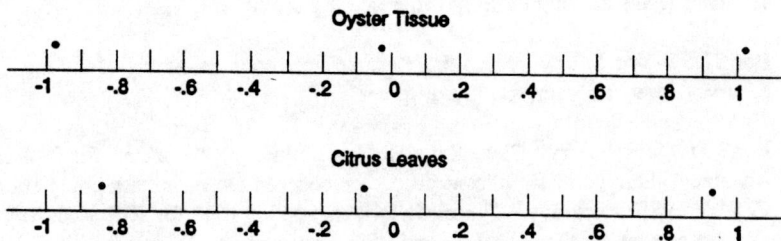

Oyster Tissue

Citrus Leaves

Bovine Liver

Human Serum

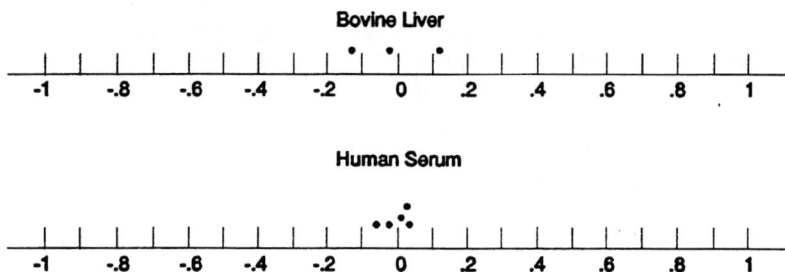

The spreads of the residuals for Oyster Tissue and Citrus Leaves appear to be much greater than the spreads of the residuals for Bovine Liver and Human Serum. The assumption of equal variances could be violated.

14.69　For a completely randomized design, the residuals are found by subtracting the treatment sample mean from each of the observations in each treatment. From Exercise 14.27, the treatment sample means are:

$$\bar{x}_1 = .0817, \; \bar{x}_2 = .11, \; \bar{x}_3 = .1217, \text{ and } \bar{x}_4 = .085$$

The residuals for each treatment are:

Location 1	Location 2	Location 3	Location 4
−.0017	.04	.0083	−.035
.0183	−.02	−.0217	.025
.0083	.00	.0283	−.015
−.0117	−.01	−.0317	.005
.0083	−.03	−.0317	.025
−.0217	.02	.0483	−.005

To check for normality, we will construct stem-and-leaf displays of the residuals for each of the treatments using MINITAB. The stem-and-leaf displays are:

```
Stem-and-Leaf of Loc 1    N = 6      Stem-and-Leaf of Loc 2    N = 6
Leaf Unit = 0.0010                   Leaf Unit = 0.010

   1    -2 1                            2    -0 32
   2    -1 1                            3    -0 1
   3    -0 1                            3     0 0
   3     0 88                           2     0 2
   1     1 8                            1     0 4

Stem-and-Leaf of Loc 3    N = 6      Stem-and-Leaf of Loc 4    N = 6
Leaf Unit = 0.010                    Leaf Unit = 0.0010

   3    -0 332                          1    -3 5
   3    -0                             1    -2
   3     0 0                            2    -1 5
   2     0 2                            3    -0 5
   1     0 4                            3     0 5
                                        2     1
                                        2     2 55
```

Since there are so few observations for each of the treatments, it is very difficult to determine if the residuals are normally distributed. From the stem-and-leaf displays, there is no strong evidence that the residuals are not normally distributed.

To check for unequal variances, we will construct residual frequency plots for each of the treatments. The frequency plots are:

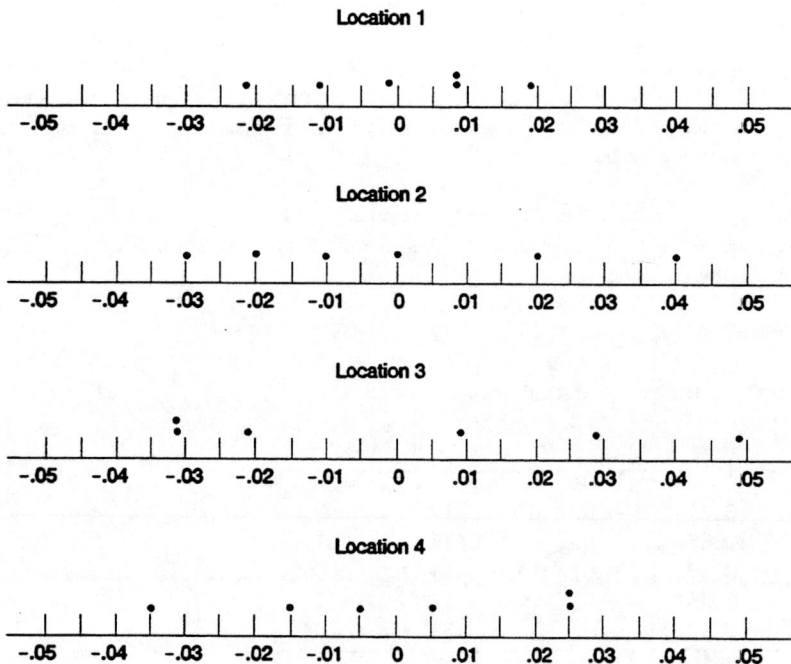

Location 1

```
  |   |   |   |   |   |   |   |   |   |   |   |   |   |   |   |   |   |   |   |   |
 -.05    -.04    -.03    -.02    -.01     0      .01     .02     .03     .04     .05
```

Location 2

```
  |   |   |   |   |   |   |   |   |   |   |   |   |   |   |   |   |   |   |   |   |
 -.05    -.04    -.03    -.02    -.01     0      .01     .02     .03     .04     .05
```

Location 3

```
  |   |   |   |   |   |   |   |   |   |   |   |   |   |   |   |   |   |   |   |   |
 -.05    -.04    -.03    -.02    -.01     0      .01     .02     .03     .04     .05
```

Location 4

```
  |   |   |   |   |   |   |   |   |   |   |   |   |   |   |   |   |   |   |   |   |
 -.05    -.04    -.03    -.02    -.01     0      .01     .02     .03     .04     .05
```

The spreads of the residuals for all four locations appear to be fairly similar. The assumption of equal variances appears to be valid.

14.71   For a factorial design, the residuals are found by subtracting the treatment sample mean from each of the observations in each treatment. From Exercise 14.41, the treatment sample means are:

$$\bar{x}_{0,W} = 18.467, \ \bar{x}_{0,O} = 22.067, \ \bar{x}_{0,A} = 20.333, \text{ and } \bar{x}_{0,F} = 19.833$$
$$\bar{x}_{3,W} = 19.033, \ \bar{x}_{3,O} = 20.433, \ \bar{x}_{3,A} = 22.233, \text{ and } \bar{x}_{3,F} = 19.933$$
$$\bar{x}_{5,W} = 20.767, \ \bar{x}_{5,O} = 21.467, \ \bar{x}_{5,A} = 21.400, \text{ and } \bar{x}_{5,F} = 18.833$$
$$\bar{x}_{10,W} = 16.300, \ \bar{x}_{10,O} = 17.833, \ \bar{x}_{10,A} = 16.733, \text{ and } \bar{x}_{10,F} = 17.200$$

The residuals for each treatment are:

Amount of Antimony	Cooling Method	Residuals		
0	WQ	−.867,	1.033,	−.167
0	OQ	−2.067,	2.233,	−.167
0	AB	−2.033,	−.533,	2.567
0	FC	−.433,	−.033,	.467
3	WQ	−.433,	.467,	−.033
3	OQ	−.433,	.467,	−.033
3	AB	−.533,	.667,	−.133
3	FC	−.933,	−.967,	−.033
5	WQ	4.533,	−1.267	−.267
5	OQ	−.567,	1.433,	−.867
5	AB	1.500,	−1.700,	.200
5	FC	.767,	−2.433,	1.667
10	WQ	−1.100,	.800,	.300
10	OQ	−1.433,	1.167,	.267
10	AB	−.933,	.567,	−.367
10	FC	−.800,	.400,	.400

To check for normality, we will construct stem-and-leaf displays of the residuals for each of the treatments using MINITAB. The stem-and-leaf displays are:

```
Stem-and-Leaf of 0, WQ    N = 3       Stem-and-Leaf of 0, OQ    N = 3
Leaf Unit = 0.10                      Leaf Unit = 1.0

  1     -0 8                            1     -0 2
 (1)    -0 1                           (1)    -0 0
  1      0                              1      0
  1      0                              1      0 2
  1      1 0
```

```
Stem-and-Leaf of 0, AB    N = 3       Stem-and-Leaf of 0, FC    N = 3
Leaf Unit = 1.0                       Leaf Unit = 0.10

  1     -0 2                            (2)   -0 40
 (1)    -0 0                            1      0 4
  1      0
  1      0 2
```

```
Stem-and-Leaf of 3, WQ    N = 3       Stem-and-Leaf of 3, OQ    N = 3
Leaf Unit = 0.10                      Leaf Unit = 0.10

  (2)   -0 40                           (2)   -0 40
  1      0 4                            1      0 4
```

```
Stem-and-Leaf of 3, AB    N = 3       Stem-and-Leaf of 3, FC    N = 3
Leaf Unit = 0.10                      Leaf Unit = 0.10

  1     -0 5                            1     -0 9
 (1)    -0 1                           (1)    -0 0
  1      0                              1      0
  1      0 6                            1      0 9
```

```
Stem-and-Leaf of 5, WQ    N = 3       Stem-and-Leaf of 5, OQ    N = 3
Leaf Unit = 0.10                      Leaf Unit = 0.10

  1     -1 2                            (2)   -0 85
 (1)    -0 2                            1     -0
  1      0                              1      0
  1      1 5                            1      0
                                       1      1 4
```

```
Stem-and-Leaf of 5, AB      N = 3        Stem-and-Leaf of 5, FC      N = 3
Leaf Unit = 0.10                         Leaf Unit = 0.10

    1     -1 7                                1     -2 4
    1     -0                                  1     -1
   (1)     0 2                                1     -0
    1      1 5                               (1)     0 7
                                             1      1 6


Stem-and-Leaf of 10, WQ     N = 3        Stem-and-Leaf of 10, OQ     N = 3
Leaf Unit = 0.10                         Leaf Unit = 0.10

    1     -1 1                                1     -1 4
    1     -0                                  1     -0
    1     -0                                 (1)     0 2
   (1)     0 3                                1      1 1
    1      0 8


Stem-and-Leaf of 10, AB     N = 3        Stem-and-Leaf of 10, FC     N = 3
Leaf Unit = 0.10                         Leaf Unit = 0.10

    1     -0 9                                1     -0 8
    1     -0                                  1     -0
   (1)     0 3                               (2)     0 44
    1      0 5
```

Since there are so few observations for each of the treatments, it is very difficult to determine if the residuals are normally distributed. From the stem-and-leaf displays, there is no strong evidence that the residuals are not normally distributed.

To check for unequal variances, we will construct frequency plots for each of the treatments. The frequency plots are:

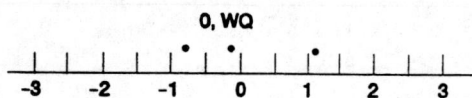

0, WQ

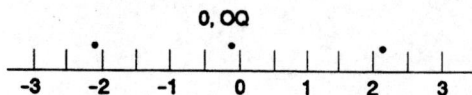

0, OQ

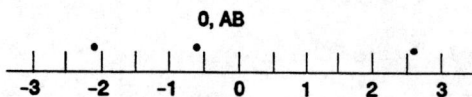

0, AB

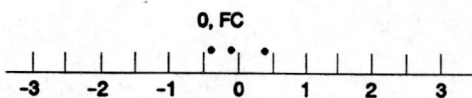

0, FC

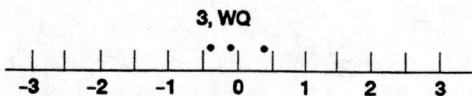

3, WQ

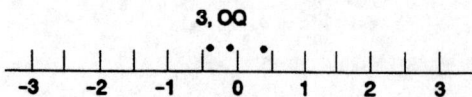

3, OQ

3, AB

```
  -3   -2   -1    0    1    2    3
```

3, FC

```
  -3   -2   -1    0    1    2    3
```

5, WQ

```
  -3   -2   -1    0    1    2    3
```

5, OQ

```
  -3   -2   -1    0    1    2    3
```

5, AB

```
  -3   -2   -1    0    1    2    3
```

5, FC

```
  -3   -2   -1    0    1    2    3
```

10, WQ

```
  -3   -2   -1    0    1    2    3
```

10, OQ

```
  -3   -2   -1    0    1    2    3
```

10, AB

```
  -3   -2   -1    0    1    2    3
```

10, FC

```
  -3   -2   -1    0    1    2    3
```

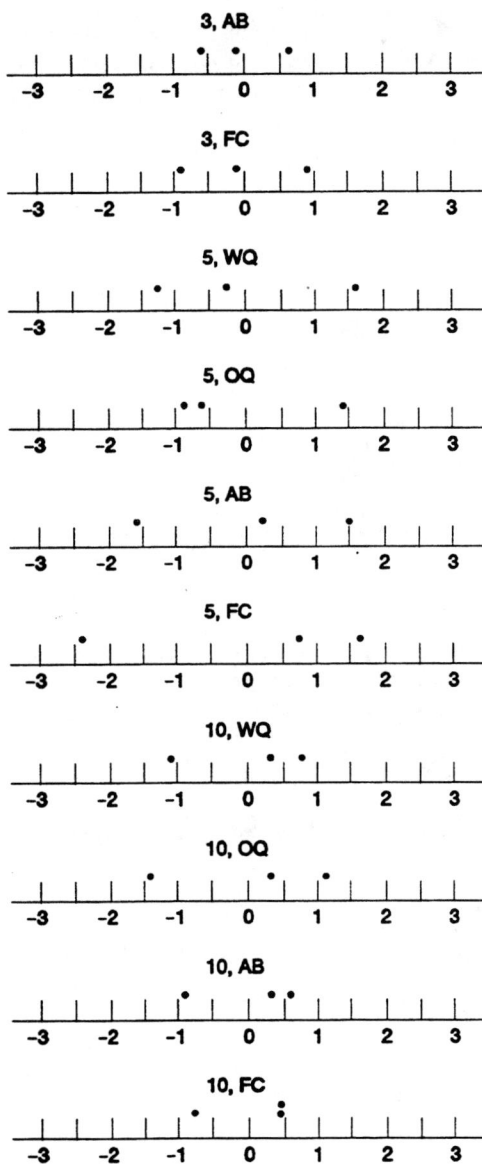

Since the spread of the residuals appears to vary from treatment to treatment, it appears that the assumption of equal variances is violated.

14.73   a.   The factors in this experiment are type of game and position.

     b.   There are three levels of type of game: aggressive video game (Missile Command), nonaggressive video game (Pac-Man), and pen-and-paper maze-solving game (control).

        There are two levels of position: player and observer.

c. There are $3 \times 2 = 6$ treatments. They are all the combinations of type of game and position. The six treatments are: (Missile Command, player), (Missile Command, observer), (Pac-Man, player), (Pac-Man, observer), (Control, player), and (Control, observer).

d. The complete model is

$$E(y) = \beta_0 + \beta_1 x_1 + \beta_2 x_2 + \beta_3 x_3 + \beta_4 x_1 x_3 + \beta_5 x_2 x_3$$

where $x_1 = \begin{cases} 1 & \text{if Missle Command} \\ 0 & \text{if not} \end{cases}$ $x_2 = \begin{cases} 1 & \text{if Pac Man} \\ 0 & \text{if not} \end{cases}$

$x_3 = \begin{cases} 1 & \text{if player} \\ 0 & \text{if not} \end{cases}$

e.

Source	df
Game	$3 - 1 = 2$
Position	$2 - 1 = 1$
Games $\times$ Position	$(3 - 1)(2 - 1) = 2$
Error	$3(2)(14 - 1) = 78$
Total	$3(2)(14) - 1 = 83$

f. The interaction between type of game and position was found to be significant. This implies that the effect of position (player or observer) on the degree of aggressive play depends on which type of game was played.

14.75 a. This is a completely randomized design with $p = 3$ treatments.

b. To determine if there is a difference among the mean times for the three methods, we test:

H_0: $\mu_A = \mu_B = \mu_C$
H_a: At least two treatment means differ

The test statistic is $F = 7.02$ (from printout).

The rejection region requires $\alpha = .01$ in the upper tail of the F distribution with $\nu_1 = p - 1 = 3 - 1 = 2$ and $\nu_2 = n - p = 15 - 3 = 12$. From Table 12, Appendix II, $F_{.01} = 6.93$. The rejection region is $F > 6.93$.

Since the observed value of the test statistic falls in the rejection region ($F = 7.02 > 6.93$), H_0 is rejected. There is sufficient evidence to indicate differences exist among the mean times for the three methods at $\alpha = .01$.

14.77 a. To determine if interaction between pH level and soil depth exists, we test:

H_0: There is no interaction between pH level and soil depth
H_a: There is interaction between pH level and soil depth

The test statistic is $t = .39$ (from printout).

The p-value from the printout is .6854. Since the p-value is greater than $\alpha = .05$, H_0 is not rejected. There is insufficient evidence to indicate interaction between pH level and soil depth exists at $\alpha = .05$.

b. To determine if blocking over time was effective, we test:

H_0: There are no differences among the block means
H_a: At least two of the block means differ

The test statistic is $t = 19.17$ (from printout).

The p-value from the printout is .0004. Since the p-value is less than $\alpha = .05$, H_0 is rejected. There is sufficient evidence to indicate a difference in the block means exists at $\alpha = .05$. Blocking was effective in removing an extraneous source of variation.

14.79 a. Some preliminary calculations are:

$$CM = \frac{\left(\sum y\right)^2}{n} = \frac{224^2}{16} = 3136$$

$$SS(Total) = \sum y^2 - CM = 3137.12 - 3136 = 1.12$$

$$SST = \frac{T_1^2}{n_1} + \frac{T_2^2}{n_2} - CM = \frac{112.5^2}{8} + \frac{111.5^2}{8} - 3136 = 3136.0625 - 3136$$
$$= .0625$$

$$SSE = SS(Total) - SST = 1.12 - .0625 = 1.0575$$

$$MST = \frac{SST}{p - 1} = \frac{.0625}{2 - 1} = .0625 \qquad MSE = \frac{SSE}{n - p} = \frac{1.0575}{16 - 2} = .0755$$

$$F = \frac{MST}{MSE} = \frac{.0625}{.0755} = .83$$

Source	df	SS	MS	F
Treatments	1	.0625	.0625	.83
Error	14	1.0575	.0755	
Total	15	1.1200		

b. To see if a difference exists between the mean strengths of the two plastics, we test:

H_0: $\mu_1 = \mu_2$
H_a: $\mu_1 \neq \mu_2$

The test statistic is $F = .83$.

The rejection region requires $\alpha = .05$ in the upper tail of the F distribution with $\nu_1 = p - 1 = 2 - 1 = 1$ and $\nu_2 = n - p = 16 - 2 = 14$. From Table 10, Appendix II, $F_{.05} = 4.60$. The rejection region is $F > 4.60$.

Since the observed value of the test statistic does not fall in the rejection region ($F = .83$ $\not> 4.60$), H_0 is not rejected. There is insufficient evidence to indicate a difference exists between the mean strengths of the two plastics at $\alpha = .05$.

14.81 a. Some preliminary calculations are:

$$CM = \frac{\left(\sum y\right)^2}{n} = \frac{1883^2}{36} = 98,491.361$$

$$SS(Total) = \sum y^2 - CM = 11,1301 - 98,491.361 = 12,809.639$$

$$SS(A) = \frac{\sum A_i^2}{br} - CM = \frac{413^2 + 614^2 + 856^2}{3(4)} - 98,491.361$$

$$= 106,691.750 - 98,491.361 = 8200.389$$

$$SS(B) = \frac{\sum B_i^2}{ar} - CM = \frac{468^2 + 623^2 + 792^2}{3(4)} - 98,491.361$$

$$= 102,868.083 - 98,491.361 = 4376.722$$

$$SS(AB) = \frac{\sum AB_{ij}^2}{r} - SS(A) - SS(B) - CM$$

$$= \frac{96^2 + 150^2 + 222^2 + 134^2 + 200^2 + 289^2 + 183^2 + 264^2 + 345^2}{4}$$

$$- 8200.389 - 4376.722 - 98,491.361$$
$$= 111,171.750 - 8200.389 - 4376.722 - 98,491.361 = 103.278$$

$$SSE = SS(Total) - SS(A) - SS(B) - SS(AB)$$
$$= 12,809.639 - 8200.389 - 4376.722 - 103.278 = 129.250$$

$$MS(A) = \frac{SS(A)}{a - 1} = \frac{8200.389}{3 - 1} = 4100.195$$

$$MS(B) = \frac{SS(B)}{b - 1} = \frac{4376.722}{3 - 1} = 2188.361$$

$$MS(AB) = \frac{SS(AB)}{(a - 1)(b - 1)} = \frac{103.278}{(3 - 1)(3 - 1)} = 25.820$$

$$MSE = \frac{SSE}{ab(r - 1)} = \frac{129.250}{3(3)(4 - 1)} = 4.787$$

$$F(A) = \frac{MS(A)}{MSE} = \frac{4100.195}{4.787} = 856.53$$

$$F(B) = \frac{MS(B)}{MSE} = \frac{2188.361}{4.787} = 457.15$$

$$F(AB) = \frac{MS(AB)}{MSE} = \frac{25.820}{4.787} = 5.39$$

Source	df	SS	MS	F
Exposure (A)	2	8200.389	4100.195	856.53
Temperature (B)	2	4376.722	2188.361	475.15
AB	4	103.278	25.820	5.39
Error	27	129.250	4.787	
Total	35	12,809.639		

b. To see if interaction between exposure and temperature exist, we test:

H_0: There is no interaction between exposure and temperature
H_a: The two factors interact

The test statistic is $F = 5.39$.

The rejection region requires $\alpha = .05$ in the upper tail of the F distribution with $\nu_1 = (a - 1)(b - 1) = (3 - 1)(3 - 1) = 4$ and $\nu_2 = ab(r - 1) = 3(3)(4 - 1) = 27$. From Table 10, Appendix II, $F_{.05} = 2.73$. The rejection region is $F > 2.73$.

Since the observed value of the test statistic falls in the rejection region ($F = 5.39 > 2.73$), H_0 is rejected. There is sufficient evidence that exposure and temperature interact at $\alpha = .05$.

14.83 a. Some preliminary calculations are:

$$CM = \frac{\left(\sum y\right)^2}{n} = \frac{6598^2}{12} = 3,627,800.333$$

$$SS(Total) = \sum y^2 - CM = 3,632,768 - 3,627,800.333 = 4967.667$$

$$SST = \frac{T_1^2}{b} + \frac{T_2^2}{b} + \frac{T_3^2}{b} + \frac{T_4^2}{b} - CM = \frac{1653^2}{3} + \frac{1702^2}{3} + \frac{1634^2}{3} + \frac{1609^2}{3}$$
$$- 3,627,800.333 = 3,629,350 - 3,627,800.333 = 1549.667$$

$$SSB = \frac{B_1^2}{p} + \frac{B_2^2}{p} + \frac{B_3^2}{p} - CM$$

$$= \frac{2265^2}{4} + \frac{2136^2}{4} + \frac{2197^2}{4} - 3,627,800.3333$$

$$= 3,629,882.5 - 3,627,800.333 = 2082.167$$

$$SSE = SS(Total) - SST - SSB = 4967.667 - 1549.667 - 2082.167$$
$$= 1,335.833$$

$$MST = \frac{SST}{p - 1} = \frac{1549.667}{4 - 1} = 516.556$$

$$MSB = \frac{SSB}{b - 1} = \frac{2082.167}{3 - 1} = 1041.084$$

$$MSE = \frac{SSE}{(p - 1)(b - 1)} = \frac{1335.833}{(4 - 1)(3 - 1)} = 222.639$$

Analysis of Variance for Designed Experiments

$$F(\text{treatments}) = \frac{\text{MST}}{\text{MSE}} = \frac{516.556}{222.639} = 2.32$$

$$F(\text{blocks}) = \frac{\text{MSB}}{\text{MSE}} = \frac{1041.084}{222.639} = 4.68$$

To determine if a difference in mean temperature exists among the four treatments, we test:

H_0: $\mu_1 = \mu_2 = \mu_3 = \mu_4$
H_a: At least two treatment means differ

The test statistic is $F = 2.32$.

The rejection region requires $\alpha = .05$ in the upper tail of the F distribution with $\nu_1 = p - 1 = 4 - 1 = 3$ and $\nu_2 = (p - 1)(b - 1) = (4 - 1)(3 - 1) = 6$. From Table 10, Appendix II, $F_{.05} = 4.76$. The rejection region is $F > 4.76$.

Since the observed value of the test statistic does not fall in the rejection region ($F = 2.32 \ngtr 4.76$), H_0 is not rejected. There is insufficient evidence to indicate a difference in the mean temperatures among the four treatments at $\alpha = .05$.

b. To determine if a difference in mean temperature exists among the three batches, we test:

H_0: There are no differences among the batch means
H_a: At least two batch means differ

The test statistic is $F = \dfrac{\text{MSB}}{\text{MSE}} = 4.68$.

The rejection region requires $\alpha = .05$ in the upper tail of the F distribution with $\nu_1 = b - 1 = 3 - 1 = 2$ and $\nu_2 = n - p - b + 1 = 12 - 4 - 3 + 1 = 6$. From Table 10, Appendix II, $F_{.05} = 5.14$. The rejection region is $F > 5.14$.

Since the observed value of the test statistic does not fall in the rejection region ($F = 4.68 \ngtr 5.14$), H_0 is not rejected. There is insufficient evidence to indicate differences among the block mean temperatures at $\alpha = .05$.

c. Since no batch differences were detected, it is recommended that no blocks be used in the future. A completely randomized design is recommended to increase degrees of freedom for error.

14.85 a. To determine if any of the factors contribute to the model, we test:

H_0: $\beta_1 = \beta_2 = \beta_3 = \cdots = \beta_{15} = 0$
H_a: At least one of the β parameters is not 0 ($i = 1, 2, \ldots, 15$)

The test statistic is $F = 40.778$.

The p-value = .0001. We will reject H_0 for any value of $\alpha > .0001$. There is sufficient evidence that at least one of the factors contributes information for the prediction of y at $\alpha > .0001$.

b. The factors which appear to affect the amount of light y are FOIL (x_1) and SHIFT*OPERATOR (x_3x_4) interaction. Both of these terms have very small p-values, .0001 and .0010, respectively.

c. The complete 2^4 factorial model is:

$$E(y) = \beta_0 + \beta_1x_1 + \beta_2x_2 + \beta_3x_3 + \beta_4x_4 + \beta_5x_1x_2 + \beta_6x_1x_3 + \beta_7x_1x_4 + \beta_8x_2x_3 \\ + \beta_9x_2x_4 + \beta_{10}x_3x_4 + \beta_{11}x_1x_2x_3 + \beta_{12}x_1x_2x_4 + \beta_{13}x_1x_3x_4 + \beta_{14}x_2x_3x_4 \\ + \beta_{15}x_1x_2x_3x_4$$

d. The degrees of freedom available for estimating σ^2 is the df for ERROR which is 16 or $n - (k + 1) = 32 - (15 + 1) = 16$.

14.87 a. To find $\sum y$, recall that $\bar{y} = \dfrac{\sum y}{n} \Rightarrow \sum y = n\bar{y}$

$$\sum y_1 = n_1\bar{y}_1 = 50(8477) = 423,850$$

$$\sum y_2 = n_2\bar{y}_2 = 50(10,404) = 520,200$$

$$\sum y = 423,850 + 520,200 = 944,050$$

$$CM = \frac{(\sum y)^2}{n} = \frac{944,050^2}{100} = 8,912,304,025$$

$$SST = \frac{T_1^2}{n_1} + \frac{T_2^2}{n_2} - CM = \frac{423,850^2}{50} + \frac{520,200^2}{50} - 8,912,304,025$$
$$= 9,005,137,250 - 8,912,304,025 = 92,833,225$$

b. Recall from Exercise 14.28

$$SSE = (n_1 - 1)s_1^2 + (n_2 - 1)s_2^2 + \cdots + (n_p - 1)s_p^2$$

Thus, $SSE = (50 - 1)820^2 + (50 - 1)928^2 = 75,145,616$

c. SS(Total) = SST + SSE = 92,833,225 + 75,145,616 = 167,978,841

Source	df	SS	MS	F
Treatment	1	92,833,225	92,833,225	121.07
Error	98	75,145,616	766,792	
Total	99	167,978,841		

14.89 **a.** To find $\sum y$, recall that $\bar{y} = \dfrac{\sum y}{n} \Rightarrow \sum y = n\bar{y}$

$$\sum y_1 = n_1\bar{y}_1 = 50(621) = 31,050$$

$$\sum y_2 = n_2\bar{y}_2 = 50(737) = 36,850$$

$$\sum y = 31,050 + 36,850 = 67,900$$

$$\text{CM} = \dfrac{(\sum y)^2}{n} = \dfrac{67,900^2}{100} = 46,104,100$$

$$\text{SST} = \dfrac{T_1^2}{n_1} + \dfrac{T_2^2}{n_2} - \text{CM} = \dfrac{31,050^2}{50} + \dfrac{36,850^2}{50} - 46,104,100$$

$$= 46,440,500 - 46,104,100 = 336,400$$

b. Recall from Exercise 14.28

$$\text{SSE} = (n_1 - 1)s_1^2 + (n_2 - 1)s_2^2 + \cdots + (n_p - 1)s_p^2$$

Thus, $\text{SSE} = (50 - 1)48^2 + (50 - 1)55^2 = 261,121$

CHAPTER FIFTEEN

· ·

Nonparametric Statistics

15.1 a. To determine if the median TCDD level in plasma for Vietnam veterans exceeds 3.0 parts per trillion, we test:

H_0: $\tau = 3.0$
H_a: $\tau > 3.0$

The test statistic is S = number of observations greater than 3 or $S = 12$.

The observed significance level is: p-value = $P(S \geq 12)$ where S has a binomial distribution with $n = 20$ and $p = .5$. From Table 1, Appendix II, p-value = $P(S \geq 12)$ = $1 - P(S \leq 11) = 1 - .7483 = .2517$.

The rejection region is to reject H_0 if $\alpha > p$-value.

Since $\alpha = .05 \not> .2517$, H_0 is not rejected. There is insufficient evidence to indicate the median TCDD level in plasma for Vietnam veterans exceeds 3.0 parts per trillion at $\alpha = .05$.

b. To determine if the median TCDD level in fat tissue for Vietnam veterans exceeds 3.0 parts per trillion, we test:

H_0: $\tau = 3.0$
H_a: $\tau > 3.0$

The test statistic is S = number of observations greater than 3 or $S = 14$.

The observed significance level is: p-value = $P(S \geq 14)$ where S has a binomial distribution with $n = 20$ and $p = .5$. From Table 1, Appendix II, p-value = $P(S \geq 14)$ = $1 - P(S \leq 13) = 1 - .9423 = .0577$.

The rejection region is to reject H_0 if $\alpha > p$-value.

Since $\alpha = .05 \not> .0577$, H_0 is not rejected. There is insufficient evidence to indicate the median TCDD level in fat tissue for Vietnam veterans exceeds 3.0 parts per trillion at $\alpha = .05$.

15.3 To determine if the median radium-226 level in soil in southern Dade County exceeds the Environmental Protection Agency limit of 4.0 pCi/L, we test:

H_0: $\tau = 4.0$
H_a: $\tau > 4.0$

The test statistic is S = number of observations greater than 4 or $S = 6$.

The observed significance level is: p-value = $P(S \geq 6)$ where S has a binomial distribution with $n = 26$ and $p = .5$. Since the binomial tables do not include $n = 26$, we will have to use the large sample Sign Test.

$$E(S) = np = 26(.5) = 13 \text{ and } V(S) = npq = 26(.5)(.5) = 6.5$$

The test statistic is $z = \dfrac{S - E(S)}{\sqrt{V(S)}} = \dfrac{6 - 13}{\sqrt{6.5}} = -2.75$

The rejection region requires $\alpha = .10$ in the upper tail of the z distribution. From Table 4, Appendix II, $z_{.10} = 1.28$. The rejection region is $z > 1.28$.

Since the observed value of the test statistic does not fall in the rejection region ($z = -2.75 \not> 1.28$), H_0 is not rejected. There is insufficient evidence to indicate the median radium-226 level in soil in southern Dade County exceeds the Environmental Protection Agency limit of 4.0 pCi/L at $\alpha = .10$.

15.5   To determine if fewer than half of the cement specimens have more than 2% coarse granules, we test:

H_0: $\tau = 2$
H_a: $\tau < 2$

The test statistic is S = number of observations less than $2 = 4$.

The observed significance level is: p-value = $P(S \geq 4)$ where S has a binomial distribution with $n = 8$ and $p = .5$. From Table 1, Appendix II, p-value = $P(S \geq 4) = 1 - P(S \leq 3)$ $= 1 - .3633 = .6367$.

The rejection region is to reject H_0 if $\alpha > p$-value.

Since $\alpha = .05 \not> .6367$, H_0 is not rejected. There is insufficient evidence to indicate that fewer than half of the specimens have more than 2% coarse granules at $\alpha = .05$.

15.7   From Exercise 15.6, we showed $P(S_1 \geq c) = P(S_2 \leq n - c)$

For a two-tailed test, the p-value will be $P(S_1 \geq c) + P(S_2 \leq n - c) = 2P(S_1 \geq c)$

15.9 First, rank the observations, then sum the ranks for each sample:

Foggy		Clear/cloudy	
Sample	Rank	Sample	Rank
.270	6.5	.618	12
.241	5	.591	11
.205	2	.225	3
.523	10	.375	9
.112	1		
.330	8		
.270	6.5		
.239	4		
$n_F = 8$	$T_F = 43$	$n_C = 4$	$T_C = 35$

To determine if the oxon/thion ratios of foggy days are different than the oxon/thion ratios on clear/cloudy days, we test:

H_0: The populations of the oxon/thion ratios are the same for foggy and clear/cloudy days

H_a: The probability distribution of the oxon/thion ratios for foggy days is shifted to the right or left of that for clear/cloudy days

The test statistic is the rank sum of the sample with the smaller sample size and is $T_C = 35$.

The rejection region is $T_C \leq T_L$ or $T_C \geq T_U$. From Table 15, Appendix II, with $\alpha = .05$ and $n_C = 4$ and $n_F = 8$, $T_L = 14$ and $T_U = 38$. The rejection region is $T_C \leq 14$ or $T_C \geq 38$.

Since the observed value of the test statistic does not fall in the rejection region ($T_C = 35 \not\leq 14$ and $T_C = 35 \not\geq 38$), H_0 is not rejected. There is insufficient evidence to indicate that the oxon/thion ratios of foggy days are different than the oxon/thion ratios on clear/cloudy days at $\alpha = .05$.

15.11 a. First, rank the observations, and then sum the ranks for each sample:

70-cm		100-cm	
Sample	Rank	Sample	Rank
6.00	1	6.80	2
7.20	3	9.20	5.5
10.20	7.5	8.80	4
13.20	13.5	13.20	13.5
11.40	11	11.20	9.5
13.60	15	14.90	16
9.20	5.5	10.20	7.5
11.20	9.5	11.80	12
$n_A = 8$	$T_A = 66$	$n_B = 8$	$T_B = 70$

To determine if there is a difference in the locations of the cracking torsion moment distributions for the two types of T-beams, we test:

H_0: The populations have identical probability distributions

H_a: The probability distribution for the 70-cm slab width is shifted to the right or left of that for the 100-cm slab width

The test statistic is either rank sum. We will use $T_A = 66$.

The rejection region is $T_A \leq T_L$ or $T_A \geq T_U$. From Table 15b, Appendix II, with $\alpha = .10$ and $n_1 = n_2 = 8$, $T_L = 52$ and $T_U = 84$. The rejection region is $T_A \leq 52$ or $T_A \geq 84$.

Since the observed value of the test statistic does not fall in the rejection region ($T_A = 66$ ≰ 52 and $T_A = 66$ ≱ 84), H_0 is not rejected. There is insufficient evidence to indicate a difference in the locations of the cracking torsion moment distributions for the two types of T-beams at $\alpha = .10$.

15.13   Since $n_1 > 10$ and $n_2 > 10$, it is appropriate to use the large sample approximation. The data are ranked and the ranks are summed:

Rural		Urban	
Sample	Rank	Sample	Rank
3.5	5	24.0	24
8.1	7	29.0	25
1.8	4	16.0	18
9.0	9	21.0	21
1.6	2	107.0	28
23.0	23	94.0	27
1.5	3	141.0	29
1.0	1	11.0	12.5
5.3	6	11.0	12.5
9.8	11	49.0	26
15.0	17	22.0	22
12.0	14.5	13.0	16
8.2	8	18.0	19.5
9.7	10	12.0	14.5
		18.0	19.5
$n_1 = 14$	$T_1 = 120.5$	$n_2 = 15$	$T_2 = 314.5$

To determine if there is a difference in the PCB levels between rural and urban areas, we test:

H_0: The probability distributions for the rural and urban areas are identical

H_a: The probability distributions for the rural and urban areas differ

The test statistic is $z = \dfrac{T_2 - \left[\dfrac{n_1 n_2 + n_2(n_2 + 1)}{2}\right]}{\sqrt{\dfrac{n_1 n_2(n_1 + n_2 + 1)}{12}}} = \dfrac{314.5 - \left[\dfrac{14(15) + 15(16)}{2}\right]}{\sqrt{\dfrac{14(15)(14 + 15 + 1)}{12}}} = 3.91$

The rejection region requires $\alpha/2 = .05/2 = .025$ in each tail of the z distribution. From Table 4, Appendix II, $z_{.025} = 1.96$. The rejection region is $z < -1.96$ or $z > 1.96$.

Since the observed value of the test statistic falls in the rejection region ($z = 3.91 > 1.96$), H_0 is rejected. There is sufficient evidence to indicate a difference exists in the PCB levels between the rural and urban areas at $\alpha = .05$.

15.15 The $4! = 24$ different ways the ranks can be assigned and the corresponding values of T_2 are:

A	B		A	B		A	B		A	B		A	B		A	B
1	3		1	4		1	2		1	4		1	2		1	3
2	4		2	3		3	4		3	2		4	3		4	2
$T_2 = 7$			$T_2 = 7$			$T_2 = 6$			$T_2 = 6$			$T_2 = 5$			$T_2 = 5$	

A	B		A	B		A	B		A	B		A	B		A	B
2	3		2	4		2	1		2	4		2	1		2	3
1	4		1	3		3	4		3	1		4	3		4	1
$T_2 = 7$			$T_2 = 7$			$T_2 = 5$			$T_2 = 5$			$T_2 = 4$			$T_2 = 4$	

A	B		A	B		A	B		A	B		A	B		A	B
3	2		3	4		3	1		3	4		3	1		3	2
1	4		1	2		2	4		2	1		4	2		4	1
$T_2 = 6$			$T_2 = 6$			$T_2 = 5$			$T_2 = 5$			$T_2 = 3$			$T_2 = 3$	

A	B		A	B		A	B		A	B		A	B		A	B
4	2		4	3		4	1		4	3		4	1		4	2
1	3		1	2		2	3		2	1		3	2		3	1
$T_2 = 5$			$T_2 = 5$			$T_2 = 4$			$T_2 = 4$			$T_2 = 3$			$T_2 = 3$	

If we assume all outcomes are equally likely, then each have a probability of 1/24.

$E(T_2) = \dfrac{7}{24} + \dfrac{7}{24} + \dfrac{6}{24} + \dfrac{6}{24} + \dfrac{5}{24} + \dfrac{5}{24} + \dfrac{7}{24} + \dfrac{7}{24} + \dfrac{5}{24} + \dfrac{5}{24} + \dfrac{4}{24} + \dfrac{4}{24}$

$\qquad + \dfrac{6}{24} + \dfrac{6}{24} + \dfrac{5}{24} + \dfrac{5}{24} + \dfrac{3}{24} + \dfrac{3}{24} + \dfrac{5}{24} + \dfrac{5}{24} + \dfrac{4}{24} + \dfrac{4}{24} + \dfrac{3}{24} + \dfrac{3}{24}$

$\quad = \dfrac{120}{24} = 5$

From the formula, $E(T_2) = \dfrac{n_2 n_1 + n_2(n_2 + 1)}{2} = \dfrac{2(2) + 2(2 + 1)}{2} = 5$

15.17   Some preliminary calculations:

Task	Differences (Human−Automated)	Rank of Absolute Values
1	5.0	2
2	−102.2	8
3	−11.1	3
4	−31.5	5
5	−29.3	4
6	4.2	1
7	−43.2	6
8	−52.1	7
$T_+ = 3$ and $T_- = 33$		

To determine if the throughput rates of tasks scheduled by a human differ from those of the automated method, we test:

H_0:   The throughput rates of humans and the automated method are the same
H_a: The probability distribution for the throughput rates of humans is shifted to the right or left of that for the automated method

The test statistic is T, the smaller of T_+ and T_-. For this problem, $T = 3$.

From Table 16, Appendix II, $T_0 = 0$ for $n = 8$ and $\alpha = .01$ in two tails. The rejection region is $T \leq 0$.

Since the observed value of the test statistic does not fall in the rejection region ($T = 3 \not\leq T_0 = 0$), H_0 is not rejected. There is insufficient evidence to indicate that the throughput rates of tasks scheduled by a human differ from those of the automated method at $\alpha = .01$.

15.19   Let population 1 correspond to 0 mg/ml concentration and population 2 correspond to .2 mg/ml concentration. Some preliminary calculations are:

Worker	Differences $(y_1 - y_2)$	Rank of Absolute Values
A	4.4	2
B	7.3	5
C	4.8	4
D	2.4	1
F	4.7	3
H	9.1	6
$T_+ = 21$ and $T_- = 0$		

To determine if a shift in the locations of the cyclic level distributions for the two GBE concentrations exists, we test:

H_0: The two populations have identical probability distributions
H_a: The probability distribution for the 0 concentration level is shifted to the right or to the left of that for the .2 concentration level

The test statistic is T, the smaller of T_+ and T_-. $T = 0$.

From Table 16, Appendix II, $T_0 = 2$ for $n = 6$ and $\alpha = .10$ in two tails. The rejection region is $T \leq 2$.

Since the observed value of the test statistic falls in the rejection region ($T = 0 \leq 2$), H_0 is rejected. There is sufficient evidence to indicate a shift in the locations of the cyclic AMP levels for the two GBE concentrations at $\alpha = .10$.

15.21   Let population 1 be the data from St. Joseph, MO, and population 2 be the data from Iowa Great Lakes. Some preliminary calculations are:

Dates	Differences $(y_1 - y_2)$	Rank of Absolute Value of Differences
Dec 21	189	4
Jan 6	293	7
Jan 21	198	6
Feb 6	193	5
Feb 21	188	3
Mar 7	171	2
Mar 21	154	1
$T_+ = 28$ and $T_- = 0$		

To determine if the irradiation levels for the two locations differ, we test:

H_0: The two populations have identical probability distributions
H_a: The distribution for the irradiation levels differ for the two locations

The test statistic is the smaller of the rank sums and is $T_- = 0$.

The rejection region is $T_- \leq T_0$ where T_0 is based on $n = 7$ and $\alpha = .02$ in two tails. From Table 16, Appendix II, $T_0 = 0$. The rejection region is $T_- \leq 0$. Since the observed value of the test statistic falls in the rejection region ($T_- = 0 \leq 0$), H_0 is rejected. There is sufficient evidence to indicate the irradiation levels differ for the two locations at $\alpha = .02$.

15.23   For the Wilcoxon signed ranks test, show that

$$T_+ + T_- = \frac{n(n + 1)}{2}$$ where n is the number of non-zero differences that are ranked.

There are n non-zero differences. These are ranked by their magnitude, ignoring their signs. Thus, the sum of the ranks is $\dfrac{n(n+1)}{2}$

T_+ is the sum of the ranks associated with the positive differences and T_- is the sum of the ranks associated with the negative differences. Together, $T_+ + T_-$ must sum to the sum of all the ranks or $\dfrac{n(n+1)}{2}$

Thus, $T_+ + T_- = \dfrac{n(n+1)}{2}$

15.25 The eight ways of ranking $n = 2$ differences and the corresponding values of T_+ are:

-1	-2	-1	-2	$+1$	$+2$	$+1$	$+2$
-2	-1	$+2$	$+1$	-2	-1	$+2$	$+1$

$T_+ = 0 \quad T_+ = 0 \quad T_+ = 2 \quad T_+ = 1 \quad T_+ = 1 \quad T_+ = 2 \quad T_+ = 3 \quad T_+ = 3$

If each case is equally likely, then each has a probability of occurring of 1/8.

$$E(T_+) = 0\left[\frac{1}{8}\right] + 0\left[\frac{1}{8}\right] + 2\left[\frac{1}{8}\right] + 1\left[\frac{1}{8}\right] + 1\left[\frac{1}{8}\right] + 2\left[\frac{1}{8}\right] + 3\left[\frac{1}{8}\right] + 3\left[\frac{1}{8}\right]$$

$$= \frac{12}{8} = 1.5$$

For $n = 2$, $E(T_+) = \dfrac{n(n+1)}{4} = \dfrac{2(2+1)}{4} = \dfrac{6}{4} = 1.5$

15.27 Some preliminary calculations are:

No Training Sample	Rank	Computer Assisted Sample	Rank	Computer + Workshop Sample	Rank
16	6.5	19	11.5	12	2
18	9	22	15.5	19	11.5
11	1	13	3	18	9
14	4	15	5	22	15.5
23	17	20	13	16	6.5
		18	9	25	18
		21	14		
$n_1 = 5$ $R_1 = 37.5$		$n_2 = 7$ $R_2 = 71$		$n_3 = 6$ $R_2 = 62.5$	

a.   To determine if the scores differ in location, we test:

H_0:   The three population probability distributions are identical
H_a: At least two of the three population probability distributions differ in location

The test statistic is $H = \dfrac{12}{n(n+1)}\sum\dfrac{R_i^2}{n_i} - 3(n+1)$

$$= \dfrac{12}{18(18+1)}\left[\dfrac{37.5^2}{5} + \dfrac{71^2}{7} + \dfrac{62.5^2}{6}\right] - 3(18+1) = .9802$$

From Table 8, Appendix II, $\chi_{.01}^2 = 9.21034$ with df $= k - 1 = 3 - 1 = 2$. The rejection region is $H > 9.21034$.

Since the observed value of the test statistic does not fall in the rejection region ($H = .9802 \not> 9.21034$), H_0 is not rejected. There is insufficient evidence to indicate the distributions of scores differ in location for the three types of training at $\alpha = .01$.

15.29 Some preliminary calculations are:

Bizcomp 1012		Cermetek 212A		Smartmodem 1200		Vadic 3451	
Sample	Rank	Sample	Rank	Sample	Rank	Sample	Rank
87	15	81	12	69	4	98	20
80	10.5	66	3	72	6.5	78	9
91	18	52	1	70	5	94	19
63	2	90	16.5	83	13	90	26.5
72	6.5	75	8	80	10.5	86	14
$n_1 = 5$	$T_1 = 52$	$n_2 = 5$	$T_2 = 40.5$	$n_3 = 5$	$T_3 = 39$	$n_4 = 5$	$T_4 = 78.5$

To determine if a difference among the performance ratings of the four smart modems exists, we test:

H_0: The four population distributions are identical
H_a: At least two of the four population distributions differ in location

The test statistic is $H = \dfrac{12}{n(n+1)}\sum\dfrac{T_i^2}{n_i} - 3(n+1)$

$$= \dfrac{12}{20(20+1)}\left[\dfrac{52^2}{5} + \dfrac{40.5^2}{5} + \dfrac{39^2}{5} + \dfrac{78.5^2}{5}\right] - 3(20+1)$$

$$= 5.7286$$

From Table 8, Appendix II, $\chi_{.10}^2 = 6.25139$ with df $= k - 1 = 4 - 1 = 3$. The rejection region is $H > 6.25139$.

Since the observed value of the test statistic does not fall in the rejection region ($H = 5.7286 \not> 6.25139$), H_0 is not rejected. There is insufficient evidence to indicate a difference among the four smart modems at $\alpha = .10$.

. . . .
Nonparametric Statistics   309

15.31 Some preliminary calculations are:

South Sample	Rank	Central Sample	Rank	North Sample	Rank
40.6	14	41.1	15	25.6	2
42.0	17	38.3	12	36.4	9
37.5	10	40.2	13	28.2	4
38.1	11	33.5	7	31.3	6
41.9	16	35.7	8	29.5	5
				22.8	1
				27.5	3
$n_1 = 5$ $T_1 = 68$		$n_2 = 5$ $T_2 = 55$		$n_3 = 7$ $T_3 = 30$	

To determine if a difference in reactivity exists among the three locations, we test:

H_0: The reactivity distributions for the three locations are identical
H_a: At least two of the distributions differ in location

The test statistic is $H = \dfrac{12}{n(n + 1)} \sum \dfrac{T_i^2}{n_i} - 3(n + 1)$

$$= \frac{12}{17(17 + 1)} \left[\frac{68^2}{5} + \frac{55^2}{5} + \frac{30^2}{7} \right] - 3(17 + 1) = 11.0342$$

From Table 8, Appendix II, $\chi_{.05}^2 = 5.99147$ with df $= k - 1 = 3 - 1 = 2$. The rejection region is $H > 5.99147$.

Since the observed value of the test statistic falls in the rejection region ($H = 11.0342 > 5.99147$), H_0 is rejected. There is sufficient evidence to indicate a difference in the reactivity of phosphoric rock mined at the three locations at $\alpha = .05$.

15.33 Some preliminary calculations are:

Level	Ranks—Computer Programs STAAD-III (1)	STAAD-III (2)	Drift
1	3	1.5	1.5
2	3	1	2
3	3	1	2
4	3	1	2
5	3	1	2
	$T_1 = 15$	$T_2 = 5.5$	$T_3 = 9.5$

To determine if the distributions of lateral displacement estimated by the three computer programs differ, we test:

H_0: The probability distributions for the three computer programs are identical
H_a: At least two of the probability distributions differ in location

The test statistic is $F_r = \dfrac{12}{bk(k+1)}\sum T_i^2 - 3b(k+1)$

$\qquad = \dfrac{12}{5(3)(3+1)}[15^2 + 5.5^2 + 9.5^2] - 3(5)(3+1) = 9.1$

From Table 8, Appendix II, $\chi^2_{.05} = 5.99147$ with df $= k - 1 = 3 - 1 = 2$. The rejection region is $F_r > 5.99147$.

Since the observed value of the test statistic falls in the rejection region ($F_r = 9.1 > 5.99147$), H_0 is rejected. There is sufficient evidence to indicate that the distributions of lateral displacement estimated by the three computer programs differ at $\alpha = .05$.

15.35   Some preliminary calculations are:

| | | | Ranks−Colors | | | | |
Subject	Green/ Black	White/ Black	Yellow/ Black	Orange/ White	Yellow	Yellow/ Amber	Yellow/ Orange
1	4.5	3	4.5	1	6	7	2
2	4.5	3	6.5	2	6.5	4.5	1
3	3.5	3.5	6	1	5	7	2
4	5	6	3.5	1.5	3.5	7	1.5
5	6	3.5	3.5	2	6	6	1
6	6	3	4	2	6	6	1
7	3.5	5	6	1	3.5	7	2
8	4	3	5.5	1.5	5.5	7	1.5
9	6	6	3.5	2	6	3.5	1
10	5.5	3.5	3.5	2	5.5	7	1
	$T_1 = 48.5$	$T_2 = 39.5$	$T_3 = 46.5$	$T_4 = 16$	$T_5 = 53.5$	$T_6 = 62$	$T_7 = 14$

a.   To determine if there is a difference among the color combinations, we test:

H_0: The probability distributions for the 7 color combinations are identical
H_a: At least two of the probability distributions differ in location

The test statistic is $F_r = \dfrac{12}{bk(k+1)}\sum T_i^2 - 3b(k+1)$

$\qquad = \dfrac{12}{10(7)(7+1)}[48.5^2 + 39.5^2 + 46.5^2 + 16^2 + 53.5^2 + 62^2 + 14^2]$

$\qquad\qquad - 3(10)(7+1) = 43.5643$

From Table 8, Appendix II, $\chi^2_{.05} = 12.5916$ with df $= k - 1 = 7 - 1 = 6$. The rejection region is $F_r > 12.5916$

Since the observed value of the test statistic falls in the rejection region ($F_r = 43.5643 > 12.5916$), H_0 is rejected. There is sufficient evidence to indicate a difference among the preference scores for the seven display colors at $\alpha = .05$.

Nonparametric Statistics

b. To determine if yellow/amber is preferred over green/black, we use a Friedman test with only the two specified treatments. The observations within each subject must be reranked:

Ranks

Subject	Green/Black	Yellow/Amber
1	1	2
2	1.5	1.5
3	1	2
4	1	2
5	1.5	1.5
6	1.5	1.5
7	1	2
8	1	2
9	2	1
10	1	2
	$T_1 = 12.5$	$T_2 = 17.5$

To determine if users prefer yellow/amber over green/black, we test:

H_0: The probability distributions for the 2 color combinations are identical

H_a: The two probability distributions differ in location

The test statistic is $F_r = \dfrac{12}{bk(k + 1)} \sum T_i^2 - 3b(k + 1)$

$$= \dfrac{12}{10(2)(2 + 1)}[12.5^2 + 17.5^2] - 3(10)(2 + 1) = 2.5$$

From Table 8, Appendix II, $\chi_{.05}^2 = 3.84146$ with df $= k - 1 = 2 - 1 = 1$. The rejection region is $F_r > 3.84146$.

Since the observed value of the test statistic does not fall in the rejection region ($F_r = 2.5 \ngtr 3.84146$), H_0 is not rejected. There is insufficient evidence to indicate that users prefer yellow/amber over green/black at $\alpha = .05$.

c. To determine if yellow/amber is preferred over yellow, we use a Friedman test with only the two specified treatments. The observations within each subject must be reranked:

Ranks

Subject	Yellow	Yellow/Amber
1	1	2
2	2	1
3	1	2
4	1	2
5	1.5	1.5
6	1.5	1.5
7	1	2
8	1	2
9	2	1
10	1	2
	$T_1 = 13$	$T_2 = 17$

To determine if users prefer yellow/amber over yellow, we test:

H_0: The probability distributions for the 2 color combinations are identical
H_a: The two probability distributions differ in location

The test statistic is $F_r = \dfrac{12}{bk(k+1)}\sum T_i^2 - 3b(k+1)$

$$= \dfrac{12}{10(2)(2+1)}[13^2 + 17^2] - 3(10)(2+1) = 1.6$$

From Table 8, Appendix II, $\chi^2_{.05} = 3.84146$ with df $= k - 1 = 2 - 1 = 1$. The rejection region is $F_r > 3.84146$.

Since the observed value of the test statistic does not fall in the rejection region ($F_r = 1.6 \ngtr 3.84146$), H_0 is not rejected. There is insufficient evidence to indicate that users prefer yellow/amber over yellow at $\alpha = .05$.

15.37 Some preliminary calculations are:

Ear	Spray A	Spray B	Spray C
1	2	3	1
2	2	3	1
3	1	3	2
4	3	2	1
5	2	1	3
6	1	3	2
7	2.5	2.5	1
8	2	3	1
9	2	3	1
10	2	3	1
	$T_1 = 19.5$	$T_2 = 26.5$	$T_3 = 14$

Ranks−Sprays

To determine if there are differences among the probability distributions of the amounts of aflatoxin present for the three sprays, we test:

H_0: The probability distributions for the three sprays are identical
H_a: At least two of the probability distributions differ in location

The test statistic is $F_r = \dfrac{12}{bk(k+1)}\sum T_i^2 - 3b(k+1)$

$$= \dfrac{12}{10(3)(3+1)}[19.5^2 + 26.5^2 + 14^2] - 3(10)(3+1) = 7.85$$

From Table 8, Appendix II, $\chi^2_{.05} = 5.99147$ with df $= k - 1 = 3 - 1 = 2$. The rejection region is $F_r > 5.99147$.

Since the observed value of the test statistic falls in the rejection region ($F_r = 7.85 >$ 5.99147), H_0 is rejected. There is sufficient evidence to indicate there are differences among the probability distributions of the amounts of aflatoxin present for the three sprays at $\alpha = .05$.

15.39   a.   Some preliminary calculations:

Annealing Time, x	Passivation Potential, y	x-Rank	y-Rank	Difference d_i	d_i^2
10	−408	1	1	0	0
20	−400	2	2	0	0
45	−392	3	3	0	0
90	−379	4	5	−1	1
120	−385	5	4	1	1
					$\sum d_i^2 = 2$

$$r_s = 1 - \frac{6\sum d_i^2}{n(n^2 - 1)} = 1 - \frac{6(2)}{5(5^2 - 1)} = .9$$

Since r_s is close to 1, there is a strong positive correlation between annealing time and passivation potential.

b.   To determine if there is a significant correlation between annealing time and passivation potential, we test:

H_0: $\rho_s = 0$
H_a: $\rho_s \neq 0$

The test statistic is $r_s = .9$.

From Table 17, Appendix II, $r_{s,\alpha} = .9$ with $\alpha = .10$ and $n = .5$.

The rejection region is $|r_s| > .90$.

Since the observed value of the test statistic does not fall in the rejection region ($r_s = .90 \not> .9$), H_0 is not rejected. There is insufficient evidence to indicate that a significant correlation between annealing time and passivation potential at $\alpha = .10$.

15.41   Some preliminary calculations are:

$$\sum u = 561 \qquad \sum v = 561$$
$$\sum u^2 = 12,521 \qquad \sum v^2 = 12,435.5 \qquad \sum uv = 11,991.5$$
$$SS_{uv} = \sum uv - \frac{(\sum u)(\sum v)}{n} = 11,991.5 - \frac{561(561)}{33} = 2454.5$$

$$SS_{uu} = \sum u^2 - \frac{(\sum u)^2}{n} = 12{,}521 - \frac{561^2}{33} = 2984$$

$$SS_{vv} = \sum v^2 - \frac{(\sum v)^2}{n} = 12{,}435.5 - \frac{561^2}{33} = 2898.5$$

Spearman's rank correlation coefficient is

$$r_s = \frac{SS_{uv}}{\sqrt{SS_{uu}SS_{vv}}} = \frac{2454.5}{\sqrt{2984(2898.5)}} = .835$$

To determine if the AUTOMARK and instructor grade assignments are positively correlated, we test:

H_0: $\rho_s = 0$
H_a: $\rho_s > 0$

The test statistic is $r_s = .835$.

Notice that Table 17, Appendix II, does not include $n = 33$. We will use $n = 30$ to form the rejection region. If H_0 is rejected at $n = 30$, H_0 would also be rejected at $n = 33$ as it becomes 'easier' to reject H_0 as the sample size increases. If H_0 is not rejected at $n = 30$, no conclusion can be made for $n = 33$. From Table 17, $r_{s,\alpha} = .305$ for $\alpha = .05$ and $n = 30$. The rejection region is $r_s > .305$.

Since the observed value of the test statistic falls in the rejection region ($r_s = .835 > .305$), H_0 is rejected. There is sufficient evidence to indicate a significant positive correlation between the AUTOMARK instructor grade assignments at $\alpha = .05$.

15.43 Some preliminary calculations are:

Integral-Fin Tube	Transfer Enhancement, y	Unflooded Area Ratio, x	Differences, # negatives	$y_j - y_i \ (i < j)$ # positives
1	2.9	1.21	—	—
2	3.2	1.26	0	1
3	2.8	1.32	2	0
4	3.5	1.32	0	3
5	3.7	1.54	0	4
6	4.1	1.58	0	5
7	4.2	1.62	0	6
8	4.5	1.64	0	7
9	4.9	1.70	0	8
10	4.7	1.77	1	8
11	4.5	1.78	2	8
12	4.6	1.88	2	9
13	4.9	1.88	0	12
14	4.4	1.93	6	7
15	5.3	1.95	0	14
16	5.2	2.00	1	14
17	5.1	2.04	2	14
18	6.1	2.12	0	17
19	5.2	2.24	2	16
20	5.3	2.26	1	18
21	6.7	2.37	0	20
22	5.8	2.47	2	19
23	7.0	2.47	0	22
24	6.0	2.77	3	20
		Totals:	24	252

To determine if a linear relationship exists, we test:

H_0: $\beta_1 = 0$
H_a: $\beta_1 > 0$

The test statistic is $C = (-1)(\text{\# negatives}) + (1) \, (\text{\# positives})$
$= (-1)(24) + (1)(252) = 228$

The observed significance level is $P(x \geq C)$. From Table 18, Appendix II, $P(x \geq C) = P(x \geq 228) \approx 0$.

Since $\alpha = .10$ and the p-value ≈ 0, we reject H_0. There is sufficient evidence to indicate a positive linear relationship between unflooded area ratio and heat transfer enhancement at $\alpha = .10$.

15.45 The 36 arrangements along with the corresponding r_s values are:

u	v		u	v		u	v		u	v		u	v		u	v
1	1		1	1		1	2		1	2		1	3		1	3
2	2		2	3		2	1		2	3		2	1		2	2
3	3		3	2		3	3		3	1		3	2		3	1
$r_s = 1$			$r_s = .5$			$r_s = .5$			$r_s = -.5$			$r_s = -.5$			$r_s = -1$	

u	v		u	v		u	v		u	v		u	v		u	v
1	1		1	1		1	2		1	2		1	3		1	3
3	2		3	3		3	1		3	3		3	1		3	2
2	3		2	2		2	3		2	1		2	2		2	1
$r_s = .5$			$r_s = 1$			$r_s = -.5$			$r_s = .5$			$r_s = -1$			$r_s = -.5$	

u	v		u	v		u	v		u	v		u	v		u	v
2	1		2	1		2	2		2	2		2	3		2	3
1	2		1	3		1	1		1	3		1	1		1	2
3	3		3	2		3	3		3	1		3	2		3	1
$r_s = .5$			$r_s = -.5$			$r_s = 1$			$r_s = -1$			$r_s = .5$			$r_s = -.5$	

u	v		u	v		u	v		u	v		u	v		u	v
2	1		2	1		2	2		2	2		2	3		2	3
3	2		3	3		3	1		3	3		3	1		3	2
1	3		1	2		1	3		1	1		1	2		1	1
$r_s = -.5$			$r_s = .5$			$r_s = -1$			$r_s = 1$			$r_s = -.5$			$r_s = .5$	

u	v		u	v		u	v		u	v		u	v		u	v
3	1		3	1		3	2		3	2		3	3		3	3
1	2		1	3		1	1		1	3		1	1		1	2
2	3		2	2		2	3		2	1		2	2		2	1
$r_s = -.5$			$r_s = -1$			$r_s = .5$			$r_s = -.5$			$r_s = 1$			$r_s = .5$	

u	v		u	v		u	v		u	v		u	v		u	v
3	1		3	1		3	2		3	2		3	3		3	3
2	2		2	3		2	1		2	3		2	1		2	2
1	3		1	2		1	3		1	1		1	2		1	1
$r_s = -1$			$r_s = -.5$			$r_s = -.5$			$r_s = .5$			$r_s = .5$			$r_s = 1$	

Since each of the 36 arrangements are equally likely, each has a probability of 1/36.

$$E(r_s) = \sum r_s p(r_s) = 1\left[\frac{1}{36}\right] + .5\left[\frac{1}{36}\right] + .5\left[\frac{1}{36}\right] + \cdots + 1\left[\frac{1}{36}\right] = 0$$

15.47   Some preliminary calculations are:

	Ranks	
A	**B**	**C**
3.5	27	1.5
11	21	7
19	17	12
15	30	8
16	23	10
1.5	25	3.5
13	29	6
18	26	14
24	22	9
20	28	5
$n_1 = 10$	$n_2 = 10$	$n_3 = 10$
$T_1 = 141$	$T_2 = 248$	$T_3 = 76$

To determine if there is a difference in location among the distributions of damage rates corresponding to the three treatments, we test:

H_0:   The three population probability distributions are identical
H_a: At least two of the three population probability distributions differ in location

The test statistic is $H = \dfrac{12}{n(n + 1)}\sum \dfrac{T_i^2}{n_i} - 3(n + 1)$

$$= \frac{12}{30(30 + 1)}\left[\frac{141^2}{10} + \frac{248^2}{10} + \frac{76^2}{10}\right] - 3(30 + 1) = 19.4658$$

From Table 8, Appendix II, $\chi_{.05}^2 = 5.99147$ with df $= k - 1 = 3 - 1 = 2$. The rejection region is $H > 5.99147$.

Since the observed value of the test statistic falls in the rejection region ($H = 19.4568 > 5.99147$), H_0 is rejected. There is sufficient evidence to indicate a difference among the damage rates for the three treatments at $\alpha = .05$.

15.49 The following information is computed from the data:

Data Set	u_i	v_i	u_i^2	v_i^2	$u_i v_i$
1	10.5	10	110.25	100	105
2	6.5	6	42.25	36	39
3	14	14	196	196	196
4	1.5	1	2.25	1	1.5
5	12.5	12	156.25	144	150
6	4.5	4	20.25	16	18
7	10.5	11	110.25	121	115.5
8	9	9	81	81	81
9	4.5	7	20.25	49	31.5
10	15	15	225	225	225
11	3	3	9	9	9
12	12.5	13	156.25	169	162.5
13	6.5	5	42.25	25	32.5
14	1.5	2	2.25	4	3
15	8	8	64	64	64
	120	120	1237.5	1240	1233.5

a. $$SS_{uv} = \sum uv - \frac{(\sum u)(\sum v)}{n} = 1233.5 - \frac{120(120)}{15} = 273.5$$

$$SS_{uu} = \sum u^2 - \frac{(\sum u)^2}{n} = 1237.5 - \frac{120^2}{15} = 277.5$$

$$SS_{vv} = \sum v^2 - \frac{(\sum v)^2}{n} = 1240 - \frac{120^2}{15} = 280$$

$$r_s = \frac{SS_{uv}}{\sqrt{SS_{uu}SS_{vv}}} = \frac{273.5}{\sqrt{277.5(280)}} = .9812$$

Since the value of r_s is very close to 1, there is a very strong positive correlation between number of records and number of disk I/O's.

b. To determine if the number of records and the number of disk I/O's are positively correlated, we test:

$$H_0: \rho_s = 0$$
$$H_a: \rho_s > 0$$

The test statistic is $r_s = .9812$.

From Table 17, Appendix II, $r_{s,\alpha} = .623$ with $\alpha = .01$ and $n = 15$.

The rejection region is $r_s > .623$.

Since the observed value of the test statistic falls in the rejection region ($r_s = .981 > .623$), H_0 is rejected. There is sufficient evidence to indicate a significant positive correlation between the number of records and the number of disk I/O's at $\alpha = .01$.

15.51   Some preliminary calculations are:

		Ranks		
Week	Orlando	Turkey Point	St. Lucie	Crystal River
1	4	1.5	1.5	3
2	1	3	2	4
3	1.5	4	1.5	3
4	1	2	4	3
5	1	4	3	2
6	4	3	1	2
7	1	3	2	4
8	1	2.5	2.5	4
9	4	3	1	2
10	1	3	2	4
	$T_1 = 19.5$	$T_2 = 29$	$T_3 = 20.5$	$T_4 = 31$

a.   To determine if a difference in the radioactivity of air particulates at the four Florida sites exists, we test:

H_0: The probability distributions for the four cities are identical
H_a: At least two of the probability distributions differ in location

The test statistic is $F_r = \dfrac{12}{bk(k+1)} \sum T_i^2 - 3b(k+1)$

$$= \frac{12}{10(4)(4+1)}[19.5^2 + 29^2 + 20.5^2 + 31^2] - 3(10)(4+1)$$

$$= 6.15$$

From Table 8, Appendix II, $\chi^2_{.05} = 7.81473$ with df $= k - 1 = 4 - 1 = 3$. The rejection region is $F_r > 7.81473$.

Since the observed value of the test statistic does not fall in the rejection region ($F_r = 6.15 \ngtr 7.81473$), H_0 is not rejected. There is insufficient evidence to indicate radioactivity of air particulates differ among the four Florida cities at $\alpha = .05$.

b.   To compare only Crystal River and Orlando, the observations must be reranked:

Ranks

Week	Orlando	Crystal River
1	2	1
2	1	2
3	1	2
4	1	2
5	1	2
6	2	1
7	1	2
8	1	2
9	2	1
10	1	2
	$T_1 = 13$	$T_2 = 17$

To determine if a difference in the radioactivity of air particulates at the two Florida sites exists, we test:

H_0: The probability distributions for the two cities are identical
H_a: The two probability distributions differ in location

The test statistic is $F_r = \dfrac{12}{bk(k+1)}\sum T_i^2 - 3b(k+1)$

$$= \dfrac{12}{10(2)(2+1)}[13^2 + 17^2] - 3(10)(2+1) = 1.6$$

From Table 8, Appendix II, $\chi^2_{.05} = 3.84146$ with df $= k - 1 = 2 - 1 = 1$. The rejection region is $F_r > 3.84146$.

Since the observed value of the test statistic does not fall in the rejection region ($F_r = 1.6 \not> 3.84146$), H_0 is not rejected. There is insufficient evidence to indicate radioactivity of air particulates differ among the two Florida cities at $\alpha = .05$.

15.53 Some preliminary calculations are:

Ranks

Plane 1		Plane 2	
3	7	9	19
16	1	5	8
14	20	2	17
4	13	15	12
18	10	11	6
$n_1 = 10$		$n_2 = 10$	
$T_1 = 106$		$T_2 = 104$	

To determine if there is a shift in location of the lifelength distributions for the two planes, we test:

H_0: The populations have identical probability distributions
H_a: The probability distribution for plane 1 is shifted to the right or left of that for plane 2

The test statistic is either rank sum, T_1 or T_2. Let $T = 106$.

From Table 15a, Appendix II, $T_L = 79$ and $T_U = 131$ for $n_1 = n_2 = 10$ and $\alpha = .05$ in two tails. The rejection region is $T \leq 79$ or $T \geq 131$.

Since the observed value of the test statistic does not fall in the rejection region, ($T = 106 \nleq 79$ and $T = 106 \ngeq 131$), H_0 is not rejected. There is insufficient evidence to indicate a shift in the location of the lifelength distributions for the two planes at $\alpha = .05$.

15.55   Some preliminary calculations are:

		Ranks		
5	**10**	**15**	**20**	**40**
1	18.5	5	11	21
16.5	22.5	9	22.5	25
2	13.5	4	18.5	16.5
6	7.5	3	15	24
10	12	7.5	13.5	20
$n_1 = 5$	$n_2 = 5$	$n_3 = 5$	$n_4 = 5$	$n_5 = 5$
$T_1 = 35.5$	$T_2 = 74$	$T_3 = 28.5$	$T_4 = 80.5$	$T_5 = 106.5$

To determine if there is a difference among the probability distributions of the depths of penetration of oxygen for the five water depths, we test:

H_0: The five population probability distributions are identical
H_a: At least two of the probability distributions differ in location

The test statistic is $H = \dfrac{12}{n(n+1)} \sum \dfrac{T_i^2}{n_i} - 3(n+1)$

$$= \frac{12}{25(25+1)} \left[\frac{35.5^2}{5} + \frac{74^2}{5} + \frac{28.5^2}{5} + \frac{80.5^2}{5} + \frac{106.5^2}{5} \right]$$

$$- 3(25+1) = 15.6775$$

From Table 8, Appendix II, $\chi_{.05}^2 = 9.48773$ with df $= k - 1 = 5 - 1 = 4$. The rejection region is $H > 9.48773$.

Since the observed value of the test statistic falls in the rejection region ($H = 15.6775 >$ 9.48773), H_0 is rejected. There is sufficient evidence to indicate a difference among the probability distributions of the depths of penetration of oxygen for the five water depths at $\alpha = .05$.

15.57 a.   Some preliminary calculations are:

	Ranks		
u_i	v_i	d_i	d_i^2
1	1	0	0
2	2	0	0
3	4	−1	1
4	3	1	1
5	5	0	0
			2

Since there are no ties in the ranks, we can use the shortcut formula to compute r_s.

$$r_s = 1 - \frac{6\sum d_i^2}{n(n^2 - 1)} = 1 - \frac{6(2)}{5(5^2 - 1)} = .90$$

b.   To determine if a positive correlation exists, we test:

$H_0: \rho_s = 0$
$H_a: \rho_s > 0$

The test statistic is $r_s = .90$.

Table 17, Appendix II, does not have the appropriate α-level for this problem. However, if H_0 can be rejected at $\alpha = .05$, it can also be rejected at $\alpha = .10$. If H_0 is not rejected at $\alpha = .05$, no conclusion can be made at $\alpha = .10$. From Table 17, $r_{s,\alpha} = .90$ for $\alpha = .05$ and $n = 5$. The rejection region is $r_s > .90$.

Since the observed value of the test statistic is not in the rejection region ($r_s = .90 \not>$ .90), H_0 is not rejected at $\alpha = .05$. However, since the tabled value for $\alpha = .10$ would be smaller than that for $\alpha = .05$, it is reasonable to assume that H_0 would be rejected at $\alpha = .10$. There is sufficient evidence to indicate a positive correlation between the number of users and the response time at $\alpha = .10$.

15.59 Some preliminary calculations are:

38°F	Rank	42°F	Rank	46°F	Rank	50°F	Rank
22	16	15	3	14	2	17	6.5
24	18.5	21	14	28	22	18	8.5
16	4.5	26	21	21	14	13	1
18	8.5	16	4.5	19	10.5	20	12
19	10.5	25	20	24	18.5	21	14
		17	6.5	23	17		
$n_1 = 5$ $T_1 = 58$		$n_2 = 6$ $T_2 = 69$		$n_3 = 6$ $T_3 = 84$		$n_4 = 5$ $T_4 = 42$	

To determine if the weights produced by the four temperatures differ, we test:

H_0: The four population probability distributions are identical
H_a: At least two of the four probability distributions differ

The test statistic is $H = \dfrac{12}{n(n + 1)} \sum_i \dfrac{T_i^2}{n_i} - 3(n + 1)$

$$= \frac{12}{22(22 + 1)}\left[\frac{58^2}{5} + \frac{69^2}{6} + \frac{84^2}{6} + \frac{42^2}{5}\right] - 3(22 + 1) = 2.03$$

From Table 8, Appendix II, $\chi^2_{.10} = 6.25139$ with df $= k - 1 = 4 - 1 = 3$.

The rejection region is $H > 6.25139$.

Since the observed value of the test statistic does not fall in the rejection region ($H = 2.03 \ngtr$ 6.25139), H_0 cannot be rejected. There is insufficient evidence to indicate that at least two of the temperatures tend to produce larger weight increases that differ at $\alpha = .10$.

CHAPTER SIXTEEN

· ·

Statistical Process and Quality Control

16.1   First, we computer $\bar{x}$ and s for the sample.

$$\sum x = 29.97 \quad \sum x^2 = 44.911408$$

$$\bar{x} = \frac{\sum x}{n} = \frac{29.97}{20} = 1.4985$$

$$s^2 = \frac{\sum x^2 - \frac{(\sum x)^2}{n}}{n - 1} = \frac{44.911408 - \frac{29.97^2}{20}}{20 - 1} = .000071736$$

$$s = \sqrt{.000071736} = .0085$$

Now, compute the lower and upper control limits:

$$\text{LCL} = \bar{x} - 3s = 1.4985 - 3(.0085) = 1.473$$
$$\text{UCL} = \bar{x} + 3s = 1.4985 + 3(.0085) = 1.524$$

The control chart is:

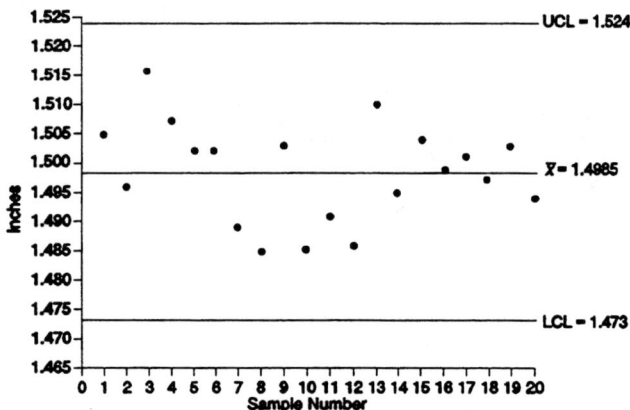

The process appears to be in control because all observations are within the control limits.

16.3   a.   First, we compute $\bar{x}$ and s for the sample.

$$\sum x = 117.9 \qquad \sum x^2 = 697.39$$

$$\bar{x} = \frac{\sum x}{n} = \frac{117.9}{20} = 5.895$$

$$s^2 = \frac{\sum x^2 - \frac{(\sum x)^2}{n}}{n - 1} = \frac{697.39 - \frac{117.9^2}{20}}{20 - 1} = .1247105$$

$$s = \sqrt{.1247105} = .35314$$

Now, compute the lower and upper control limits:

$$\text{LCL} = \bar{\bar{x}} - 3s = 5.895 - 3(.35314) = 4.83558$$
$$\text{UCL} = \bar{\bar{x}} + 3s = 5.895 + 3(.35314) = 6.95442$$

The control chart is:

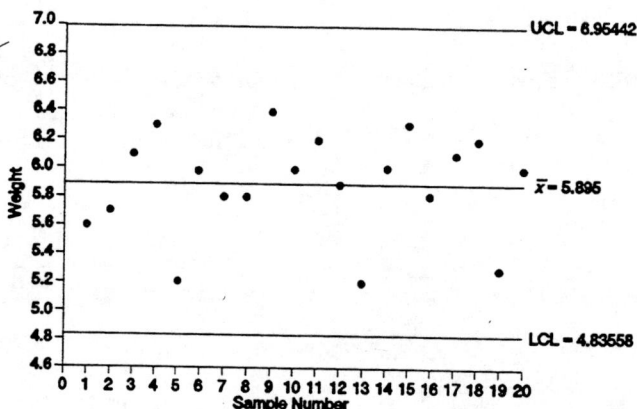

b.   The process appears to be in control because all observations are within the control limits.

16.5   For each of the sampling times, we will compute the sample mean pin lengths and the sample ranges.

30-Minute Interval	Sample Mean $\bar{x}_i$	Range R_i
1	1.020	.06
2	.970	.08
3	.996	.04
4	.994	.05
5	1.020	.09
6	.984	.14
7	.974	.09
8	1.012	.08
9	.988	.04
10	1.000	.07

a.   The center line for an $\bar{x}$-chart is $\bar{\bar{x}} = \dfrac{\sum\limits_{i=1}^{10} \bar{x}_i}{10} = \dfrac{1.02 + .907 + \cdots + 1.000}{10} = .9958$

b.   The upper and lower control limits are found using:

UCL: $\bar{\bar{x}} + A_2\bar{R}$
LCL: $\bar{\bar{x}} - A_2\bar{R}$

where A_2 is given in Table 19, Appendix II and $\bar{R}$ is the average range for the samples.

From Table 19, Appendix II, $A_2 = .577$ for $n = 5$.

$$\bar{R} = \frac{\sum\limits_{i=1}^{k} R_i}{k} = \frac{.06 + .08 + \cdots + .07}{10} = .074$$

Thus, $\text{UCL} = \bar{\bar{x}} + A_2\bar{R} = .9958 + .577(.074) = 1.0385$
         $\text{LCL} = \bar{\bar{x}} - A_2\bar{R} = .9958 - .577(.074) = .9531$

c.   The $\bar{x}$-chart is:

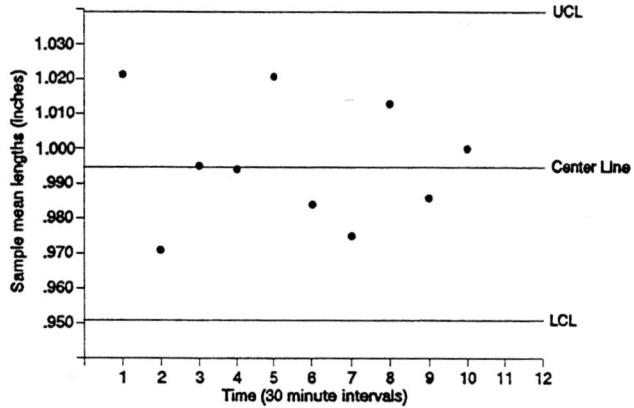

d.   The Defense Department's specification is that the firing pins be $1.00 \pm .08$ or from .92 to 1.08. Since all the sample means are well within this range, the manufacturing process appears to be in control.

16.7   a.   For each of the days, we will compute the sample mean weights and the sample range.

Day	Sample Mean $\bar{x}_i$	Range R_i	Day	Sample Mean $\bar{x}_i$	Range R_i
1	5.73333	.2	11	5.86667	.6
2	6.00000	.6	12	5.83333	.2
3	5.80000	.8	13	5.46667	.5
4	6.00000	.5	14	6.03333	.1
5	5.80000	1.1	15	5.96667	.6
6	5.96667	1.5	16	6.03333	.4
7	5.86667	.4	17	6.36667	.5
8	6.00000	.4	18	5.86667	.5
9	5.96667	.8	19	5.40000	.2
10	5.93333	.4	20	6.03333	.1

The center line for the $\bar{x}$-chart is

$$\bar{\bar{x}} = \frac{\sum\limits_{i=1}^{20} \bar{x}_i}{20} = \frac{5.73333 + 6 + \cdots + 6.03333}{20} = \frac{117.93333}{20} = 5.89667$$

The upper and lower control limits are found using:

UCL: $\bar{\bar{x}} + A_2\bar{R}$
LCL: $\bar{\bar{x}} - A_2\bar{R}$

where A_2 is given in Table 19, Appendix II and $\bar{R}$ is the average range for the samples.

From Table 19, Appendix II, $A_2 = 1.023$ for $n = 3$.

$$\bar{R} = \frac{\sum_{i=1}^{k} R_i}{k} = \frac{.2 + .6 + \cdots + .1}{20} = \frac{10.4}{20} = .52$$

Thus,   UCL $= \bar{\bar{x}} + A_2\bar{R} = 5.89667 + 1.023(.52) = 6.42863$
LCL $= \bar{\bar{x}} - A_2\bar{R} = 5.89667 - 1.023(.52) = 5.36471$

The $\bar{x}$-chart is:

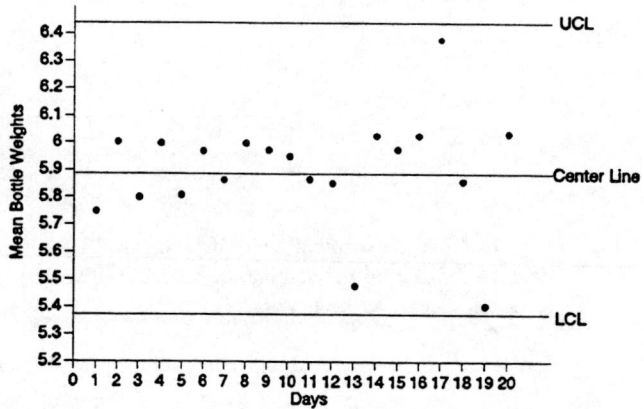

b.   Yes, the process appears to be in control because none of the sample means lie outside the control limits.

16.9 a. For each of the hours, we will compute the sample mean distance and the sample range.

Hour	Sample Mean $\bar{x}_i$	Range R_i	Hour	Sample Mean $\bar{x}_i$	Range R_i
1	.1378	.009	15	.1418	.010
2	.1430	.008	16	.1404	.013
3	.1412	.015	17	.1400	.006
4	.1398	.006	18	.1418	.008
5	.1400	.010	19	.1404	.008
6	.1392	.008	20	.1388	.006
7	.1412	.010	21	.1414	.006
8	.1400	.008	22	.1414	.007
9	.1420	.007	23	.1406	.008
10	.1392	.013	24	.1414	.013
11	.1396	.012	25	.1404	.008
12	.1414	.009	26	.1420	.007
13	.1412	.003	27	.1410	.008
14	.1406	.008			

The center line for the $\bar{x}$-chart is

$$\bar{\bar{x}} = \frac{\sum\limits_{i=1}^{27} \bar{x}_i}{27} = \frac{.1378 + .1430 + \cdots + .1410}{27} = \frac{3.7976}{27} = .14065$$

b. The upper and lower control limits are found using:

UCL: $\bar{\bar{x}} + A_2\bar{R}$

LCL: $\bar{\bar{x}} - A_2\bar{R}$

where A_2 is found in Table 19, Appendix II and $\bar{R}$ is the average range for the samples.

$$\bar{R} = \frac{\sum\limits_{i=1}^{k} R_i}{k} = \frac{.009 + .008 + \cdots + .008}{27} = \frac{.234}{27} = .00867$$

From Table 19, $A_2 = .577$ for $n = 5$.

Thus, UCL $= \bar{\bar{x}} + A_2\bar{R} = 1.4065 + .577(.00867) = .14565$

LCL $= \bar{\bar{x}} - A_2\bar{R} = 1.4065 - .577(.00867) = .13565$

The $\bar{x}$-chart is:

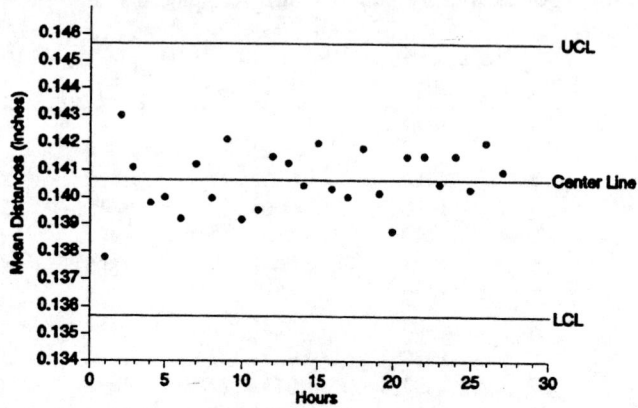

c. The process appears to be in control since all sample means are between the upper and lower control limits.

16.11 Refer to Exercise 16.6. For each hour, the sample mean and sample range are computed.

Hour	Sample Mean $\bar{x}_i$	Range R_i	Hour	Sample Mean $\bar{x}_i$	Range R_i
1	5.00000	.15	7	4.99875	.13
2	4.97625	.22	8	4.98625	.12
3	4.99375	.11	9	5.01250	.09
4	4.98250	.13	10	4.98000	.10
5	5.00750	.18	11	5.00250	.04
6	4.94875	.20	12	5.00500	.20

The center line of the R-chart is

$$\bar{R} = \frac{\sum\limits_{i=1}^{12} R_i}{12} = \frac{.15 + .22 + \cdots + .20}{12} = \frac{1.67}{12} = .13917$$

b. The upper and lower control limits are found using:

UCL: $\bar{R}D_4$
LCL: $\bar{R}D_3$

where D_3 and D_4 are found in Table 19, Appendix II. For $n = 8$, $D_3 = .136$ and $D_4 = 1.864$.

Thus, UCL $= \bar{R}D_4 = .13917(.1864) = .25941$
LCL $= \bar{R}D_3 = .13917(.136) = .01893$

The *R*-chart is:

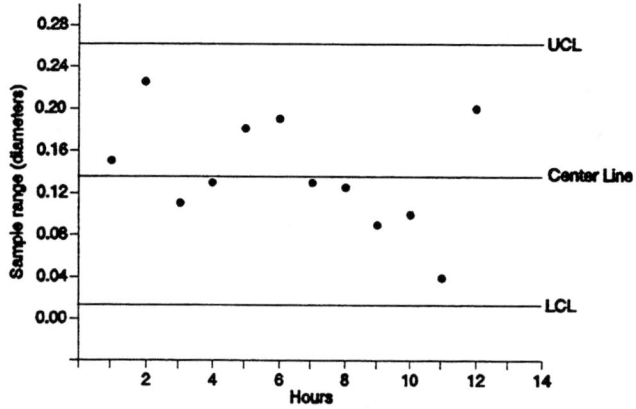

The process appears to be in control since none of the ranges fall outside the control limits.

16.13 Refer to Exercise 16.8. For each hour, the sample mean and sample range are computed.

Hour	Sample Mean $\bar{x}_i$	Range R_i	Hour	Sample Mean $\bar{x}_i$	Range R_i
1	34.0	4	11	35.8	4
2	31.6	4	12	38.4	4
3	30.8	2	13	34.0	14
4	33.0	3	14	35.0	4
5	35.0	5	15	33.8	7
6	32.2	2	16	31.6	5
7	33.0	5	17	33.0	5
8	32.6	13	18	28.2	3
9	33.8	19	19	31.8	9
10	37.8	6	20	35.6	6

The center line of the *R*-chart is $\bar{R} = \dfrac{\displaystyle\sum_{i=1}^{k} R_i}{12} = \dfrac{124}{20} = 6.2$

Since the measurements are expressed in units of .0001, the actual mean range is 6.2(.0001) = .00062.

The upper and lower control limits are found using:

UCL: $\bar{R}D_4$
LCL: $\bar{R}D_3$

where D_3 and D_4 are found in Table 19, Appendix II.

For $n = 5$, $D_3 = 0$ and $D_4 = 2.115$.

Thus,   UCL $= \bar{R}D_4 = .00062(2.115) = .00131$
           LCL $= \bar{R}D_3 = .00062(0) = 0$

The R-chart is:

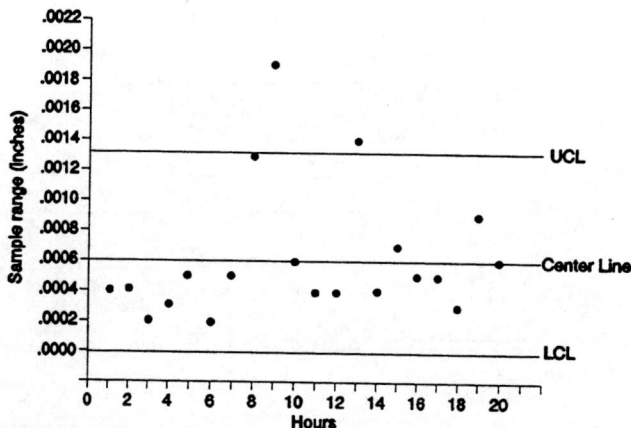

No, the process does not appear to be in control. Two of the ranges, for hours 9 and 13, fall outside the control limits.

16.15 Refer to the solution to Exercise 16.9.

$$\bar{R} = .00867$$

The upper and lower control limits are found using:

   UCL: $\bar{R}D_4$
   LCL: $\bar{R}D_3$

where D_3 and D_4 are found in Table 19, Appendix II.

For $n = 5$, $D_3 = 0$ and $D_4 = 2.115$.

Thus,   UCL $= \bar{R}D_4 = .00867(2.115) = .01834$
           LCL $= \bar{R}D_3 = .00867(0) = 0$

The R-chart is:

The process appears to be in control because none of the sample ranges are outside the control limits.

16.17 From the $\bar{x}$-chart in Exercise 16.5, the runs are

```
 +  -  +  -  +  --  +  -  +
 ⌣  ⌣  ⌣  ⌣  ⌣  ⌣   ⌣  ⌣  ⌣
 1  2  3  4  5  6   7  8  9
```

The longest run is two points. There is no evidence of a trend. From the R-chart in Exercise 16.10, the runs are

```
 -  +  --  ++++  --
 ⌣  ⌣  ⌣   ⌣⌣⌣⌣  ⌣
 1  2  3    4     5
```

The longest run is 4 points. There is no evidence of a trend.

16.19 From the $\bar{x}$-chart in Exercise 16.7, the runs are

```
 -  +  -  +  -  +  -  +++  ---  ++++  --  +
 ⌣  ⌣  ⌣  ⌣  ⌣  ⌣  ⌣  ⌣    ⌣    ⌣    ⌣  ⌣
 1  2  3  4  5  6  7  8    9    10   11 12
```

The longest run is 4 points. There is no evidence of a trend.

From the R-chart in Exercise 16.12, the runs are

```
 -  ++  -  ++  --  +  -  +  ---  +  -----
 ⌣  ⌣   ⌣  ⌣   ⌣   ⌣  ⌣  ⌣  ⌣    ⌣   ⌣
 1  2   3  4   5   6  7  8  9    10  11
```

The longest run is 5 points. There is no evidence of a trend.

16.21 From the $\bar{x}$-chart in Exercise 16.9, the runs are

```
 -  ++  ---  +  -  +  --  ++  -  +  --  +  --  ++  -  +  -  ++
 ⌣  ⌣   ⌣    ⌣  ⌣  ⌣  ⌣   ⌣   ⌣  ⌣  ⌣   ⌣  ⌣   ⌣   ⌣  ⌣  ⌣  ⌣
 1  2   3    4  5  6  7   8   9  10 11  12 13  14  15 16 17 18
```

The longest run is three points. There is no evidence of a trend.

From the *R*-chart in Exercise 16.15, the runs are

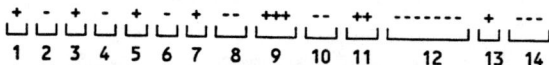

```
  +  -  +  -  +  -  +  --  +++  --  ++  -------  +  ---
  └┘ └┘ └┘ └┘ └┘ └┘ └┘ └┘  └┘  └────┘ └─────┘  └┘ └─┘
  1  2  3  4  5  6  7  8   9   10 11    12     13  14
```

The longest run is 7 points. Since there are seven consecutive points on the same side of the center line, there is evidence of a trend.

16.23 a. The 25 sample proportions are computed by dividing the number of defectives each hour by the sample size, 100. The sample proportions are:

HOUR	$\hat{p}_i$	HOUR	$\hat{p}_i$	HOUR	$\hat{p}_i$
1	6/100 = .06	10	.02	18	.01
2	.04	11	.01	19	0
3	.09	12	.03	20	.03
4	.03	13	.04	21	.07
5	0	14	.05	22	.09
6	.06	15	.05	23	.02
7	.04	16	.02	24	.10
8	.02	17	.01	25	.03
9	.01				

$$\bar{p} = \frac{\sum \hat{p}_i}{k} = \frac{.06 + .04 + .09 + \cdots + .03}{25} = \frac{.93}{25} = .0372$$

The lower and upper control limits are:

$$LCL = \bar{p} - 3\sqrt{\frac{p(1-p)}{n}} = .0372 - 3\sqrt{\frac{.0372(.9628)}{100}} = .0372 - .0568$$
$$= -.0916 = 0$$

$$UCL = \bar{p} + 3\sqrt{\frac{p(1-p)}{n}} = .0372 + 3\sqrt{\frac{.0372(.9628)}{100}} = .0372 + .0568$$
$$= .0940$$

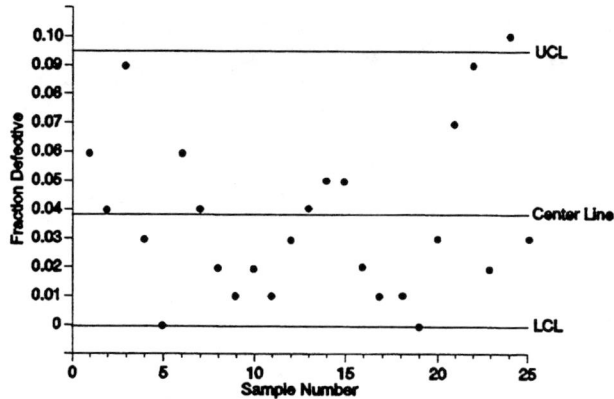

b. See the *p*-chart in part a for the center line, which is $\bar{p} = .0372$.

c. See the *p*-chart in part a for the upper and lower control limits, which are LCL = 0 and UCL = .0940. Since the sample proportion for sample 24 falls outside the control limits, this indicates the process is out of control.

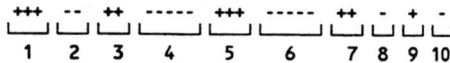

d. The runs are

```
  +++   --   ++   -----   +++   -----   ++   -   +   -
  └─┘  └─┘  └─┘   └───┘   └─┘   └───┘   └─┘ └┘ └┘ └┘
   1    2    3      4      5      6      7  8  9 10
```

The longest run is 5 points. There is no evidence of a trend.

16.25 The 21 sample proportions are computed by dividing the number of defectives each week by the sample size, 50. The sample proportions are:

Day	$\hat{p}_i$	Day	$\hat{p}_i$	Day	$\hat{p}_i$
1	11/50 = .22	8	.28	15	.22
2	.30	9	.18	16	.22
3	.24	10	.26	17	.32
4	.20	11	.30	18	.30
5	.18	12	.46	19	.20
6	.24	13	.30	20	.26
7	.24	14	.24	21	.24

$$6\bar{p} = \frac{\sum \hat{p}_i}{k} = \frac{.22 + .30 + .24 + \cdots + .24}{21} = \frac{5.40}{21} = .2571$$

The lower and upper control limits are:

$$LCL = \bar{p} - 3\sqrt{\frac{p(1 - p)}{n}} = .2571 - 3\sqrt{\frac{.2571(1 - .2571)}{50}} = .2571 - .1854 = .0717$$

$$UCL = \bar{p} + 3\sqrt{\frac{p(1 - p)}{n}} = .2571 + 3\sqrt{\frac{.2571(1 - .2571)}{50}} = .2571 + .1854 = .4425$$

a.

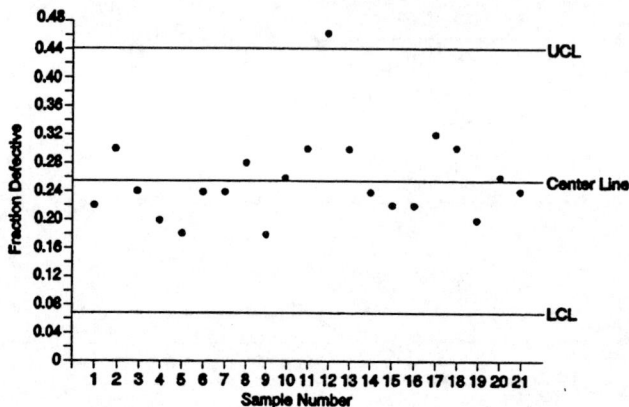

b. See the p-chart in part a for the center line, which is $\bar{p} = .2571$.

c. See the p-chart in part a for the lower and upper control limits, which are LCL = .0717 and UCL = .4425.

d. The process does not appear to be in control as sample proportion 12 falls outside the control limits. Dropping this sample out and recalculating the control limits, we get

$$\bar{p} = \frac{\sum \hat{p}_i}{k} = \frac{4.94}{20} = .247$$

The modified control limits are:

$$\text{LCL} = \bar{p} - 3\sqrt{\frac{p(1-p)}{n}} = .247 - 3\sqrt{\frac{.247(1-.247)}{50}} = .247 - .1830 = .064$$

$$\text{UCL} = \bar{p} + 3\sqrt{\frac{p(1-p)}{n}} = .247 + 3\sqrt{\frac{.247(1-.247)}{50}} = .247 + .1830 = .430$$

e. The runs are

$$
\begin{array}{cccccccccc}
- & + & ----- & + & - & ++++ & --- & ++ & - & + & - \\
\underline{} & \underline{} & \underline{} & \underline{} & \underline{} & \underline{} & \underline{} & \underline{} & \underline{} & \underline{} & \underline{} \\
1 & 2 & 3 & 4 & 5 & 6 & 7 & 8 & 9 & 10 & 11
\end{array}
$$

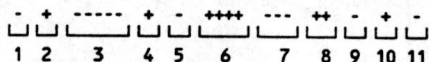

Since the longest run is 5 points, there is no evidence of a trend.

16.27 **a.** $\bar{c} = \dfrac{\sum c_i}{k} = \dfrac{130}{20} = 6.5$

The lower and upper control limits are:

$$\text{LCL} = \bar{c} - 3\sqrt{\bar{c}} = 6.5 - 3\sqrt{6.5} = -1.149 \approx 0$$

$$\text{UCL} = \bar{c} + 3\sqrt{\bar{c}} = 6.5 + 3\sqrt{6.5} = 14.149$$

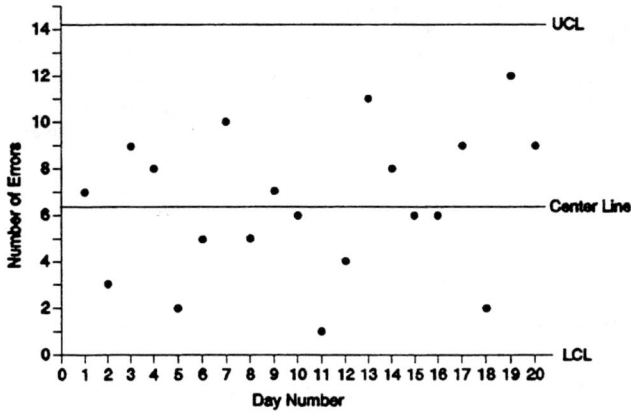

b.   See the c-chart in part **a** for the center line, which is $\bar{c} = 6.5$.

c.   See the c-chart in part **a** for the control limits, which are LCL = 0 and UCL = 14.149. Since no days have error numbers that lie outside the control limits, the process appears to be in control.

d.   The runs are

```
  +   -   ++   --   +   -   +   ---   ++   --   +   -   ++
 └┘  └┘  └┘   └┘   └┘  └┘  └┘  └──┘  └┘   └┘   └┘  └┘  └┘
  1   2   3    4    5   6   7    8     9   10   11  12  13
```

Since the longest run is 3 points, there is no evidence of a trend.

16.29 a. $\bar{c} = \dfrac{\sum c_i}{k} = \dfrac{436}{50} = 8.72$

The control limits are:

$$\text{LCL} = \bar{c} - 3\sqrt{\bar{c}} = 8.72 - 3\sqrt{8.72} = 8.72 - 8.86 = -.14 \approx 0$$
$$\text{UCL} = \bar{c} + 3\sqrt{\bar{c}} = 8.72 + 3\sqrt{8.72} = 8.72 + 8.86 = 17.58$$

Since the number of errors for aircraft 36 falls outside the control limits, the process appears out of control.

b.   The runs for the all 50 aircraft are:

```
 -------   +++   ---   +   -   +   --   +   --   +   --   +
 └─────┘  └──┘  └──┘  └┘  └┘  └┘  └┘  └┘  └┘  └┘  └┘  └┘
    7       3     3    1   1   1   2   1   2   1   2   1
```

Since the longest run is 7 points, there is evidence of a trend.

16.31   **a.**   From Exercise 16.8, $\bar{x} = .40336$.

$$s^2 = \frac{\sum x^2 - \frac{(\sum x)^2}{n}}{n - 1} = \frac{16.26953793 - \frac{40.3355^2}{100}}{99} = .000000124$$

$$s = \sqrt{.000000124} = .0003529$$

The tolerance interval is $\bar{x} \pm Ks$

From Table 20, Appendix II, with $n = 100$, $\alpha = .05$, and $\gamma = .99$, $K = 2.934$. The tolerance interval is

$$.40336 \pm 2.934(.0003529) \Rightarrow .40336 \pm .00104 \Rightarrow (.40232, .40440)$$

b.   The specification interval is $.4037 \pm .0013 \Rightarrow (.4024, .4050)$. The lower limit of the specification interval is just smaller than the lower tolerance limit, while the upper limit of the specification interval is larger than the upper tolerance limit. This implies the specifications are being met.

c.   From Table 21, Appendix II, with $\alpha = .05$ and $\gamma = .95$, and $n = 93$. Since $n = 100$ in this example, we can form the tolerance interval. It is

$$(x_{min}, x_{max}) \Rightarrow (.4023, .4043)$$

16.33   **a.**   Since μ and σ are known, the tolerance interval for 99% of the complaint rates is

$$\mu \pm z_{.005}\sigma \Rightarrow 26 \pm 2.576(11.3) \Rightarrow 26 \pm 29.1088 \Rightarrow (-3.1088, 55.1088)$$

This is a 100% tolerance interval. Since μ and σ are known and the population is assumed normal, we know that 99% of the data will fall within 2.576 standard deviations of the mean.

b.   Since 93.12 falls outside the tolerance interval, the observed rate is probably due to a specific cause.

16.35   **a.**   To sketch the operating characteristic curve, we need to compute $P(A) = P($accepting the lot$)$ for different values of p. Let $y = $ number of defectives in the n trials. Then y has a binomial distribution.

Let $n = 5$, and $a = 0$. Using Table 1, Appendix II,

For $p = .1$, $P(A) = P(y \le 0) = p(0) = .5905$
For $p = .3$, $P(A) = P(y \le 0) = p(0) = .1681$
For $p = .5$, $P(A) = P(y \le 0) = p(0) = .0313$

The operating characteristic curve is:

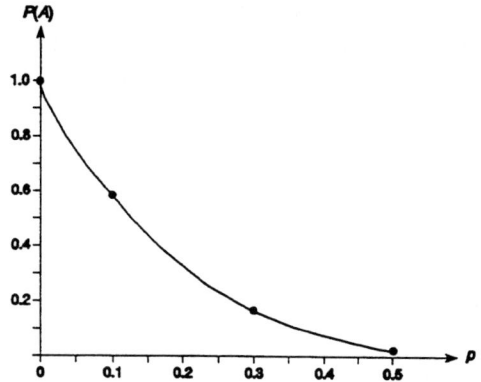

b. The producer's risk is $\alpha = 1 - P(A)$. When AQL $= .01$, $\alpha = 1 - P(A) = 1 - .9510$ $= .0490$ (from Table 1, Appendix II).

c. The consumer's risk is $\beta = P(A)$. When $p_1 = .10$, $\beta = P(A) = .5905$ (from Table 1, Appendix II).

16.37 a. Using Table 22 of Appendix II, for a lot of 5000 items, the code for a reduced inspection level is J.

Now using Table 23, we find the row labelled J and see the recommended size sample of $n = 80$. We follow this row across to the 4% AQL and see we will accept the lot if 7 or fewer are defective.

b. Using the same procedure as above:

Code $=$ L $n = 200$ $a = 14$
Code $=$ M $n = 315$ $a = 21$

16.39 Using Table 22 in Appendix II, for a lot of 400 items, the code for the lot size is H.

Using Table 23, we find the recommended sample size is $n = 50$. For an AQL $= 2.5\%$, we should accept the lot if there are 3 or fewer defects in the sample.

16.41 a. To sketch operating characteristic curves, we need to compute $P(A) = P$(accepting the lot) for different values of p. Let $y =$ number of defective items in n trials. Then y has a binomial distribution.

Let $n = 5$ and $a = 1$. Using Table 1, Appendix II,

For $p = .05$, $P(A) = P(y \leq 1) = .9774$
For $p = .10$, $P(A) = P(y \leq 1) = .9185$
For $p = .20$, $P(A) = P(y \leq 1) = .7373$
For $p = .30$, $P(A) = P(y \leq 1) = .5282$
For $p = .40$, $P(A) = P(y \leq 1) = .3370$

The operating characteristic curve is:

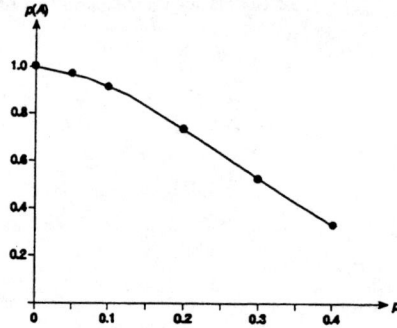

Let $n = 25$ and $a = 5$. Using Table 1, Appendix II,

For $p = .05$, $P(A) = P(y \leq 5) = .9988$
For $p = .1$, $P(A) = P(y \leq 5) = .9666$
For $p = .2$, $P(A) = P(y \leq 5) = .6167$
For $p = .3$, $P(A) = P(y \leq 5) = .1935$
For $p = .4$, $P(A) = P(y \leq 5) = .0294$

The operating characteristics curve is:

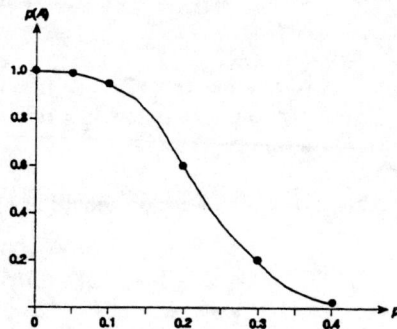

b. As a seller, one would want the producer's risk, α, to be as small as possible. The producer's risk is $\alpha = 1 - P(A)$ when $p = $ AQL $= .10$.

When $n = 5$, $\alpha = 1 - P(A) = 1 - .9816 = .0814$
When $n = 25$, $\alpha = 1 - P(A) = 1 - .9666 = .0334$

As a seller, you would want α to be as small as possible. Thus, the second plan with $\alpha = .0334$ would be preferred.

c. As a buyer, one would want the consumer's risk, β, to be as small as possible. The consumer's risk is:

$\beta = P(A)$ when $p_1 = .3$

When $n = 5$, $\beta = P(A) = .5282$
When $n = 25$, $\beta = P(A) = .1935$

As a buyer, you would want β to be as small as possible. Thus, the second plan with $\beta = .1935$ would be preferred.

16.43   a.   To find the MIL-STD-105D general inspection sampling for a lot size of 250 and acceptance level of 10%, we first use Table 22, Appendix II.

For the normal inspection level (II), and lot size 250, the code letter is G from Table 22.

From Table 23, with a code letter of G, the sample size should be 32. To find the acceptance number, move across the top row to 10%. The acceptance (Ac) number is $a = 7$, the intersection of the 10% column and the G row.

b.   For the tightened acceptance level (III) with a lot size of 250 and acceptance level of 10%, we again start with Table 22. For $n = 250$ and tightened acceptance level III, the code letter is H from Table 22.

From Table 23, with a code letter of H, the sample size should be 50. To find the acceptance number, move across the top row to 10%. The acceptance (Ac) number is $a = 10$, the intersection of the 10% column and H row.

16.45   a.   Let $y =$ number of defectives in $n = 20$ trials. The distribution of y is binomial. For $a = 2$ and AQL $= .05$, the producer's risk is $\alpha = 1 - P(A) = 1 - P(y \le 2) = 1 - .9245 = .0755$ (from Table 1, Appendix II).

b.   The consumer's risk is $\beta = P(A)$. For $p_1 = .10$,

$$\beta = P(A) = P(y \le 2) = .6769 \text{ (from Table 1, Appendix II)}$$

c.   To find the operating characteristic curve, we must find $P(A)$ for several values of p.

For $p = .1$, $P(A) = P(y \le 2) = .6769$ (from Table 1, Appendix II)
For $p = .2$, $P(A) = P(y \le 2) = .2061$
For $p = .3$, $P(A) = P(y \le 2) = .0355$
For $p = .4$, $P(A) = P(y \le 2) = .0036$
For $p = .5$, $P(A) = P(y \le 2) = .0002$

The operating characteristic curve is:

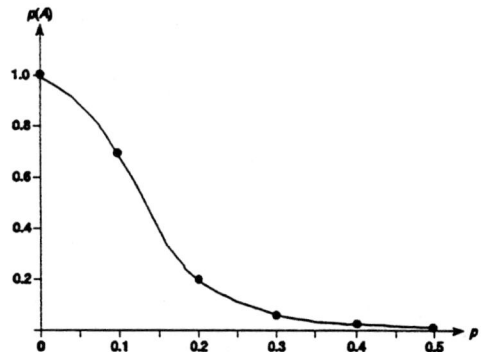

d. Using Table 22 in Appendix II, the lot code for a lot size of 1500 is K.

Using Table 23, we find a recommended sample size of $n = 125$. Using an AQL = 5%, we should accept the lot if 10 or fewer are defective (use AQL = 4% for a conservative value).

e. For $n = 125$ and $a = 10$.

Let y = number of defective items in $n = 125$ trials.

Then y has a binomial distribution with $n = 125$ and $p = $ AQL = .05. The producer's risk is $\alpha = 1 - P(A) = 1 - P(y \leq 10)$.

Since $n = 125$, we will use the normal approximation to the binomial with

$$\mu = np = 125(.05) = 6.25 \text{ and}$$
$$\sigma = \sqrt{npq} = \sqrt{125(.05)(.95)} = 2.4367$$

Thus, $\alpha = 1 - P(y \leq 10) \approx P\left(z \leq \dfrac{10.5 - 6.25}{2.4367}\right)$

$$= 1 - P(z \leq 1.74)$$
$$= 1 - (.5 + P(0 \leq z \leq 1.74))$$
$$= 1 - .5 - .4591$$
$$= .0409$$

f. For $n = 125$, and $p_1 = .08$, the consumer's risk is $\beta = P(y \leq 10)$. Using the normal approximation to the binomial,

$$\mu = np = 125(.08) = 10 \text{ and}$$
$$\sigma = \sqrt{npq} = \sqrt{125(.08)(.92)} = 3.03315$$

Thus, $\beta = P(y \leq 10) \approx P\left(z \leq \dfrac{10.5 - 10}{3.03315}\right) = P(z \leq .16)$

$$= .5 + P(0 \leq z \leq .16) = .5 + .0636 = .5636$$

CHAPTER SEVENTEEN

. . . .

Product and System Reliability

17.1 a.   The density function for the Weibull distribution is

$$f(t) = \begin{cases} \dfrac{\alpha}{\beta} t^{\alpha-1} e^{-t^{\alpha}/\beta} & 0 \le t \le \infty,\ \alpha > 0,\ \beta > 0 \\ 0 & \text{elsewhere} \end{cases}$$

This cumulative distribution function of the Weibull distribution is

$$F(t) = 1 - e^{-t^{\alpha}/\beta}$$

The hazard rate is $z(t) = \dfrac{f(t)}{1 - F(t)} = \dfrac{\dfrac{\alpha}{\beta} t^{\alpha-1} e^{-t^{\alpha}/\beta}}{1 - \left(1 - e^{-t^{\alpha}/\beta}\right)} = \dfrac{\alpha}{\beta}\left(t^{\alpha-1}\right)$

For $\alpha = 1,\ \beta = 4,\ z(t) = \dfrac{1}{4}t^{0} = \dfrac{1}{4}$

The graph of $z(t)$ is:

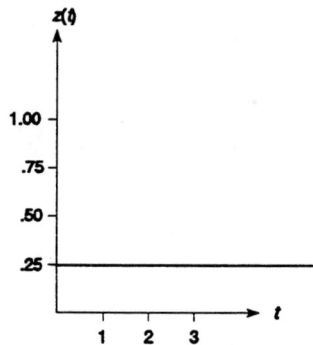

b.   For $\alpha = 2,\ \beta = 2,\ z(t) = \dfrac{2}{2}t^{2-1} = t$

The graph of $z(t)$ is:

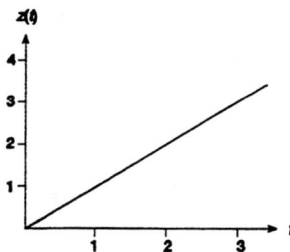

. . . .
Product and System Reliability   343

c.   For $\alpha = 3$, $\beta = 6$, $z(t) = \dfrac{3}{6}t^{3-1} = \dfrac{t^2}{2}$

The graph of $z(t)$ is:

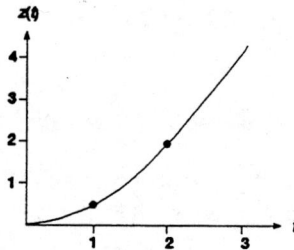

17.3   The density function for t is

$$f(t) = \frac{1}{\sqrt{2\pi}}e^{-\frac{(t-3)^2}{2}} \qquad -\infty < t < \infty$$

For $t = 0, 1, 2, \ldots, 6$,

$$f(0) = \frac{1}{\sqrt{2\pi}}e^{-\frac{(0-3)^2}{2}} = \frac{1}{\sqrt{2\pi}}e^{-9/2} = .004431848 \approx .0044$$

$$f(1) = \frac{1}{\sqrt{2\pi}}e^{-\frac{(1-3)^2}{2}} = \frac{1}{\sqrt{2\pi}}e^{-4/2} = .053990966 \approx .0540$$

$$f(2) = \frac{1}{\sqrt{2\pi}}e^{-\frac{(2-3)^2}{2}} = \frac{1}{\sqrt{2\pi}}e^{-1/2} = .241970724 \approx .2420$$

$$f(3) = \frac{1}{\sqrt{2\pi}}e^{-\frac{(3-3)^2}{2}} = \frac{1}{\sqrt{2\pi}}e^{0} = .39894228 \approx .3989$$

$$f(4) = \frac{1}{\sqrt{2\pi}}e^{-\frac{(4-3)^2}{2}} = \frac{1}{\sqrt{2\pi}}e^{-1/2} = .241970724 \approx .2420$$

$$f(5) = \frac{1}{\sqrt{2\pi}}e^{-\frac{(5-3)^2}{2}} = \frac{1}{\sqrt{2\pi}}e^{-4/2} = .053990966 \approx .0540$$

$$f(6) = \frac{1}{\sqrt{2\pi}}e^{-\frac{(6-3)^2}{2}} = \frac{1}{\sqrt{2\pi}}e^{-9/2} = .004431843 \approx .0044$$

$F(t_0) = P(t \le t_0)$ where t has a normal distribution with a mean of 3 and a standard deviation of 1.

$$F(0) = P(t \leq 0) = P\left[z \leq \frac{0-3}{1}\right] = P(z \leq -3) = .5 - P(0 \leq z \leq 3)$$
$$= .5 - .4987 = .0013$$
(using Table 4, Appendix II)

$$F(1) = P(t \leq 1) = P\left[z \leq \frac{1-3}{1}\right] = P(z \leq -2) = .5 - P(0 \leq z \leq 2)$$
$$= .5 - .4772 = .0228$$

$$F(2) = P(t \leq 2) = P\left[z \leq \frac{2-3}{1}\right] = P(z \leq -1) = .5 - P(0 \leq z \leq 1)$$
$$= .5 - .3413 = .1587$$

$$F(3) = P(t \leq 3) = P\left[z \leq \frac{3-3}{1}\right] = P(z \leq 0) = .5$$

$$F(4) = P(t \leq 4) = P\left[z \leq \frac{4-3}{1}\right] = P(z \leq 1) = .5 + P(0 \leq z \leq 1)$$
$$= .5 + .3413 = .8413$$

$$F(5) = P(t \leq 5) = P\left[z \leq \frac{5-3}{1}\right] = P(z \leq 2) = .5 + P(0 \leq z \leq 2)$$
$$= .5 + .4772 = .9772$$

$$F(6) = P(t \leq 6) = P\left[z \leq \frac{6-3}{1}\right] = P(z \leq 3) = .5 + P(0 \leq z \leq 3)$$
$$= .5 + .4987 = .9987$$

We know that $z(t) = \dfrac{f(t)}{1 - F(t)}$

Thus,
$$z(0) = \frac{f(0)}{1 - F(0)} = \frac{.00443184}{1 - .0013} = .0044$$

$$z(1) = \frac{f(1)}{1 - F(1)} = \frac{.053990966}{1 - .0228} = .0553$$

$$z(2) = \frac{f(2)}{1 - F(2)} = \frac{.241970724}{1 - .1587} = .2876$$

$$z(3) = \frac{f(3)}{1 - F(3)} = \frac{.39894228}{1 - .5} = .7979$$

$$z(4) = \frac{f(4)}{1 - F(4)} = \frac{.241970724}{1 - .8413} = 1.5247$$

$$z(5) = \frac{f(5)}{1 - F(5)} = \frac{.053990966}{1 - .9772} = 2.3680$$

$$z(6) = \frac{f(6)}{1 - F(6)} = \frac{.00443184}{1 - .9987} = 3.4091$$

b.

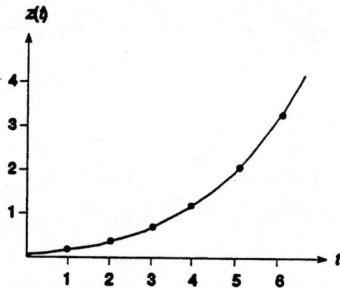

17.5　a.　$F(t) = \int_0^t \dfrac{2ye^{-y^2/100}}{100}dy = -e^{-y^2/100}\Big]_0^t = -e^{-t^2/100} - \left(-e^{-0^2/100}\right) = 1 - e^{-t^2/100}$

　　b.　$R(t) = 1 - F(t) = 1 - \left(1 - e^{-t^2/100}\right) = e^{-t^2/100}$

$$z(t) = \dfrac{f(t)}{R(t)} = \dfrac{\dfrac{2te^{-t^2/100}}{100}}{e^{-t^2/100}} = \dfrac{2t}{100} = \dfrac{t}{50}$$

　　c.　$R(8) = e^{-8^2/100} = e^{-.64} = .5273$

　　　　$z(8) = 8/50 = .16$

17.7　If y has an exponential distribution with parameter β, then $\dfrac{2y}{\beta}$ has a chi-square distribution with 2 degrees of freedom (see Exercise 8.89).

Thus, $\sum \dfrac{2y_i}{\beta} = \dfrac{2\sum y_i}{\beta}$ has a chi-square distribution with $2n$ degrees of freedom.

Using the pivotal statistic, $\dfrac{2\sum y_i}{\beta}$, the confidence interval is

$$P\left[\chi^2_{1-\alpha/2} \le \dfrac{2\sum y_i}{\beta} \le \chi^2_{\alpha/2}\right] = 1 - \alpha$$

For confidence coefficient .90, $\alpha = 1 - .90 = .10$ and $\alpha/2 = .10/2 = .05$. From Table 8, Appendix II, $\chi^2_{.95} = 6.57063$ and $\chi^2_{.05} = 23.6848$ with degrees of freedom $= 2n = 2(7) = 14$.

$$P\left[\chi^2_{1-\alpha/2} \leq \frac{2\sum y_i}{\beta} \leq \chi^2_{\alpha/2} \right] = P\left[\frac{1}{\chi^2_{1-\alpha/2}} \geq \frac{\beta}{2\sum y_i} \geq \frac{1}{\chi^2_{\alpha/2}} \right]$$

$$= P\left[\frac{2\sum y_i}{\chi^2_{1-\alpha/2}} \geq \beta \geq \frac{2\sum y_i}{\chi^2_{\alpha/2}} \right] = 1 - \alpha$$

$$\text{or } P\left[\frac{2\sum y}{\chi^2_{\alpha/2}} \leq \beta \leq \frac{2\sum y_i}{\chi^2_{1-\alpha/2}} \right] = 1 - \alpha$$

From the data, $\sum y_i = 281$. The 90% confidence interval is

$$\left[\frac{2(281)}{23.6848} \leq \beta \leq \frac{2(281)}{6.57063} \right] \Rightarrow (23.728 \leq \beta \leq 85.532)$$

17.9 The approximate confidence interval for the mean miles to failure, based on censored sampling with T fixed at 135 is:

$$\frac{2(\text{Total life})}{\chi^2_{\alpha/2}} \leq \beta \leq \frac{2(\text{Total life})}{\chi^2_{1-\alpha/2}}$$

where Total life $= \sum_{i=1}^{r} t_i + (n - r)T = 3307.5 + (96 - 37)(135) = 11{,}272.5$

For confidence coefficient .95, $\alpha = 1 - .95 = .05$ and $\alpha/2 = .05/2 = .025$. From Table 8, Appendix II, $\chi^2_{.975} \approx 57.1532$ and $\chi^2_{.025} \approx 106.629$, with df $= 2r + 2 = 2(37) + 2 = 76$. The 95% confidence interval is:

$$\left[\frac{2(11{,}272.5)}{106.629} \leq \beta \leq \frac{2(11{,}272.5)}{57.1532} \right] \Rightarrow (211.4 \leq \beta \leq 394.5)$$

17.11 The approximate confidence interval for the mean time between failures based on censored sampling with fixed $T = 2000$ is:

$$\frac{2(\text{Total life})}{\chi^2_{\alpha/2}} \leq \beta \leq \frac{2(\text{Total life})}{\chi^2_{1-\alpha/2}}$$

where Total life $= \sum_{i=1}^{3} t_i + (n - r)T = 4048 + (100 - 3)(2000) = 198{,}048$

For confidence coefficient .99, $\alpha = 1 - .99 = .01$ and $\alpha/2 = .01/2 = .005$. From Table 8, Appendix II, $\chi^2_{.995} = 1.344419$ and $\chi^2_{.005} = 21.955$ with df $= 2r + 2 = 2(3) + 2 = 8$. The 99% confidence interval is:

$$\left[\frac{2(198{,}048)}{21.955} \leq \beta \leq \frac{2(198{,}048)}{1.344419} \right] \Rightarrow (18{,}041.27 \leq \beta \leq 294{,}622.44)$$

We are 99% confident the mean time between failures of the capacitors is between 18,041.27 and 294,622.44 hours.

17.13   Some preliminary calculations are:

Time (i)	$x_i = \ln i$	Number of Survivors (n_i)	$\hat{R}(i) = \dfrac{n_i}{n}$	$-\ln \hat{R}(i)$	$y = \ln[-\ln \hat{R}(i)]$
1	0	47	.94	.061875	−2.782633
2	.69315	39	.78	.248461	−1.392468
3	1.09861	29	.58	.544727	−.607470
4	1.38629	18	.36	1.021651	.021420
5	1.60944	11	.22	1.514128	.414840
6	1.79176	5	.10	2.302585	.834032
7	1.94591	3	.06	2.813411	1.034398
8	2.07944	1	.02	3.912023	1.364055

For the model $y_i = \beta_0 + \beta_1 x_i + \epsilon_i$, we need to estimate β_0 and β_1 by the method of least squares. The preliminary calculations are:

$$\sum x_i = 10.6046 \qquad \sum y_i = -1.113826 \qquad \sum x_i y_i = 5.408493728$$

$$\sum x_i^2 = 17.52053827 \qquad \sum y_i^2 = 13.84981903$$

$$SS_{xy} = \sum x_i y_i - \frac{\sum x_i \sum y_i}{n} = 5.408493728 - \frac{10.6046(-1.113826)}{8} = 6.884953628$$

$$SS_{xx} = \sum x_i^2 - \frac{\left(\sum x_i\right)^2}{n} = 17.520538 - \frac{10.6046^2}{8} = 3.46334562$$

$$SS_{yy} = \sum y_i^2 - \frac{\left(\sum y_i\right)^2}{n} = 13.84981903 - \frac{(-1.113826)^2}{8} = 13.69474299$$

$$\hat{\beta}_1 = \frac{SS_{xy}}{SS_{xx}} = \frac{6.884953628}{3.46334562} = 1.987948759 \approx 1.987949$$

$$\hat{\beta}_0 = \bar{y} - \hat{\beta}_1 \bar{x} = \frac{-1.113826}{8} - 1.987948759\left(\frac{10.6046}{8}\right) = -2.774403426$$

$$\approx -2.77440$$

$$SSE = SS_{yy} - \frac{SS_{xy}^2}{SS_{xx}} = 13.69474299 - \frac{6.884953628^2}{3.46334562} = .00780797$$

$$MSE = \frac{SSE}{n-2} = \frac{.00780797}{8-2} = .001301$$

a.   Using the method of least squares:

$$\hat{\alpha} = \hat{\beta}_1 = 1.987949$$

$$\hat{\beta} = e^{-\hat{\beta}_0} = e^{-(-2.77440)} = 16.029$$

b. The confidence interval for α is the same as the confidence interval for β_1 which is:

$$\hat{\beta}_1 \pm t_{\alpha/2}s(\hat{\beta}_1) \quad \text{where } s(\hat{\beta}_1) = \sqrt{\frac{MSE}{SS_{xx}}} = \sqrt{\frac{.001301}{3.46334562}} = .019382$$

For confidence coefficient .95, $\alpha = 1 - .95 = .05$ and $\alpha/2 = .05/2 = .025$. From Table 7, Appendix II, $t_{.025} = 2.447$ with $n - 2 = 8 - 2 = 6$ degrees of freedom. The 95% confidence interval is:

$$1.987949 \pm 2.447(.019382) \Rightarrow 1.987949 \pm .04743 \Rightarrow (1.9405, 2.0354)$$

c. To find a confidence interval for β, we first find a confidence interval for β_0. Then the confidence interval for β is found by raising e to the negative endpoints of the confidence interval for β_0.

The form of the confidence interval for β_0 is:

$$\hat{\beta}_0 \pm t_{\alpha/2}s(\hat{\beta}_0) \quad \text{where } s(\hat{\beta}_0) = \sqrt{MSE\left[\frac{1}{n} + \frac{\bar{x}^2}{SS_{xx}}\right]}$$

$$= \sqrt{.001301\left[\frac{1}{8} + \frac{1.325575^2}{3.46334562}\right]} = .02868$$

From part b, $t_{.025} = 2.447$. The confidence interval for β_0 is

$$-2.77440 \pm 2.447(.02688) \Rightarrow -2.77446 \pm .0702 \Rightarrow (-2.84458, -2.70422)$$

Thus, the confidence interval for β is

$$\left(e^{2.70422}, e^{2.84458}\right) \Rightarrow (14.943, 17.194)$$

17.15 a. For the Weibull distribution, the density function is

$$f(t) = \begin{cases} \dfrac{\alpha}{\beta}t^{\alpha-1}e^{-t^{\alpha}/\beta} & 0 \le t \le \infty, \alpha > 0, \beta > 0 \\ 0 & \text{elsewhere} \end{cases}$$

and the cumulative distribution function is

$$F(t) = 1 - e^{-t^{\alpha}/\beta}$$

Using the least squares estimators for α and β from Exercise 17.13, $\hat{\alpha} = 1.987949$ and $\hat{\beta} = 16.029$.

Thus, $\quad f(t) = \dfrac{1.987949}{16.029}t^{.987949}e^{-t^{1.987949}/16.029}$

or $\quad f(t) = .124022t^{.987949}e^{-t^{1.987949}/16.029}$

and $F(t) = 1 - e^{-t^{1.987949}/16.029}$

The hazard rate, $z(t)$, is

$$z(t) = \frac{f(t)}{1 - F(t)} = \frac{.124022t^{.987949}e^{-t^{1.987949}/16.029}}{e^{-t^{1.987949}/16.029}} = .124022t^{.987949}$$

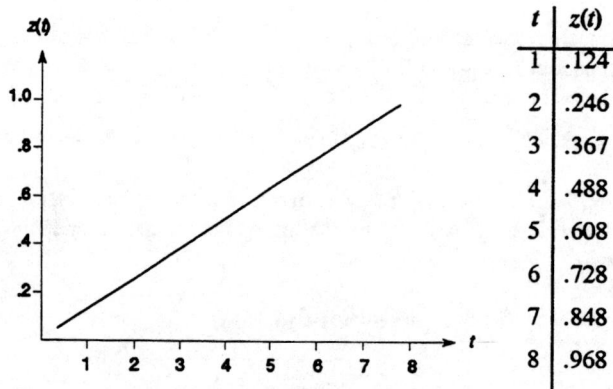

t	$z(t)$
1	.124
2	.246
3	.367
4	.488
5	.608
6	.728
7	.848
8	.968

b. $z(4) = .124015(4^{.987949}) = .488$

17.17 a.

Time (i)	Number of Survivors (n_i)
1	$12 - 3 = 9$
2	$12 - 8 = 4$
3	$12 - 10 = 2$

b. Expanding the table in a, we get:

Time (i)	$x_i = \ln i$	Number of Survivors (n_i)	$\hat{R}(i) = \dfrac{n_i}{n}$	$-\ln \hat{R}(i)$	$y = \ln[-\ln \hat{R}(i)]$
1	0	9	.75	.28768	-1.2459
2	.69315	4	.333	1.09861	.094048
3	1.09861	2	.167	1.79176	.583198

For the model $y_i = \beta_0 + \beta_1 x_i + \epsilon_i$, we need to estimate β_0 and β_1 by the method of least squares. The preliminary calculations are:

$$\sum x_i = 1.79176 \qquad \sum y_i = -.568654 \qquad \sum x_i y_i = .705897$$

$$\sum x_i^2 = 1.6874 \qquad \sum y_i^2 = 1.901232$$

$$SS_{xy} = \sum x_i y_i - \frac{\sum x_i \sum y_i}{n} = .705897 - \frac{1.79176(-.568654)}{3} = 1.045527$$

$$SS_{xx} = \sum x_i^2 - \frac{\left(\sum x_i\right)^2}{n} = 1.6874 - \frac{1.79176^2}{3} = .617266$$

$$SS_{yy} = \sum y_i^2 - \frac{\left(\sum y_i\right)^2}{n} = 1.901232 - \frac{(-5.68654)^2}{3} = 1.793443$$

$$\hat{\beta}_1 = \frac{SS_{xy}}{SS_{xx}} = \frac{1.045527}{.617266} = 1.693802996 \approx 1.69380$$

$$\hat{\beta}_0 = \bar{y} - \hat{\beta}_1 \bar{x} = \frac{-.568654}{3} - 1.693802996 \left(\frac{1.79176}{3}\right) = -1.201180819$$
$$\approx -1.201181$$

$$SSE = SS_{yy} - \frac{SS_{xy}^2}{SS_{xx}} = 1.793443 - \frac{1.045527^2}{.617266} = .022526235$$

$$MSE = \frac{SSE}{n-2} = \frac{.022526235}{3-2} = .022526235$$

a. Using the method of least squares:
$$\hat{\alpha} = \hat{\beta}_1 = 1.6938$$
$$\hat{\beta} = e^{-\hat{\beta}_0} = e^{-(-1.201181)} = 3.324$$

b. The confidence interval for α is the same as the confidence interval for β_1 which is:

$$\hat{\beta}_1 \pm t_{\alpha/2}s(\hat{\beta}_1) \quad \text{where } s(\hat{\beta}_1) = \sqrt{\frac{MSE}{SS_{xx}}} = \sqrt{\frac{.022526235}{.617266}} = .19103$$

For confidence coefficient .95, $\alpha = 1 - .95 = .05$ and $\alpha/2 = .05/2 = .025$. From Table 7, Appendix II, $t_{.025} = 12.706$ with $n - 2 = 3 - 2 = 1$ degrees of freedom. The 95% confidence interval is:

$$1.6938 \pm 12.706(.19103) \Rightarrow 1.6938 \pm 2.4272 \Rightarrow (-.7334, 4.1210)$$

c. To find a confidence interval for β, we first find a confidence interval for β_0. The confidence interval for β is found by raising e to the negative endpoints of the confidence interval for β_0.

The form of the confidence interval for β_0 is:

$$\hat{\beta}_0 \pm t_{\alpha/2}s(\hat{\beta}_0) \quad \text{where } s(\hat{\beta}_0) = \sqrt{MSE\left[\frac{1}{n} + \frac{\bar{x}^2}{SS_{xx}}\right]}$$
$$= \sqrt{.022526235\left[\frac{1}{3} + \frac{.597253333^2}{.617266}\right]} = .1433$$

From above, $t_{.025} = 12.706$. The confidence interval for β_0 is

$$-1.201181 \pm 12.706(.1433) \Rightarrow -1.201181 \pm 1.8208 \Rightarrow (-3.0220, .6196)$$

Thus, the confidence interval for β is

$$\left(e^{-.6196},\ e^{3.0220}\right) \Rightarrow (.5382,\ 20.5323)$$

d.   Let y = time until repair.

$$F(y) = 1 - e^{-y^{\alpha}/\beta}$$

Substituting the estimates of α and β,

$$F(y) = 1 - e^{-y^{1.6938}/3.324}$$

$$P(y \le 2) = F(2) = 1 - e^{-2^{1.6938}/3.324} = 1 - .37785 = .62215$$

17.19   For a series system consisting of 4 independently operating components, A, B, C, and D is:

$$P(\text{series system functions}) = p_A p_B p_C p_D = .88(.95)(.90)(.80) = .60192$$

17.21   To find the reliability of the system, we must first find the reliability of several subsystems:

The reliability of the subsystem (parallel)

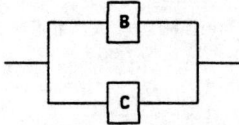

is $P(\text{subsystem B and C functions}) = p_{BC} = 1 - (1 - p_B)(1 - p_C)$
$$= 1 - (1 - .95)(1 - .85)$$
$$= 1 - .0075 = .9925$$

The reliability of the subsystem (parallel)

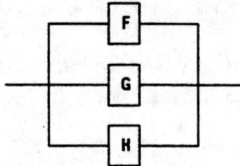

is $P(\text{subsystem F, G, and H}) = p_{FGH} = 1 - (1 - p_F)(1 - p_G)(1 - p_H)$
$$= 1 - (1 - .80)(1 - .95)(1 - .95)$$
$$= 1 - .0005 = .9995$$

The reliability of the subsystem (series)

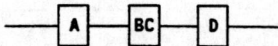

is $p_{ABCD} = p_A p_{BC} p_D = .90(.9925)(.85) = .7592625$

The reliability of the subsystem (series)

is $p_{EFGH} = p_E p_{FGH} = .98(.9995) = .97951$

The reliability of the entire system (parallel)

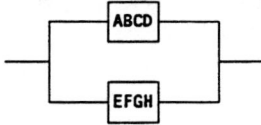

is $P(\text{system functions}) = 1 - (1 - P_{ABCD})(1 - P_{EFGH})$
$$= 1 - (1 - .7592625)(1 - .97951)$$
$$= 1 - .00493 = .99507$$

17.23 **a.** $P(\text{system functions}) = .95$

If the system has 3 identical components connected in series,

$$P(\text{system functions}) = p_A p_B p_C = p^3 = .95$$

Thus, $p = \sqrt[3]{.95} = .983$

b. If the components are connected in parallel,

$$P(\text{system functions}) = 1 - (1 - p_A)(1 - p_B)(1 - p_C)$$
$$= 1 - (1 - p)^3 = .95$$
$$\Rightarrow (1 - p)^3 = 1 - .95 = .05$$
$$\Rightarrow 1 - p = \sqrt[3]{.05} = .3684$$
$$\Rightarrow p = 1 - .3684 = .632$$

17.25 **a.** For the Weibull distribution, the cumulative distribution function is $F(y) = 1 - e^{-t^\alpha/\beta}$.
If $\alpha = .05$ and $\beta = .70$, then $F(y) = 1 - e^{-t^{.05}/.70}$.

The probability the light fails before time $t = 1000$ is

$$P(y \le 1000) = F(1000) = 1 - e^{-1000^{.05}/.70} = 1 - .1329 = .8671$$

b. The reliability of the light is

$$R(t) = 1 - F(t) = 1 - \left(1 - e^{-t^{.05}/.7}\right) = e^{-t^{.05}/.7}$$

Thus, $R(500) = e^{-500^{.05}/.7} = .1424$

c. The hazard rate of the light is $z(t) = \dfrac{f(t)}{R(t)}$

For the Weibull distribution, $f(t) = \dfrac{\alpha}{\beta} t^{\alpha-1} e^{-t^\alpha/\beta}$

With $\alpha = .05$, $\beta = .7$,

$$f(t) = \frac{.05}{.7} t^{.05-1} e^{-t^{.05}/.7} = .0714 t^{-.95} e^{-t^{.05}/.7}$$

Thus, $z(t) = \dfrac{.0714t^{-.95}e^{-t^{.05}/.7}}{e^{-t^{.05}/.7}} = .0714t^{-.95}$

$z(500) = .0714(500)^{-.95} = .000195$

17.27   a. $F(t) = P(y \le t) = \displaystyle\int_0^t \frac{1}{\beta}\, dy = \frac{y}{\beta}\Big]_0^t = \frac{t}{\beta} - \frac{0}{\beta} = \frac{t}{\beta}$

$R(t) = 1 - F(t) = 1 - \dfrac{t}{\beta}$

$z(t) = \dfrac{f(t)}{R(t)} = \dfrac{\dfrac{1}{\beta}}{1 - \dfrac{t}{\beta}} = \dfrac{1}{\beta - t}$

   b.   For $\beta = 10$, $z(t) = \dfrac{1}{10 - t}$

t	$z(t)$
0	.1
1	.111
2	.125
3	.143
4	.167
5	.2

$z(0) = \dfrac{1}{10 - 0} = .1$

$z(1) = \dfrac{1}{10 - 1} = .111$

$z(2) = \dfrac{1}{10 - 2} = .125$

$z(3) = \dfrac{1}{10 - 3} = .143$

$z(4) = \dfrac{1}{10 - 4} = .167$

$z(5) = \dfrac{1}{10 - 5} = .2$

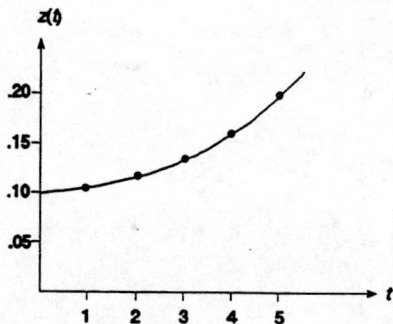

   c.   The reliability of the system when $t = 4$ and $\beta = 10$ is

$R(4) = 1 - \dfrac{4}{10} = .6$

17.29 a. First, we must fill out the table.

Time (i)	$x_i = \ln i$	Number of Survivors (n_i)	$\hat{R}(i) = \dfrac{n_i}{n}$	$-\ln \hat{R}(i)$	$y = \ln[-\ln \hat{R}(i)]$
1	0	438	.876	.132389	-2.02201
2	.69315	280	.560	.579818	$-.54504$
3	1.09861	146	.292	1.231001	.207828
4	1.38629	51	.102	2.282782	.825395
5	1.60943	15	.030	3.506558	1.254635

For the model $y_i = \beta_0 + \beta_1 x_i + \epsilon_i$, we need to estimate β_0 and β_1 by the method of least squares. The preliminary calculations are:

$$\sum x_i = 4.78748 \qquad \sum y_i = -.279192 \qquad \sum x_i y_i = 3.01401$$

$$\sum x_i^2 = 6.19947 \qquad \sum y_i^2 = 6.68417$$

$$SS_{xy} = \sum x_i y_i - \frac{\sum x_i \sum y_i}{n} = 3.0140 - \frac{4.78748(-.279192)}{5} = 3.281337$$

$$SS_{xx} = \sum x_i^2 - \frac{\left(\sum x_i\right)^2}{n} = 6.19947 - \frac{4.78748^2}{5} = 1.615473$$

$$SS_{yy} = \sum y_i^2 - \frac{\left(\sum y_i\right)^2}{n} = 6.68417 - \frac{(-.279192)^2}{5} = 6.668582$$

$$\hat{\beta}_1 = \frac{SS_{xy}}{SS_{xx}} = \frac{3.281337}{1.615473} = 2.031193$$

$$\hat{\beta}_0 = \bar{y} - \hat{\beta}_1 \bar{x} = \frac{-.279192}{5} - 2.031193\left(\frac{4.78748}{5}\right) = -2.000698$$

$$SSE = SS_{yy} - \frac{SS_{xy}^2}{SS_{xx}} = 6.668582 - \frac{3.281337^2}{1.615473} = .003554168$$

$$MSE = \frac{SSE}{n-2} = \frac{.003554168}{5-2} = .001184722$$

Using the method of least squares:

$$\hat{\alpha} = \hat{\beta}_1 = 2.031193 \quad \text{and}$$

$$\hat{\beta} = e^{-\hat{\beta}_0} = e^{2.000698} = 7.394215$$

b. For the Weibull distribution, the density function is

$$f(t) = \frac{\alpha}{\beta} t^{\alpha-1} e^{-t^\alpha/\beta} = \frac{2.031193}{7.394215} t^{1.031193} e^{-t^{2.031193}/7.394215}$$

$$= .2747 t^{1.031193} e^{-t^{2.031193}/7.394215}$$

The cumulative distribution function is:

$$F(t) = 1 - e^{-t^{\alpha}/\beta} = 1 - e^{-t^{2.031193}/7.394215}$$

The hazard rate $z(t)$ is

$$z(t) = \frac{f(t)}{1 - F(t)} = \frac{.2747t^{1.031193}e^{-t^{2.031193}/7.394215}}{1 - \left(1 - e^{-t^{2.031193}/7.394215}\right)} = .2747t^{1.031193}$$

The reliability $R(t)$ is

$$R(t) = 1 - F(t) = 1 - \left(1 - e^{-t^{2.031193}/7.394215}\right) = e^{-t^{2.031193}/7.394215}$$

c. $P(t \geq 1) = 1 - F(1) = 1 - \left(1 - e^{-1^{2.031193}/7.394215}\right) = e^{-1/7.394215} = .8735$

17.31 The density function for the exponential distribution is

$$f(t) = \begin{cases} \dfrac{1}{\beta}e^{-t/\beta} & 0 \leq t \leq \infty \\ 0 & \text{elsewhere} \end{cases}$$

For $\beta = 1000$, $f(t) = \begin{cases} \dfrac{1}{1000}e^{-t/1000} & 0 \leq t \leq \infty \\ 0 & \text{elsewhere} \end{cases}$

The cumulative distribution function for the exponential distribution is:

$$F(t) = 1 - e^{-t/1000}$$

The reliability of a single component is

$$R(t) = 1 - F(t) = 1 - \left(1 - e^{-t/1000}\right) = e^{-t/1000}$$

At time = 1400, the reliability of a component is:

$$R(1400) = e^{-1400/1000} = .2466$$

The reliability of a system composed of two resistors connected in series is

$$P(\text{system functions}) = p_A p_B = .2466(.2466) = .0608$$

17.33 a. The confidence interval for the mean time β based on censored sampling with 4 failures is

$$\frac{2(\text{Total life})}{\chi^2_{\alpha/2}} \leq \beta \leq \frac{2(\text{Total life})}{\chi^2_{1-\alpha/2}}$$

where Total life $= \sum_{i=1}^{4} t_i + (n - r)t_r = 5283 + (10 - 4)(2266) = 18879$

For confidence coefficient .95, $\alpha = 1 - .95 = .05$ and $\alpha/2 = .05/2 = .025$. From Table 8, Appendix II, $\chi^2_{.025} = 17.5346$ and $\chi^2_{.975} = 2.17973$ with $2r = 2(4) = 8$ degrees of freedom. The 95% confidence interval is:

$$\left[\frac{2(18879)}{17.5346} \leq \beta \leq \frac{2(18879)}{2.17973} \right] \Rightarrow (2153.343 \leq \beta \leq 17322.329)$$

b. The cumulative distribution function for the exponential distribution is:

$$F(t) = 1 - e^{-t/\beta}$$

The reliability is $R(t) = 1 - F(t) = 1 - \left(1 - e^{-t/\beta}\right) = e^{-t/\beta}$

The point estimator for β is $\hat{\beta} = \dfrac{\sum_{i=1}^{4} t_i + (n - r)t_r}{r} = \dfrac{5283 + (10 - 4)(2266)}{4}$
$$= 4719.75$$

Thus, an estimate of the reliability function is

$$R(t) = e^{-t/4719.75}$$

The probability the semiconductor will still be in operation after 4000 hours is:

$$R(4000) = e^{-4000/4719.75} = .4285$$

The 95% confidence interval for reliability is found by substituting the endpoints of the confidence interval for β into the equation for the reliability. The 95% confidence interval is

$$e^{-4000/2153.343} \leq e^{-4000/\beta} \leq e^{-4000/17322.329} \Rightarrow (.1560, .7938)$$

c. The hazard rate is $z(t) = \dfrac{f(t)}{R(t)} = \dfrac{\frac{1}{\beta}e^{-t/\beta}}{e^{-t/\beta}} = \dfrac{1}{\beta}$

For $\hat{\beta} = 4719.75$, the hazard rate is estimated by

$$z(t) = \frac{1}{4719.75} = .000212$$

Because the hazard rate is a constant, it is just as likely the component will fail in one unit interval as any other.

The 95% confidence interval for the hazard rate is found by substituting the endpoints of the confidence interval for β into the hazard function. The 95% confidence interval is:

$$\left(\frac{1}{17322.329}, \frac{1}{2153.343} \right) \Rightarrow (.0000577, .0004644)$$

17.35   To find the reliability of the system, we must first find the reliability of the subsystems.

The reliability of the subsystem (parallel)

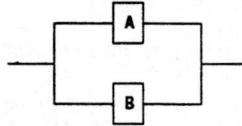

is $p_{AB} = 1 - (1 - p_A)(1 - p_B) = 1 - (1 - .85)(1 - .25) = 1 - .0375 = .9625$

The reliability of the subsystem (parallel)

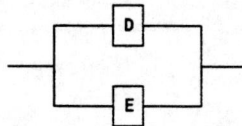

is $p_{DE} = 1 - (1 - p_D)(1 - p_E) = 1 - (1 - .90)(1 - .95) = 1 - .005 = .995$

The reliability of the subsystem (parallel)

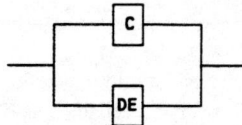

is $p_{CDE} = 1 - (1 - p_C)(1 - p_{DE}) = 1 - (1 - .75)(1 - .995) = 1 - .00125 = .99875$

The reliability of the entire system (series)

is: $p_{ABCD} = p_{AB}p_{CDE} = .9625(.99875) = .9613$

APPENDIX I

· ·

Matrix Algebra

I.1   a. $\mathbf{AB} = \begin{bmatrix} 3 & 0 \\ -1 & 4 \end{bmatrix} \begin{bmatrix} 2 & 1 \\ 0 & -1 \end{bmatrix} = \begin{bmatrix} 3(2) + 0(0) & 3(1) + 0(-1) \\ -1(2) + 4(0) & -1(1) + 4(-1) \end{bmatrix} = \begin{bmatrix} 6 & 3 \\ -2 & -5 \end{bmatrix}$

     b. $\mathbf{AC} = \begin{bmatrix} 3 & 0 \\ -1 & 4 \end{bmatrix} \begin{bmatrix} 1 & 0 & 3 \\ -2 & 1 & 2 \end{bmatrix} = \begin{bmatrix} 3(1) + 0(-2) & 3(0) + 0(1) & 3(3) + 0(2) \\ -1(1) + 4(-2) & -1(0) + 4(1) & -1(3) + 4(2) \end{bmatrix}$

           $= \begin{bmatrix} 3 & 0 & 9 \\ -9 & 4 & 5 \end{bmatrix}$

     c. $\mathbf{BA} = \begin{bmatrix} 2 & 1 \\ 0 & -1 \end{bmatrix} \begin{bmatrix} 3 & 0 \\ -1 & 4 \end{bmatrix} = \begin{bmatrix} 2(3) + 1(-1) & 2(0) + 1(4) \\ 0(3) - 1(-1) & 0(0) - 1(4) \end{bmatrix} = \begin{bmatrix} 5 & 4 \\ 1 & -4 \end{bmatrix}$

I.3   a. $\mathbf{AB}$ is a 3×4 matrix $(3 \times 2)(2 \times 4) \Rightarrow 3 \times 4$

     b.   No, it is not possible to find $\mathbf{BA}$. In order to multiply two matrices, the inner dimension numbers must be equal $\Rightarrow (2 \times 4)(3 \times 2) \Rightarrow 4 \neq 3$.

I.5   a. $\mathbf{AB} = \begin{bmatrix} 1 & 0 & 0 \\ 0 & 3 & 0 \\ 0 & 0 & 2 \end{bmatrix} \begin{bmatrix} 2 & 3 \\ -3 & 0 \\ 4 & -1 \end{bmatrix} = \begin{bmatrix} 1(2) + 0(-3) + 0(4) & 1(3) + 0(0) + 0(-1) \\ 0(2) + 3(-3) + 0(4) & 0(3) + 3(0) + 0(-1) \\ 0(2) + 0(-3) + 2(4) & 0(3) + 0(0) + 2(-1) \end{bmatrix}$

           $= \begin{bmatrix} 2 & 3 \\ -9 & 0 \\ 8 & -2 \end{bmatrix}$

     b. $\mathbf{CA} = \begin{bmatrix} 3 & 0 & 2 \end{bmatrix} \begin{bmatrix} 1 & 0 & 0 \\ 0 & 3 & 0 \\ 0 & 0 & 2 \end{bmatrix}$

           $= \begin{bmatrix} 3(1) + 0(0) + 2(0) & 3(0) + 0(3) + 2(0) & 3(0) + 0(0) + 2(2) \end{bmatrix} = \begin{bmatrix} 3 & 0 & 4 \end{bmatrix}$

     c. $\mathbf{CB} = \begin{bmatrix} 3 & 0 & 2 \end{bmatrix} \begin{bmatrix} 2 & 3 \\ -3 & 0 \\ 4 & -1 \end{bmatrix} = \begin{bmatrix} 3(2) + 0(-3) + 2(4) & 3(3) + 0(0) + 2(-1) \end{bmatrix} = \begin{bmatrix} 14 & 7 \end{bmatrix}$

· ·

I.7   **a.** $\begin{bmatrix} 1 & 0 \\ 0 & 1 \end{bmatrix}$

b. $\mathbf{IA} = \begin{bmatrix} 1 & 0 \\ 0 & 1 \end{bmatrix} \begin{bmatrix} 3 & 0 & 2 \\ -1 & 1 & 4 \end{bmatrix} = \begin{bmatrix} 3 & 0 & 2 \\ -1 & 1 & 4 \end{bmatrix} = \mathbf{A}$

c. $\begin{bmatrix} 1 & 0 & 0 \\ 0 & 1 & 0 \\ 0 & 0 & 1 \end{bmatrix}$

d. $\mathbf{AI} = \begin{bmatrix} 3 & 0 & 2 \\ -1 & 1 & 4 \end{bmatrix} \begin{bmatrix} 1 & 0 & 0 \\ 0 & 1 & 0 \\ 0 & 0 & 1 \end{bmatrix} = \begin{bmatrix} 3 & 0 & 2 \\ -1 & 1 & 4 \end{bmatrix} = \mathbf{A}$

I.9 $\mathbf{AA}^{-1} = \mathbf{A}^{-1}\mathbf{A} = \mathbf{I}$

$\mathbf{AA}^{-1} = \begin{bmatrix} 12 & 0 & 0 & 8 \\ 0 & 12 & 0 & 0 \\ 0 & 0 & 8 & 0 \\ 8 & 0 & 0 & 8 \end{bmatrix} \begin{bmatrix} 1/4 & 0 & 0 & -1/4 \\ 0 & 1/12 & 0 & 0 \\ 0 & 0 & 1/8 & 0 \\ -1/4 & 0 & 0 & 3/8 \end{bmatrix} = \begin{bmatrix} 1 & 0 & 0 & 0 \\ 0 & 1 & 0 & 0 \\ 0 & 0 & 1 & 0 \\ 0 & 0 & 0 & 1 \end{bmatrix}$

$\mathbf{A}^{-1}\mathbf{A} = \begin{bmatrix} 1/4 & 0 & 0 & -1/4 \\ 0 & 1/12 & 0 & 0 \\ 0 & 0 & 1/8 & 0 \\ -1/4 & 0 & 0 & 3/8 \end{bmatrix} \begin{bmatrix} 12 & 0 & 0 & 8 \\ 0 & 12 & 0 & 0 \\ 0 & 0 & 8 & 0 \\ 8 & 0 & 0 & 8 \end{bmatrix} = \begin{bmatrix} 1 & 0 & 0 & 0 \\ 0 & 1 & 0 & 0 \\ 0 & 0 & 1 & 0 \\ 0 & 0 & 0 & 1 \end{bmatrix}$

I.11 To verify Theorem I.1, show $DD^{-1} = D^{-1}D = I$

$$DD^{-1} = \begin{bmatrix} d_{11} & 0 & 0 & \cdots & 0 \\ 0 & d_{22} & 0 & \cdots & 0 \\ 0 & 0 & d_{33} & \cdots & 0 \\ . & . & . & \cdots & . \\ . & . & . & \cdots & . \\ . & . & . & \cdots & . \\ 0 & 0 & 0 & \cdots & d_{nn} \end{bmatrix} \begin{bmatrix} 1/d_{11} & 0 & 0 & \cdots & 0 \\ 0 & 1/d_{22} & 0 & \cdots & 0 \\ 0 & 0 & 1/d_{33} & \cdots & 0 \\ . & . & . & \cdots & . \\ . & . & . & \cdots & . \\ . & . & . & \cdots & . \\ 0 & 0 & 0 & \cdots & 1/d_{nn} \end{bmatrix}$$

$$= \begin{bmatrix} 1 & 0 & 0 & \cdots & 0 \\ 0 & 1 & 0 & \cdots & 0 \\ 0 & 0 & 1 & \cdots & 0 \\ . & . & . & \cdots & . \\ . & . & . & \cdots & . \\ . & . & . & \cdots & . \\ 0 & 0 & 0 & \cdots & 1 \end{bmatrix}$$

$$D^{-1}D = \begin{bmatrix} 1/d_{11} & 0 & 0 & \cdots & 0 \\ 0 & 1/d_{22} & 0 & \cdots & 0 \\ 0 & 0 & 1/d_{33} & \cdots & 0 \\ . & . & . & \cdots & . \\ . & . & . & \cdots & . \\ . & . & . & \cdots & . \\ 0 & 0 & 0 & \cdots & 1/d_{nn} \end{bmatrix} \begin{bmatrix} d_{11} & 0 & 0 & \cdots & 0 \\ 0 & d_{22} & 0 & \cdots & 0 \\ 0 & 0 & d_{33} & \cdots & 0 \\ . & . & . & \cdots & . \\ . & . & . & \cdots & . \\ . & . & . & \cdots & . \\ 0 & 0 & 0 & \cdots & d_{nn} \end{bmatrix}$$

$$= \begin{bmatrix} 1 & 0 & 0 & \cdots & 0 \\ 0 & 1 & 0 & \cdots & 0 \\ 0 & 0 & 1 & \cdots & 0 \\ . & . & . & \cdots & . \\ . & . & . & \cdots & . \\ . & . & . & \cdots & . \\ 0 & 0 & 0 & \cdots & 1 \end{bmatrix}$$

I.13   a.   Rewrite the linear equation as:

$$10v_1 + 0v_2 + 20v_3 = 60$$
$$0v_1 + 20v_2 + 0v_3 = 60$$
$$20v_1 + 0v_2 + 63v_3 = 176$$

$$A = \begin{bmatrix} 10 & 0 & 20 \\ 0 & 20 & 0 \\ 20 & 0 & 68 \end{bmatrix} \qquad V = \begin{bmatrix} v_1 \\ v_2 \\ v_3 \end{bmatrix} \qquad G = \begin{bmatrix} 60 \\ 60 \\ 176 \end{bmatrix}$$

b.   Show $A^{-1}A = AA^{-1} = I$

$$A^{-1}A = \begin{bmatrix} 17/70 & 0 & -1/14 \\ 0 & 1/20 & 0 \\ -1/14 & 0 & 1/28 \end{bmatrix} \begin{bmatrix} 10 & 0 & 20 \\ 0 & 20 & 0 \\ 20 & 0 & 68 \end{bmatrix} = \begin{bmatrix} 1 & 0 & 0 \\ 0 & 1 & 0 \\ 0 & 0 & 1 \end{bmatrix}$$

$$AA^{-1} = \begin{bmatrix} 10 & 0 & 20 \\ 0 & 20 & 0 \\ 20 & 0 & 68 \end{bmatrix} \begin{bmatrix} 17/70 & 0 & -1/14 \\ 0 & 1/20 & 0 \\ -1/14 & 0 & 1/28 \end{bmatrix} = \begin{bmatrix} 1 & 0 & 0 \\ 0 & 1 & 0 \\ 0 & 0 & 1 \end{bmatrix}$$

c. $V = A^{-1}G = \begin{bmatrix} 17/70 & 0 & -1/14 \\ 0 & 1/20 & 0 \\ -1/14 & 0 & 1/28 \end{bmatrix} \begin{bmatrix} 60 \\ 60 \\ 176 \end{bmatrix} = \begin{bmatrix} 2 \\ 3 \\ 2 \end{bmatrix}$